THE
LIVES
OF
BEES

———

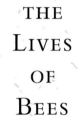

THE LIVES OF BEES

The Untold Story
of the Honey Bee
in the Wild

Thomas D. Seeley

PRINCETON UNIVERSITY PRESS
PRINCETON AND OXFORD

Copyright © 2019 by Princeton University Press

Published by Princeton University Press

41 William Street, Princeton, New Jersey 08540

6 Oxford Street, Woodstock, Oxfordshire OX20 1TR

press.princeton.edu

All Rights Reserved

LCCN 2019930923

ISBN 978-0-691-16676-6

British Library Cataloging-in-Publication Data is available

Editorial: Alison Kalett and Kristin Zodrow

Production Editorial: Brigitte Pelner

Text and Jacket/Cover Design: Carmina Alvarez

Jacket/Cover Credit: Bees/iStock

Production: Jacqueline Poirier

Publicity: Sara Henning-Stout

Copyeditor: Amy K. Hughes

This book has been composed in Perpetua and Bembo

Printed on acid-free paper ∞

Printed in Canada

1 3 5 7 9 10 8 6 4 2

Dedicated to Roger A. Morse (1927–2000), who for more than forty years as a scientist, writer, and teacher at Cornell University informed and entertained students about the honey bee, and whose support of the author laid the foundation for this book

Contents

	Preface	vii
1	Introduction	1
2	Bees in the Forest, Still	17
3	Leaving the Wild	57
4	Are Honey Bees Domesticated?	79
5	The Nest	99
6	Annual Cycle	140
7	Colony Reproduction	155
8	Food Collection	187
9	Temperature Control	215
10	Colony Defense	243
11	Darwinian Beekeeping	277
	Notes	293
	References	317
	Acknowledgments	337
	Illustration Credits	341
	Index	349

Preface

We humans have always been fascinated by the honey bee—*Apis mellifera*, the "honey-carrying bee." For hundreds of thousands of years, our earliest ancestors in Africa, Europe, and Asia surely marveled at this bee's astonishing industry in storing honey and making beeswax, two substances of great value. More recently, during the last 10,000 years, we invented the intricate craft of beekeeping and we began our scientific studies of honey bees. It was, for example, the ancient philosopher Aristotle who first described this bee's practice of "flower constancy": a worker bee generally sticks to one type of flower throughout a foraging trip to boost the efficiency of her food collection. And within the past few hundred years, we have written tens of thousands of scientific articles about the honey bee, many on the practical stuff of apiculture but also countless others on the fundamental biology of this endlessly enchanting bee. We have also authored thousands of books on bees and beekeeping. In the United States alone, nearly 4,000 titles on beekeeping, honey bee science, children's tales about bees, and the like have been published between the 1700s and 2010.

Given humanity's enduring fascination with the honey bee, it is a curious thing that until recently we have known rather little about the true natural history of *Apis mellifera*—that is, about how colonies of this species live in the wild. What explains our long delay in making a broad reconnaissance of the natural lives of honey bees? I think the answer is simple: the most ardent students of this industrious and intriguing insect—beekeepers and biologists—have almost always worked with managed colonies inhabiting man-made hives crowded in apiaries, not wild colonies occupying hollow trees and rock crevices dispersed across the landscape. Managed colonies are the ones that produce our honey and pollinate our crops, so it is not surprising that beekeepers have focused their attention on the colonies living in their hives. Managed colonies are also the ones best suited for scientific investigations, which require controlled experiments, so it is also not surprising that biologists too have worked primarily with colonies living in artificial homes. The Nobel laureate Karl von Frisch, for example, would never have discovered the meaning of the honey bee's waggle dance if he had not worked with a colony living in a glass-walled observation hive, labeled some of its foragers with paint marks for individual identification, and then watched how these bees behaved inside his hive when they returned from an artificial food source, a little dish of sugar syrup that he had set up in the courtyard just outside his laboratory.

Humankind's long-standing focus of interest on honey bees housed in hives—whether clay cylinders, woven baskets, wooden boxes, or (most recently) polystyrene containers—continues to this day. Over the last few decades, however, beekeepers and biologists have begun to investigate how these engaging insects live when they do so without our supervision, and this "return to nature" has opened our eyes to many new mysteries in the lives of honey bees. This book is my attempt to review what has been learned about how colonies of honey bees live in their natural world. We will see that the free-living colonies residing in tree cavities and rock crevices lead lives that differ substantially from the lives of beekeepers' colonies inhabiting the white boxes that we see parked in apple orchards and blueberry fields, jam-packed in apiaries, and nestled in backyards. Perhaps most remarkably, we will see that the wild colonies are surviving and are maintaining their numbers, while at the same time some 40 percent of the colonies managed by beekeepers are dying each year.

The story of the wild honey bees is an important one, because it can expand how we view ourselves in relation to *Apis mellifera* and how we conduct the craft of beekeeping. We might, for example, start to think of the honey bee not only as a compliant, hardworking insect that we can manipulate to produce honey crops and fulfill pollination contracts but also as a wondrous insect that we can admire, respect, and treat in truly bee-friendly ways. In the coming chapters, I will show how numerous strands of work on honey bee colonies living in the wild—their nest architecture, nest spacing, foraging range, mating system, disease resistance, colony genetics, and more—have come together to reveal how these colonies living all on their own are thriving. In the final chapter, "Darwinian Beekeeping," I will discuss how we can use this growing body of knowledge to address a critically important issue: how to be better partners with *Apis mellifera*, a species that has sweetened the lives of humans for many thousands of years, and on which humanity's food supply depends more and more every year.

My fascination with honey bees living in the wild began in the spring of 1963, when I was not quite 11 years old. I lived then, as I still do today, in a little valley called Ellis Hollow, which lies a few miles east of Ithaca, New York. It is a stream valley barely 1.6 kilometers (1 mile) wide and 3.2 kilometers (2 miles) long, and it lies between two steep-sided hills, Mount Pleasant and Snyder Hill. These are parallel units of an ancient escarpment of sandstone that runs through the Finger Lakes region of central New York State and adds rugged beauty to the area. Ellis Hollow was a good place to grow up, because the wooded hillsides and the valley bottom—with its sloping fields bordered by dark groves of hemlock trees, sunny swamps patrolled by dragonflies, and the gently flowing Cascadilla Creek meandering through it all—seemed endless. It is where I first observed a magnificent pileated woodpecker chiseling into a tree for carpenter ants, first watched a steely-eyed snapping turtle laying eggs deep in moist soil, and first showed my pet raccoon how to hunt for crayfish under rocks in little streams. Fortunately, there were no "No Trespassing" signs to limit my explorations of this ever-fascinating place. Even today, when I drive home along the Ellis Hollow Creek Road, I take note of spots that I still need to investigate.

One day, back in early June 1963, I was walking along Ellis Hollow Road, when I heard a loud buzzing sound and saw a bread-truck-size cloud of honey bees circling the ancient black walnut tree that stands beside the road about 100 meters (330 feet) east of my family's house. I was frightened, so I crossed to the shady woods on the other side of the road and watched from what felt like a safe distance. From there, I saw that the bees were landing on a thick limb about 4 meters (12 feet) off the ground, blanketing it with their thousands of leathery-brown bodies, and streaming into a knothole the size of a golf ball. The bees were moving in! Suddenly this huge tree, which already I had prized as a good climbing tree and a rich source of black walnuts, was *super* special. It was now a bee tree! I visited it often that summer and gradually overcame my fear of the bees, eventually learning that I could watch them close up (while perched atop a stepladder) without being stung. It was a time of wonder.

My mother noted my curiosity about the bees, and for Christmas 1963 my parents gave me a beautifully illustrated children's book on honey bees, *The Makers of Honey* (1956), written by Mary Geisler Phillips. I read it closely and liked how it introduced me to the biology of honey bees. It sits on my writing desk as I type these words. I feel an especially deep connection to this little book because its author was also a Cornell University professor, in the College of Home Economics (now the College of Human Ecology), where she served as editor of the college's radio scripts and outreach publications. Moreover, the first professor of apiculture at Cornell, Everett F. Phillips, was her husband.

Given these lovely introductions to honey bees as a boy, especially the firsthand observations of a bee colony living wild in a tree, it is not surprising that, when I started graduate school in 1974 to earn a PhD in biology, and I had to choose a topic for my thesis research, I decided to investigate what honey bees seek when they (not a beekeeper) choose their living quarters. In doing so, I figured that I could apply to honey bees the "know-thy-animal-in-its-world" rule that I was learning from my thesis adviser at Harvard, the German ethologist Bert Hölldobler. I also hoped that I could foster a new approach to studying honey bees, one in which we view them as amazing wild creatures that live in hollow trees in forests, not just as the "angels of agriculture" that live in white boxes in apiaries. Furthermore, I hoped that through this thesis research I would solve the mystery I had sensed back in 1963 as I cautiously watched a swarm moving into its new home: What was it about the dark cavity in the black walnut tree near my parent's house that attracted the bees to make it their dwelling place? Watching *that* swarm take up residence in *that* tree on *that* day is the spark that ignited my long-standing passion to understand how honey bees live in the wild.

Thomas D. Seeley
Ithaca, New York

The Lives of Bees

I

INTRODUCTION

> We have never known what we were doing
> because we have never known what we were undoing.
> We cannot know what we are doing until we know
> what nature would be doing if we were doing nothing.
> —Wendell Berry, "Preserving Wildness," 1987

This book is about how colonies of the honey bee (*Apis mellifera*) live in the wild. Its purpose is to provide a synthesis of what is known about how honey bee colonies function when they are not being managed by beekeepers for human purposes and instead are living on their own and in ways that favor their survival, their reproduction, and thus their success in contributing to the next generation of colonies. Our goal is to understand the natural lives of honey bees—how they build and warm their nests, rear their young, collect their food, thwart their enemies, achieve their reproduction, and stay in tune with the seasons. Besides looking at *how* honey bee colonies live in nature, we will examine *why* they live as they do when they manage their affairs themselves. In other words, we will also explore how natural selection has shaped the biology of this important species during its long journey through the labyrinth of evolution. Doing so will reveal how *Apis mellifera* achieved a native range that includes Europe, western Asia, and most of Africa, and so became a world-class species even before beekeepers introduced it to the Americas, Australia, and eastern Asia.

Knowing how the honey bee lives in its natural world is important for a broad range of scientific studies. This is because *Apis mellifera* has become one of the model systems for investigating basic questions in biology, especially those related to behavior. Whether one is studying these bees to solve some mystery in animal cognition, behavioral genetics, or social behavior, it is critically important to become familiar with their natural biology before designing one's experimental investigations. For example, when sleep researchers used honey bees to explore the functions of sleep, they benefited greatly from knowing that it is only the elderly bees within a colony, the foragers, that get most of their sleep at night and in comparatively long bouts. If these researchers had not known which bees are a colony's soundest sleepers come nightfall, then they might have failed to design truly meaningful sleep-deprivation experiments. A good experiment with honey bees, as with all organisms, taps into their natural way of life.

Knowing how honey bee colonies function when they live in the wild is also important for improving the craft of beekeeping. Once we understand the natural lives of honey bees, we can see more clearly how we create stressful living conditions for these bees when we manage them intensively for honey production and crop pollination. We can then start to devise beekeeping practices that are better—for both the bees and ourselves. The importance of using nature as a guide for developing sustainable methods of agriculture was expressed beautifully by the author, environmentalist, and farmer Wendell Berry, when he wrote: "We cannot know what we are doing until we know what nature would be doing if we were doing nothing."

The current state of beekeeping shows us all too clearly how problems in the lives of animals under our management can arise when we fail to consider how they would be living if we were not forcing them to live in artificial ways that serve mainly our own interests. Many beekeepers—especially those in North America who practice industrial-scale beekeeping with tens of thousands of colonies of *Apis mellifera*, the species that is the focus of this book—are experiencing colony mortality rates of 40 percent or more each year. To be sure, this is not due entirely to the colony management practices of beekeepers. Changes in the crop production practices of farmers, especially the use of systemic insecticides that are absorbed by plants and contaminate their nectar and pollen, and the switch in many places to growing corn and soybeans instead of clover and alfalfa, also play roles in this sad story. But the heavy-handed manipulations of the lives of the honey bee colonies housed in beekeepers' hives certainly do contribute to the sky-high rates of colony mortality. We will see that when beekeepers force colonies to live crowded together in apiaries—where the bees' homes are less than 1 meter (ca. 3 feet) apart, rather than the hundreds of meters (at least 1,000 feet) apart in nature—beekeepers boost the efficiency of their work but they also foster the spread of the bees' diseases. Likewise, when beekeepers supersize their colonies by housing them in huge hives that are nearly as tall as themselves, rather than in smaller hives the size of the bees' natural nesting cavities, they boost the honey production of their colonies, but they also turn them into stupendous hosts for the pathogens and parasites of *Apis mellifera*, such as the deadly ectoparasitic mite *Varroa destructor*.

Given the harmful effects on the bees that arise from the standard practices of beekeeping, it is not surprising that many beekeepers are now exploring alternative approaches to this craft. These folks are keen to use nature as a model, and this requires a solid understanding of how honey bees live on their own in nature. To help readers who want to adjust their beekeeping practices to make them more bee-friendly, I have included a final chapter on what I like to call "Darwinian beekeeping," which is an approach to beekeeping that aims to give bees the opportunity to live the way they do in the wild.

FOCUS ON WILD COLONIES IN THE NORTHEASTERN UNITED STATES

This book does not attempt to provide a comprehensive account of how *Apis mellifera* lives in nature across its vast geographic range, which now includes Europe, some of Asia, all of Africa except the great desert regions,

FIG. 1.1. *Left:* Bee-tree home of a wild colony of honey bees living in the Arnot Forest, of Cornell University, in the United States. Red arrow indicates the small knothole entrance of this colony's nest. *Right:* Nest entrance of a wild colony of honey bees living in Munich, Germany.

most of North and South America, and parts of Australia and New Zealand. Instead, it focuses on how colonies of our most important pollinator are living in the wild in the deciduous forests of the northeastern United States, a place where they have thrived as an introduced species for nearly 400 years. This is also the place where, for more than 40 years, my collaborators and I have studied the behavior, social life, and ecology of honey bees living in the wild (Fig. 1.1). Although our studies are based on honey bees living outside their native range, I believe that what we have learned about how honey bees live in the woods in the northeastern corner of the United States can help us understand how these bees originally lived in nature in Europe, especially in its northern and western regions.

Until the mid-1800s, all the honey bees living in the northeastern United States were descendants of the colonies of honey bees that were brought to North America from northern Europe starting in the early 1600s. Insect taxonomists recognize some 30 subspecies (geographic variants) of *Apis mellifera*, and they refer to the honey bees native to northern Europe as members of the subspecies *Apis mellifera mellifera* Linnaeus. This subspecies of *Apis mellifera*—also called the dark European honey bee—has the distinction of being the first kind of honey bee to be described taxonomically. This was done 360 years ago, in 1758, when Carl Linnaeus, a professor of botany and zoology at Uppsala University in Sweden, published his work *Systema Naturae*, in which he presented the system of taxonomic classification that biologists have used ever since.

The dark European honey bee is so named because its body color ranges from dark brown to jet black and historically it lived throughout northern Europe, from the British Isles in the west to the Ural Mountains in the east, and from the Pyrenees and Alps in the south to the coasts of the Baltic Sea in the north (Fig. 1.2). We know from archaeological studies, which have found traces of beeswax in fragments of pottery vessels dating from 7,200 to 7,500 years before present, that this bee was living in Germany and Austria some 8,000 years ago. We also know from genetic studies of the bees themselves that as the climate of northern Europe underwent postglacial warming, starting about 10,000 years ago, this bee expanded its

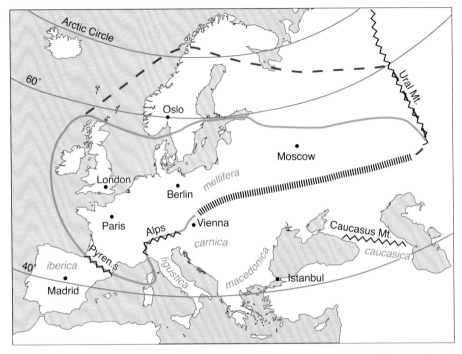

FIG. 1.2. Distribution map of the dark European honey bee, *Apis mellifera mellifera*. Green line: original distribution limits to the west, north, and east. Vertically hatched line: transition zone to the honey bee races of southern and eastern Europe (*A. m. ligustica, carnica, macedonica,* and *caucasica*). Red dashed line: northern limit of beekeeping.

range north and east from Ice Age refugia—pockets of woodlands in the mountains of southern France and Spain—as it tracked the expansion of forests populated with thermophilic trees such as willow, hazel, oak, and beech. Evidently, *Apis mellifera mellifera* flourished as it spread and eventually extended its range farther north within Europe than any other subspecies, drawing on the adaptations for winter survival that it had evolved during the glacial period. It is estimated that within the eastern, heavily forested two-thirds of its range (from eastern Germany to the Urals) there once lived millions of colonies of the dark European honey bee. And there is no doubt that tree beekeeping—cutting cavities in trees to create nest sites and then harvesting honey without killing the colonies living in the artificial hollows—provided most of the beeswax and honey traded in Europe in the Middle Ages. A vestige of this centuries-old tradition of tree beekeeping is found in the South Ural region of the republic of Bashkortostan, part of the Russian Federation. Colonies of pure *A. m. mellifera* still inhabit the forests of this region, and Bashkir tree-hive beekeepers still harvest basswood (*Tilia cordata*) honey from colonies residing in man-made nest cavities high in the trees.

The dark European honey bee is superbly adapted to living in forested regions with relatively cool summers and long, cold winters. It is not surprising, therefore, that when bees of this subspecies were brought to Massachusetts, Delaware, and Virginia in North America by English and Swedish immigrants starting in the early 1600s, they escaped (swarmed) from the beekeepers' hives and soon became an important part of the local fauna. Already in 1720, a Mr. Paul Dudley published, in *Philosophical Transactions*, a journal of the Royal Society of London, a letter titled "An account of a method lately found out in New-England for discovering where the bees hive in the woods, in order to get their honey." Analysis of letters, diaries, and accounts of travels in North America written in the 1600s and 1700s has revealed that these honey bees spread speedily across the heavily forested eastern half of North America below the Great Lakes (Fig. 1.3). Also, the journals of the Lewis and Clark Expedition document the dark European honey bee's rapid colonization of North America east of the Mississippi River. For example, on Sunday, 25 March 1804, shortly after the expedition party had left St. Louis and was camped along the Kansas River, William Clark wrote in his journal: "River rose 14 Inch last night, the men find numbers of Bee Trees, & take great quantities of honey."

These days, the honey bees living wild in the forests of the northeastern United States are no longer a genetically pure population of *Apis mellifera mellifera*. This is because, in 1859, following the advent of steamship service between Europe and the United States, American beekeepers began importing queen bees of several other subspecies of *Apis mellifera*, ones that are native to southern Europe or northern Africa. These imports continued for more than 60 years, during which time many thousands of mated

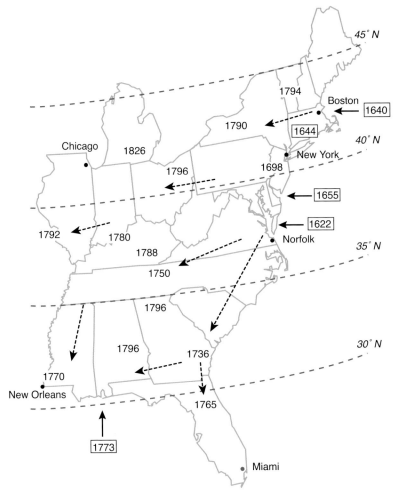

FIG. 1.3. The dispersal of the dark European honey bee across eastern North America following introductions (indicated by solid arrows) in Virginia, Massachusetts, Connecticut, and Maryland in the 1600s, and in Alabama in 1773. Dashed arrows indicate the bee's subsequent spread.

queen bees were shipped to North America, but they abruptly ceased in 1922. This was the year the U.S. Congress passed the Honey Bee Act, which prohibited further imports to protect honey bees in the United States from the Isle of Wight disease, an unspecific but supposedly highly infectious and lethal disease named for the location of its first reputed outbreak, in southern England. As we shall see, the genetic composition of the wild honey bees living in the northeastern United States is now a blend of the genes of *A. m. mellifera* and several other subspecies of *Apis mellifera*. Of those introduced in the late 1800s and early 1900s, the three most important all came from south-central Europe: *A. m. ligustica* (from Italy), *A. m. carnica* (from Slovenia), and *A. m. caucasica* (from the Caucasus Mountains). Several other subspecies were introduced from the Middle East and Africa—*A. m. lamarckii* (from Egypt), *A. m. cypria* (from Cyprus), *A. m. syriaca* (from Syria and the eastern Mediterranean region), and *A. m. intermissa* (from northern Africa)—but they did not prove popular, and it seems that they are not well represented genetically anywhere in the United States.

More recently, in 1987, a subspecies of *Apis mellifera* that is native to eastern and southern Africa, *A. m. scutellata*, entered the southern United States by way of Florida, perhaps when a beekeeper, seeking bees that would thrive in subtropical Florida, smuggled in some queens of this tropically adapted subspecies. Since then, colonies of *A. m. scutellata* have indeed thrived in Florida and have greatly influenced the genetics of the honey bees living in the southeastern United States but not of the honey bees living in the northeastern United States, probably because colonies of *A. m. scutellata* cannot survive northern winters. Bees of the African subspecies *A. m. scutellata* entered the United States a second time and in a second place in 1990, when swarms flew across the U.S.–Mexico border into Texas. Here again they mixed with the European honey bees already in residence. Since then, populations of colonies that are hybrids of African and European honey bees (so-called Africanized honey bees) have developed in the humid, subtropical parts of southern Texas and in the southernmost parts of New Mexico, Arizona, and California. As of 2013, the gene pool of the Africanized bees in southern Texas still had a small genetic contribution (ca. 10%, for both mitochondrial and nuclear genes) from European honey bees.

The complex history of countless introductions to North America of honey bees from various regions of Europe, the Middle East, and Africa raises an important question: What is the mix of subspecies of *Apis mellifera* in the wild colonies living in the northeastern United States, the main subjects of this book? Fortunately, we now have a clear answer to this question for the colonies living in the vast woodlands of southern New York State. In 1977 and then again in 2011, I collected worker bees from 32 wild colonies living in this heavily forested region. The 32 sets of bees from 1977 were stored as pinned (voucher) specimens in the Cornell University Insect Collection, and the 32 sets of bees from 2011 were stored in vials filled with ethanol, which preserves DNA quite nicely. In 2012, specimens from both groups of bees were shipped to one of my former students, Professor Alexander S. Mikheyev, who heads the Ecology and Evolution Unit at the Okinawa Institute of Science and Technology in Japan. There the DNA was extracted from one bee from each of the 64 colonies that I had sampled and an analysis based on whole-genome sequencing was performed to determine the subspecies composition of both the 1977 ("old") and the 2011 ("modern") populations of bees (Fig. 1.4).

This genetic detective work found that the bees in both the old and modern samples are primarily descendants of two of the subspecies of *Apis mellifera* that were imported from southern Europe, specifically from Italy and from Slovenia: *A. m. ligustica* and *A. m. carnica*, respectively. This finding was not surprising, because these are the two subspecies that have proven the most popular among beekeepers in North America since the 1800s. Colonies of these two subspecies tend to be good-natured (not prone to stinging) and good producers of honey. What was surprising, however, was the discovery that the bees in both the old and modern samples also possessed many genes from the dark honey bees imported from north of the Alps (*A. m. mellifera*) starting in the 1600s and from the gray mountain honey bees (*A. m. caucasica*) imported from the Caucasus Mountains starting in the late 1800s (see Fig. 1.4). This genetic sleuthing also revealed that the bees in the modern (2011) sample, but not those in the old (1977) sample, have a small percentage (less than 1%) of genes from two African subspecies: *A. m. scutellata*, native to Africa south of the Sahara, and *A. m. yemenetica*, native to the hot arid zones of Arabia (e.g., Saudi Arabia, Yemen,

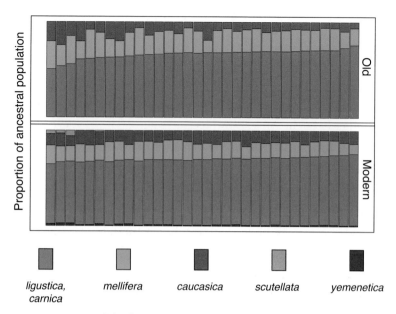

FIG. 1.4. Ancestries of the honey bees (*Apis mellifera*) living in the forests south of Ithaca, New York. Both the old (1970s) and the modern (2010s) populations are largely descendants of bees from southern and southeastern Europe: the subspecies *A. m. ligustica*, *carnica*, and *caucasica*, which have been popular with beekeepers in North America since the 1880s. Both old and modern populations also have clear ancestry from the dark European honey bees (*A. m. mellifera*) of northern Europe that were introduced to North America starting in the 1600s. The modern population also shows a small ancestry of bees from Africa (*A. m. scutellata*) and the Arabian peninsula (*A. m. yemenitica*).

and Oman) and eastern Africa (e.g., Sudan, Somalia, and Chad). This slight introgression of African genes into the modern population of wild colonies living in the forests near Ithaca, New York, is probably a result of Africanized honey bees—hybrids of African and European races of *Apis mellifera*—becoming established in parts of the southern United States in the late 1980s and early 1990s. These southern regions—which include the states of Florida, Georgia, Alabama, and Texas—have warm climates favorable to Africanized honey bees, and they are where much of the commercial queen production in the United States takes place. Evidently, for the last

25 or so years, queen producers in the southern states have been shipping queens carrying some genes of African descent to beekeepers in the northern states. Migratory beekeepers who keep their colonies in Florida over winter and then truck them north in spring to pollinate apples, cranberries, and other crops, have probably also contributed to the northward trickle of genes of the Africanized honey bees.

This new, high-tech look at the genes of the honey bees living in the woodlands south of Ithaca has revealed two important things. First, it has shown us that the arrivals of African honey bees in Florida and Texas in the 1980s and 1990s have affected only very slightly the genetic composition of the wild colonies living in the forests near Ithaca. In other words, the genetic makeup of these wild colonies still reflects mainly the nearly 400-year history of imports of honey bees from Europe. Second, it shows that the genes in this population of wild colonies have come predominantly from honey bees native to *southern* Europe, even though the introductions of honey bees from *northern* Europe started some 200 years earlier. Presumably, this reflects the greater popularity among beekeepers of bees from Italy and Slovenia (*A. m. ligustica* and *A. m. carnica*), in southern Europe, relative to the dark European honey bees (*A. m. mellifera*) from various places in northern Europe. Most beekeepers prefer bees that are calm and produce much honey, and the lighter-colored bees of southern Europe, compared to the darker-colored bees of northern Europe, tend to be less likely to run around when a hive is opened and more likely to build up large populations of worker bees and amass large stores of honey.

Given that most of the genes in the wild colonies living near Ithaca are from honey bees adapted to the relatively mild climates of southern Europe, and given that Ithaca winters are long, snowy (Fig. 1.5), and often bitterly cold (lowest temperatures about $-23°C/-10°F$), we need to ask: Are the wild honey bees living near the Ithaca area well adapted for life in this northern region of North America? We will see in the coming chapters that the answer to this question is a solid yes; multiple studies have found that the wild colonies of honey bees living in this region are impressively skilled at living here. These studies have looked at how the colonies' nest-

FIG. 1.5. Winter look of an apiary located in Ellis Hollow, near Ithaca, New York.

site preferences, seasonal patterns of brood rearing and swarming, foraging skills, overwintering abilities, and defenses against pathogens and parasites are all highly adaptive for life in this northeastern corner of the United States. Perhaps the most compelling indication that these wild colonies are well adapted to their current environment is that they possess a powerful set of behavioral defenses against the aptly named ectoparasitic mite *Varroa destructor*. In chapter 10, we will see how the population of wild colonies living near Ithaca was decimated when this mite—whose original host is an Asian honey bee species, *Apis cerana*—reached the Ithaca area in the mid-1990s but then recovered through strong selection for multiple defensive behaviors in worker bees that kill these mites. Indeed, we now know that the density of wild colonies of honey bees in the 2010s (ca. 20 years post-*Varroa* arrival) matches what it was in the 1970s (ca. 20 years pre-*Varroa* arrival).

We should not be surprised that the wild colonies of honey bees living in the forests around Ithaca are well adapted to survive and reproduce in these northern woodlands, where the winters are far longer and colder than the winters in many of the bees' ancestral homelands in Europe. After all, these colonies have been exposed to strong natural selection to adapt to the climate throughout their nearly 400-year history of living in the northeastern United States, and there are countless studies by biologists that demonstrate that evolution by natural selection can produce a population of plants or animals with a robust solution to a new problem in just a few years. Besides the rapid evolution of *Varroa* resistance in the wild colonies of honey bees living in the forests around Ithaca, there is the example of the Africanized honey bees (*Apis mellifera scutellata*) in Puerto Rico evolving docility in only 10 or so years. Evidently, this rapid evolutionary change, which occurred between 1994 and 2006, was driven by natural selection favoring genes for reduced aggression in honey bees living where there are no major predators. Another striking example of an insect's rapid behavioral adaptation to a changed circumstance is the adaptive disappearance between the late 1990s and 2003 of the calling song of male field crickets (*Teleogryllus oceanicus*) on the Hawaiian island of Kauai. This behavioral change followed the accidental introduction of parasitic flies that locate host crickets by orienting to the crickets' chirps. Male crickets with mutations for wing structures that silenced their singing were strongly favored by natural selection. Quick evolution led to quiet crickets!

ROAD MAP TO WHAT FOLLOWS

This book aims to provide you with a clear view of the natural lives of honey bee colonies, especially those living in cold climate regions of the world. To enjoy this view, you will need to work your way through some new scientific terrain, make dozens of stops along the way, and look carefully in a different direction at each stopping place. You will soon see that this book is partly a synthesis of the work of many research biologists and partly a travelogue of my personal quest to better understand this special piece of nature. Here is a road map to what follows.

Chapter 2 describes when, where, and how I became intrigued with the puzzle of how colonies of the "domestic" honey bee, *Apis mellifera*, live in the wild. This chapter introduces you to the landscape and forests south of the small city of Ithaca, in central New York State, which is where many of the investigations described in this book were conducted. It also describes how, in the late 1970s, I began studying the population of wild colonies living in one of these forests, the Arnot Forest. It further describes how, in the early 2000s, I was amazed to find wild colonies still living in this forest even though the deadly ectoparasitic mite *Varroa destructor* had spread to the Ithaca area sometime in the early 1990s. This chapter goes on to review what we know about the abundance (and persistence) of wild colonies of *Apis mellifera* in other places. By the end of chapter 2, you will see clearly the two main puzzles that are solved, step-by-step, in the rest of the book: 1) How are the wild colonies of honey bees living in the woods around Ithaca able to survive without being treated with miticides? And, more broadly, 2) How do the lives of wild and managed colonies of honey bees differ, and what can we learn from these differences to be better stewards of our most important pollinator?

Chapters 3 and 4 take a step back from the present-day biology of *Apis mellifera* to explore why, until recently, we have known so little about the natural lives of honey bees. We will see that honey bee colonies probably began living in man-made structures (hives) as soon as humans made the shift from being mobile hunters and gatherers to living as sedentary herders and farmers, some 10,000 years ago. It is likely that as soon as we humans stopped being destructive honey hunters, we began to become manipulative beekeepers. We will also see how, over thousands of years, we gradually refined our artificial housing of managed honey bee colonies to make it easier and easier for us to reach into their homes and steal their golden honey. Thus, step-by-step, we grew increasingly disconnected from how honey bees live in the wild. Meanwhile, the bees never yielded their nature to us and instead continued to follow a way of life set millions of years ago. It was not until about 70 years ago that we perfected the means—

artificial insemination of queen bees—to control this insect's mating and breed it for our purposes. Fortunately, even today, precious few queens are artificially inseminated. Most still mate with whatever drones they encounter.

Chapters 5 through 10 review what has been learned, mostly over the last 40 years, about the natural history of honey bees living in temperate regions of the world. Here we will examine the interwoven topics of nest architecture, annual cycle, colony reproduction, food collection, temperature control, and colony defense. Throughout these chapters, we will see how the marvelous inner workings of a honey bee colony have been shaped by natural selection for life in the wild, not in domestic settings, so honey bees are still perfectly able to survive and reproduce without a beekeeper's supervision. More specifically, we will see how the bees build and use their beeswax combs, time their swarming and drone rearing, operate a factory-like organization of food and water collection, maintain thermal homeostasis in their nests, and sustain an arsenal of colony defenses. These are all parts of a honey bee colony's complex suite of adaptations for passing on its genes to the next generation of colonies.

Finally, chapter 11 presents the take-home lessons from what we have learned about how *Apis mellifera* lives in its natural world. The chapter first summarizes the findings reported in the previous chapters in the form of a 21-point comparison between the lives of colonies living in the wild and those of colonies managed for apicultural purposes. It then offers 14 practical suggestions of ways beekeepers can help their bees live closer to their natural lifestyle and so enjoy less stress and better health.

2

BEES IN THE FOREST, STILL

> Reports of my death have been greatly exaggerated.
> —*Mark Twain, New York Journal, 1897*

Many of the studies that I will describe in this book are ones that my colleagues and I have conducted over the past 40 years with the wild colonies of honey bees that live in the forests near the small city of Ithaca, the home of Cornell University, in central New York State. Ithaca lies at the southern end of Cayuga Lake, an elongate, glacially deepened lake that runs north for nearly 65 kilometers (40 miles). It is one of the 11 Finger Lakes, which extend south to north across the middle of New York like the fingers on a pair of outstretched hands (Fig. 2.1). The landscape *between* these lakes is one of rolling hills with deep, rich soils lying atop a bedrock of limestone. This is a fertile and productive region with intensive agriculture, including vineyards, orchards, and dairy farms. But *south* of these lakes, and thus south of Ithaca, we find a hilly, wooded landscape whose topography and soils differ markedly from those to the north. Here we encounter narrow valleys that snake between rugged hills, some with nearly vertical slopes, and acidic soils thinly covering bedrocks of shales and sandstones. The region south of Ithaca is part of the Appalachian Highlands and has elevations exceeding 610 meters (2,000 feet). Most of it is ill-suited for farming, but it supports beautiful hardwood forests that provide prime habitat for wild animals, including black bears, beavers, bobcats, fishers, mink, porcupines, foxes, and ravens. Also, wild colonies of honey bees.

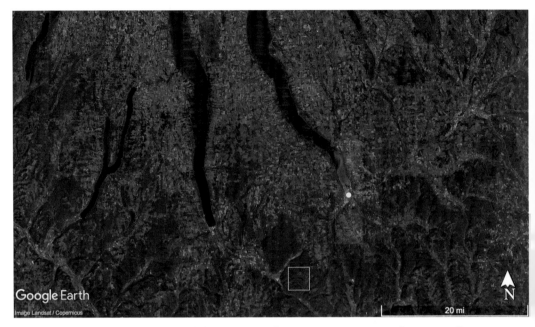

Fig. 2.1. Aerial photo of the Finger Lakes region in New York State. Yellow dot: the city of Ithaca. Yellow square: the Arnot Forest. Scale bar: 32 kilometers (20 miles).

The climate of the Finger Lakes region resembles that of northern Europe. Summers are short, hot, and humid, with temperatures rarely exceeding 32°C (90°F). Winters are long, cold, and snowy. The temperature often drops to −18°C (0°F) or lower, and the annual snowfall total averages more than 150 centimeters (5 feet). If a colony of honey bees is to thrive in this part of the world, it must be able to cope with dramatic changes in the weather across the seasons.

ECOLOGICAL HISTORY OF THE FORESTS AROUND ITHACA

The earliest inhabitants of the lands around the Finger Lakes were Paleo-Indian hunters. Carbon dating the charred remains of their campfires has shown that these seminomadic hunters arrived soon after the last glaciers disappeared, some 13,000 years ago, and were present here until about 4,000 years ago. These hunter-gatherers were followed by agricultural Native Americans who lived in villages of bark-covered longhouses and tended fields in the rich soils of the land between the lakes. They grew maize, squash, and beans—the famous "three sisters" of Native American agriculture—along with the native tobacco species *Nicotiana rustica*. They also hunted game (including deer, turkey, and passenger pigeons), fished for eels and salmon, gathered acorns and berries, and produced ceramic cooking pots. Their lifestyle lasted intact from approximately 1000 BCE to the early 1600s, when Europeans from France, England, and the Netherlands started intruding upon it. By then the Native Americans living in the Finger Lakes region were called the Iroquois (or Haudenosaunee, "people of the longhouse").

After the American Revolutionary War (1775–1783), the state of New York gained title to all the Iroquois land, except several small reservations, and in the late 1790s settlers from the states perched along the Atlantic coast—mainly New York, Pennsylvania, New Jersey, Massachusetts, and Connecticut—began moving in. The wealthier settlers built their farms on the gently sloping, verdant lands between the Finger Lakes, where the Iroquois people had previously cleared large fields, but the poorer folks settled in the wooded, hilly lands south of Ithaca, where the land was sold for bargain prices or was rented to tenant farmers. The hill farmers cleared the virgin forest, grew potatoes, various grains (wheat, oats, and barley), and fruits (especially apples), and raised sheep for wool. In the 1840s and 1850s, many of the people living on the poorest lands were Irish immigrants who had escaped the Great Famine, caused by the potato blight, and their presence is recorded in the place names found in these hills. For example, in the Arnot Forest—the rugged, 1,700-hectare (4,200-acre) forest preserve where I have conducted many of my studies of wild honey bees—the most prominent hill is called Irish Hill, and the rocky dirt road that winds up from the village of Cayuta in Pony Hollow to the abandoned hill farms atop Irish Hill is the McClary Road. The 1860s census records show that many of the settlers of Irish Hill—including William Hetherington, Abram and Azara Sealy, and Mary Pearson—were born in Ireland. By the 1870s, however, farmers working the low-grade soils in the hills south of Ithaca were deserting their farms in rapidly growing numbers, and most were gone by the 1920s. Consequently, the vast deforestation of these hills that occurred in 1800–1860 was followed by reversion to post-agricultural, second-growth forests starting in the 1870s.

Today, most of the abandoned fields and pastures in the hills south of Ithaca have grown back to closed-canopy woods that are dotted with the stone foundations of long-gone houses and barns, stone-lined wells, and wagon-size stone piles marking the boundaries of what were once plowed fields. These forestlands also shade many abandoned cemeteries, such as the small one along Irish Hill Road in the Arnot Forest. Of the 20 or so graves here, most are detectable only as coffin-shaped depressions in the soil; a few others are marked with unlettered rocks standing on edge, and just six have a costly granite gravestone bearing names and dates. The years of death chiseled into them are 1860, 1862, 1864, 1871, 1881, and 1884. It seems that the population boom on Irish Hill was over by the 1880s, whereupon the forests began their return.

The growth rings of the trees covering the hills south of Ithaca also provide us with a clear record of the rewilding of this region over the last 130 or so years. When I moved back to Ithaca in 1986, my wife, Robin, and I began buying up 40 hectares (100 acres) of forested land on a hillside southeast of Ithaca, in a corner of Ellis Hollow, along Hurd Road. It is only a few miles from where I grew up. Much of what is now our forest was once the farm of the Hurd family. They lived on it for two generations, starting in the early 1800s, when Asa Hurd came with Peleg Ellis to settle the hollow, and ending in 1883, when Asa Hurd's son, Wesley Hurd, sold the land shortly before he died at age 82. For the next several decades, a series of owners cut hay and raised stock on the farm's fields, but none lived on the farm, and bit by bit the house, the barn, and the fields were abandoned. By the 1930s, all the land was growing back to a species-rich, mostly hardwood forest with some widely spaced eastern white pines (*Pinus strobus*) towering over all. There are also dark stands of eastern hemlock trees (*Tsuga canadensis*) shading the steep, north-facing slopes, for here the soil stays cool and damp.

Like most of the hilly woodlands south of Ithaca, ours contains diverse native hardwood tree species. These include red, white, and chestnut oak

(*Quercus* spp.); sugar, red, striped, and boxelder maple (*Acer* spp.); white and green ash (*Fraxinus* spp.); shagbark, bitternut, and pignut hickory (*Carya* spp.); yellow, black, and white birch (*Betula* spp.); black and pin cherry (*Prunus* spp.); butternut and black walnut (*Juglans* spp.); American beech (*Fagus grandifolia*); American basswood (*Tilia americana*); tulip tree (*Liriodendron tulipifera*); sassafras (*Sassafras albidum*); cucumber magnolia (*Magnolia acuminata*); American hornbeam (*Carpinus caroliniana*); eastern hophornbeam (*Ostrya virginiana*); and even a few American chestnut (*Castanea dentata*) trees. Not surprisingly, there are also old apple (*Malus pumila*) and pear (*Pyrus communis*) trees around the site of the Hurd family's house, which remains clearly marked by its stone-walled cellar hole.

When I cleared the site for our house on this land in the winter of 1988, I felled several white oaks (*Quercus alba*) some 80 centimeters (32 inches) in diameter at breast height and studied their growth rings. I learned that these oaks were 100 to 110 years old. I also learned that for the first 50 years of their lives these trees had grown rapidly, quickly reaching diameters of about 56 centimeters (22 inches), but that for the rest of their lives they had grown more slowly, so that over the last 50 to 60 years their diameters had enlarged by only another 25 centimeters (10 inches). I inferred from the growth rings of these oaks, and from the stubs of rusty barbed wire poking out of large eastern hemlock and sugar maple (*Acer saccharum*) trees marking a nearby property boundary, that these white oaks had sprung up in an abandoned pasture in the 1880s and, with little competition for light, had continued growing rapidly until the 1930s, and then had grown more slowly as the forest canopy closed in. Today our forest, which is typical for the area, is dominated by large trees ranging in age up to 140 or so years. These trees are large enough to provide homes for the various animals that need good-size nesting cavities, including raccoons, pileated woodpeckers, barred owls, and honey bees. Indeed, in August 2016, my wife, Robin Hadlock Seeley, heard the roar of a swarm of bees flying over the treetops and succeeded in following it to a red maple (*Acer rubrum*) tree, where she found the bees flying in and out a dark knothole, the entrance to their new home (Fig. 2.2).

Fig. 2.2. Robin's bee tree. The entrance to the nest cavity in this red maple (*Acer rubrum*) tree is the dark knothole near top of photo. The entrance is 5.9 meters (19.4 feet) up from the base of the tree.

What plants provide the forage for the honey bees living in these woodlands? Among the trees growing in the forests, the maple, cherry, basswood, tulip, cucumber magnolia, and chestnut trees are all rich sources of nectar or pollen or both. Likewise, various shrubs and herbaceous plants, found in the forest understory or in sunny places along streams and marshy areas, provide the bees with excellent food. The shrubs include common alder (*Alnus incana*), pussy willow (*Salix discolor*), staghorn sumac (*Rhus typhina*), spicebush (*Lindera benzoin*), serviceberry (*Amelanchier* spp.), hawthorns (*Crataegus* spp.), and northern shrub honeysuckle (*Lonicera canadensis*). Among the herbaceous plants growing in sunny spots in these woods, the most important as food sources for bees are the brambles (*Rubus* spp.), goldenrods (*Solidago* spp.), and asters (*Aster* spp.). And because honey bee foragers can fly to food sources 10 kilometers (6 miles) or more from their homes (discussed further in chapter 8), the colonies living up in the forested hills can also exploit the food-rich flowers growing in the farms, gardens, roadways, and waste areas down in the valleys. These sites include apple orchards, stands of black locust trees (*Robinia pseudoacacia*), fields of buckwheat (*Fagopyrum esculentum*), hayfields seeded with white and sweet clovers (*Trifolium repens* and *Melilotus alba*) and alfalfa (*Medicago sativa*), and boggy places that contain many excellent sources of pollen and nectar, including cattails (*Typha latifolia*), jewelweed (*Impatiens capensis*), and purple loosestrife (*Lythrum salicaria*). In the flower gardens and roadsides, the bees find many native and introduced plants that provide rich forage. Those found most commonly are crocuses (*Crocus vernus*), dandelions (*Taraxacum officinale*), chicory (*Cichorium intybus*), milkweeds (e.g., *Asclepias syriaca*), Japanese knotweed (*Fallopia japonica*), and various herbs, including catnip (*Nepeta cataria*), borage (*Borago officinalis*), and mints (*Mentha* spp.).

HOW ABUNDANT ARE WILD HONEY BEE COLONIES IN THE FORESTS AROUND ITHACA AND BEYOND?

It has long been known that honey bees were introduced to North America in the mid-1600s, and that wild colonies of honey bees were thriving throughout the forests east of the Mississippi River by the late 1700s. It was not until the 1970s, however, that we began to get reliable information about the abundance of honey bee colonies living in the wild anywhere across this whole continent. Since then, the question of the density (and thus the spacing) of honey bee colonies living in natural areas has been investigated in multiple sites in North America, Europe, and Australia. These explorations have been undertaken to enable us to better understand this important element of the natural history of *Apis mellifera*.

I began wondering about the abundance of wild colonies back in the 1970s, when I started searching for whatever information was available about how honey bees live when they inhabit natural cavities rather than beekeepers' hives. My richest find was a little book with the intriguing title *Survey of a Thousand Years of Beekeeping in Russia* (1971), written by Dorothy Galton, a scholar in the School of Slavonic and East European Studies of the University College London. In it, Galton describes the activities of tree beekeepers (*bortniki*) in Russian forests in the 1100s to the 1600s. This was when Russian princes owned vast bee forests in which the largest trees were protected by law to provide homes for honey bees. The people living in these forests served their lords (to whom they belonged) by gouging out nesting cavities for the bees high in the biggest trees, fitting them with sturdy doors, and periodically inspecting them to see which ones were occupied. In late summer, they would return to the trees with bees, climb each one while carrying wooden pails, open the access door to expose the bees' nest, and remove some of the honeycombs, leaving enough to sustain the colony over winter (Fig. 2.3). The honey and wax were separated by crushing the honeycombs in water. The honey water was turned into mead, and the wax bits floating in the honey water were skimmed off and purified to supply the churches and monasteries in Russia and the Byzantine Empire with beeswax candles.

Galton explains that the several hundred tons of beeswax exports each year were hugely important to the economy of medieval Russia, so Russian tree beekeeping was a well-organized activity. She also provides information on the abundance of colonies in the Russian bee forests by citing records from the late 1600s for the Morozov estate near the city of Nizhny

FIG. 2.3. Bashkir tree beekeeper extracting honey from a colony nesting in a manmade tree hollow. Photo taken in the South Ural region of the republic of Bashkortostan, Russian Federation.

Novgorod, which is east of Moscow. This grand estate possessed four bee forests that ranged in size from 10 to 88 square kilometers (4 to 28 square miles) and that contained from 3 to 50 trees with nesting cavities occupied by honey bees. The average density of *known* colonies in these four forests was 0.5 colonies per square kilometer (1.3 colonies per square mile). I wondered, however: Might the total density of colonies residing in these forests have been higher than these records indicate? Might there have been many cryptic colonies living in natural dens high in the trees that the *bortniki* had failed to find?

Galton's estimate of the density of honey bee colonies in the medieval Russian forests—*at least* one per square mile—made sense to me, because over the summers of 1975, 1976, and 1977, I had studied the nest-site preferences of honey bees and from this work had acquired evidence that the density of wild colonies in the woods near Ithaca was at least one per square mile. My study of the bees' housing preferences involved setting up pairs of experimental nest boxes that differed in just one property, such as cavity size or entrance size, and then seeing which box in a pair was occupied first by a wild swarm of bees (see the section on nest-site selection in chapter 5). I nailed the boxes in each pair onto trees standing at least 10 meters (33 feet) apart along roads that ran through heavily forested areas south of Ithaca in the towns of Dryden and Caroline. Because I spaced the pairs of nest boxes about a mile apart, the density of my nest-box pairs was approximately one pair per square mile. On average, I caught 0.5 swarms per pair of boxes each summer, which told me that there were enough wild colonies living in the area to generate 0.5 swarms per square mile (0.2 swarms per square kilometer). (To the best of my knowledge, there were few managed colonies in the areas where I conducted this study.) It seemed likely, however, that more than 0.5 swarms had been produced per square mile, because it was unlikely that my nest boxes had attracted *all* the swarms in the places where I had put them. So, I wondered, might there actually be several wild colonies per square mile in the woods around Ithaca? To answer this question, I knew that I must find a large forest and then try to locate all the wild honey bee colonies living within it.

Fortunately, Cornell University owns a sprawling research forest, the 17-square-kilometer (6.6-square-mile) Arnot Forest, which lies only about 25 kilometers (15 miles) southwest of Ithaca (see Fig. 2.1). This forest extends over parts of the towns of Newfield (in Tompkins County) and Cayuta (in Schuyler County) (Fig. 2.4). Moreover, the woodland habitat found in the Arnot Forest doesn't end at its boundary lines. Instead, it continues beyond the Arnot Forest's northern, western, and southern bor-

FIG. 2.4. *Top:* Aerial photo of the Arnot Forest with forest boundaries indicated by yellow lines. Red bars: upper, 1 kilometer (upper); lower, 1 mile. North is up. The complex of buildings in the valley to the west is a sawmill that processes hardwood logs harvested from the vast woodlands in southern New York and northern Pennsylvania. *Bottom:* View of the Arnot Forest, looking southeast from Irish Hill Road. Photo taken in late September, as trees began to show their autumn colors.

ders into the steeply sloped Cliffside and Newfield State Forests and beyond its eastern border into several private forests. Furthermore, on the far side of the narrow valley with good farmland (Pony Hollow) that skirts the northwestern corner of the Arnot Forest lies the rugged, 47-square-kilometer (18-square-mile) Connecticut Hill Wildlife Management Area, a state-owned preserve that is also mostly woodland. Agriculture was abandoned in all these places more than 100 years ago, and today they comprise mostly hardwood forests but also some wetlands and old field habitat. The whole area provides a superb setting for the study of wildlife, including wild honey bees.

In July 1978, a friend and fellow student of the bees, Kirk Visscher, and I began to explore for wild colonies of honey bees living in the Arnot Forest. We did so using the tools and methods of bee hunting (also called "bee lining"), an outdoor pursuit that has been practiced for centuries in Europe and North America. Most bee hunters search for wild colonies to get some honey and have some fun. Our aim was purely scientific—to discover how many wild colonies of honey bees were living in the Arnot Forest—but we had some fun, too.

When Kirk and I began our bee hunting, we were rank novices and had nobody to teach us the craft, but we had come across a first-rate guidebook on the subject that was published in 1949, *The Bee Hunter*. It was written by George H. Edgell, a bee hunter (and professor of architectural history at Harvard University) with decades of experience finding wild colonies of honey bees living in the mountains around his summer home in Newport, New Hampshire. Edgell explains how you start a hunt for a wild colony by going to a good-size clearing (the bigger the better) that is well stocked with flowers attractive to honey bees. Here, you use a small, two-chambered apparatus called a bee box to capture some bees that are foraging on the flowers. Once you have imprisoned a half dozen or so bees in your bee box, you tuck inside it a small square of beeswax comb filled with sugar syrup. The bees in the box will find the syrup-filled comb, tank up on its delicious contents, and make ready to fly home. After giving the bees about five minutes to stumble upon the bait in your bee box, you let them out and watch closely to see in which direction they fly. At this point you wait, anxiously, for some of them to return to your little feeding station. Usually some do, and if the plants in bloom at the time are providing only meager forage, then your first customers will be excited by your syrup-filled comb and will recruit hive mates to help them exploit its treasure. After an hour or so, the bees will be familiar with the flight route between your feeder and their home, and many will fly a direct course—a bee-line—back to their nest. At this point, you determine the direction to their home by measuring their vanishing bearings with a magnetic compass, and you estimate the distance to their home by measuring the round-trip times of a half dozen bees that you have labeled with paint marks. If your bees need only two to three minutes to fly home, unload, and fly back to you, then you know that their nest is only about 100 meters (330 feet) away, but if they are gone for six or seven minutes, then it is probably about 1 kilometer (0.6 miles) away. Now you will want to move your whole operation down the beeline. To do so, you trap in your bee box as many of the bees as possible, then you move your gear 100–200 meters (ca. 300–600 feet) down the beeline to another clearing, and here you release your bees. Now you again note the bees' vanishing bearings, to check that you are moving in the right direction, and you again record their round-trip times, to update your estimate of the distance. By patiently making a series of moves down the beeline, you will find your way to the stand of trees in which the bees reside, then to the one tree that is their dwelling place, and ultimately to the knothole or fissure that is the entryway to their home. Discovering it is always a huge thrill!

Kirk and I began our survey of the wild colonies in the Arnot Forest by driving to a small clearing near its center and searching there for flowers being visited by honey bees. We had difficulty finding even one honey bee in this spot, but eventually Kirk spied a bee on a multiflora rose (*Rosa multiflora*) in full bloom, and he managed to capture her in his bee box. He then slid into the bee box, without releasing the bee, a small square of beeswax comb filled with anise-scented sugar syrup. The bee loaded up on this bait and upon release flew off to the east, which revealed to us the

Fig. 2.5. Crowd of forager bees recruited to help exploit a square of comb filled with sugar syrup, at the start of a hunt for a wild colony's home.

general direction to her home. Nine minutes and 20 seconds later, she arrived back at our feeding station and landed on the comb for a refill. As she calmly drank in more of our syrup, Kirk applied a dot of green paint to her abdomen, so we could identify her. Green-abdomen became a steady visitor, and after an hour or so she had recruited several dozen nest mates to help exploit the amazing food source we were offering (Fig. 2.5). Meanwhile, Kirk and I had labeled about 10 of the recruits and had measured the bearings of their beelines home: nearly due east (see the red line in Fig. 2.6). We then began moving our feeding station, together with the bees, in a series of steps down their flight path home. Each move took more than an hour and brought us only about 100–200 meters (330–660 feet) closer to the bees' residence, but we were determined to find their bee-tree home, so we persisted. By carefully noting at each stopping place the vanishing bearings of the departing bees, and moving ahead in this direction, we managed to zero in on the entryway of their residence: a knothole about 6 meters (20 feet) up in an eastern hemlock tree some 800 meters (0.5 miles) east of our starting point.

Because we had difficulty finding honey bees on flowers in the Arnot Forest in early July, Kirk and I postponed further bee hunting until late August. We knew that by then the endless stands of goldenrod plants lining the roadsides and filling the clearings in the forest would be in bloom, attracting droves of foraging bees. This should make it easy to find the honey bee foragers needed to establish beelines leading to the other colonies living in this forest.

When I returned to the Arnot Forest on 26 August 1978, to begin several weeks of intense bee hunting, I did indeed find seas of goldenrod plants (mostly *Solidago canadensis*) in bloom, and they were alive with honey bees bobbing on their brilliant yellow inflorescences. Beautiful! Figuring that I could not survey the whole Arnot Forest in the three weeks I had available, I decided to focus my search on the forest's southern and western sectors. I chose these places because there is a narrow, flat-bottomed valley below the forest's southern and western boundaries (see Fig. 2.6), and within it there were abandoned pastures and a derelict railroad track brimming with patches of goldenrod. It was delightfully easy to find foraging honey bees here. Working in this valley, I also found it remarkably easy to get readings of the bees' vanishing bearings, because after they lifted off from my feeding station, laden with heavy payloads of thick syrup, they would fly slowly as they struggled up the steep hillsides en route to their homes high in the woods. Figure 2.6 shows that these homeward flights by the bees guided me to nine more wild colonies, eight within the Arnot Forest and one just outside its western boundary. And one more thing delighted me about this bee hunting: it hadn't led to any beekeepers' colonies! It was becoming clear that there were only wild colonies living in and around the Arnot Forest.

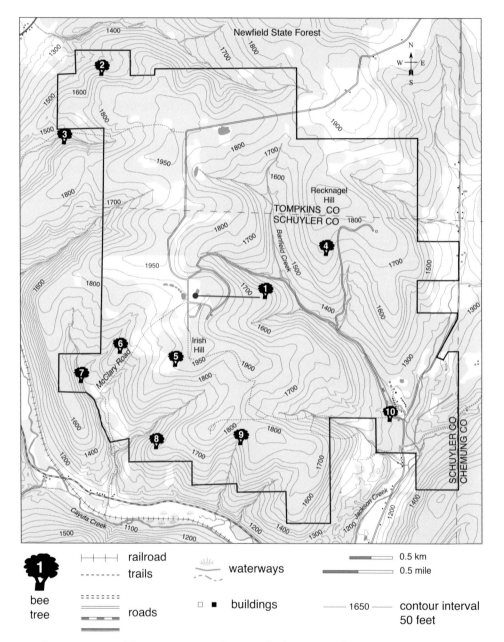

FIG. 2.6. Map of the Arnot Forest showing the locations of the 10 bee trees found there in 1978. The site of each bee tree is marked by the base of a bee-tree symbol. Red line denotes the path of the author's first bee hunt.

I knew, of course, that the nine bee-tree colonies that Kirk and I found in this forest were not all the wild colonies living in it. After all, no bee-lines had been followed from flower patches in the northern and eastern regions of the forest, so about half of the Arnot Forest was terra incognita. Moreover, I could not even be confident that I had located all the colonies in the southern and western parts of the forest. I concluded, therefore, that the nine colonies that had been found were *at most* about one half of all the colonies residing in this forest; hence there were 18 or more colonies living in this 17-square-kilometer (6.6-square-mile) forest. So, I figured that the density of wild colonies living in the Arnot Forest in September 1978 was at least one colony per square kilometer, hence 2.5 or more per square mile.

HOW ABUNDANT ARE WILD COLONIES OF HONEY BEES ELSEWHERE?

Building on the 1978 study of the density of honey bee colonies living within the Arnot Forest, other biologists have investigated this matter at various sites in North America, Europe, and Australia. The first of these additional studies was led by Roger A. Morse, the entomology professor at Cornell University who generously let me start working in his honey bee laboratory when I was still a high school student back in 1969. He and a team of seven graduate students conducted their study in the spring of 1990, in the small port city of Oswego, on Lake Ontario in northern New York State. Their investigation was triggered by the discovery of a colony of Africanized honey bees—a hybrid between European subspecies and the African subspecies *A. m. scutellata* (discussed below)—nesting in a shipment of pipes from Brazil. The presence of these exotic honey bees raised concerns that Africanized bees, and the fearsome ectoparasitic mite (*Varroa destructor*) that these bees could carry, might have been introduced to North America, so attempts were made to locate all the honey bee colonies living near the port so they could be checked for Africanized bees and *Varroa* mites. Newspaper and radio advertisements were run offering a $35 reward for information on honey bee colonies living in the semicircular area within 1.6 kilometers (1 mile) of the port. Eleven wild colonies living in trees and buildings, and one managed colony residing in a backyard beehive, were found. This work revealed that in this small city, the density of the wild colonies was 2.7 colonies per square kilometer (7 colonies per square mile), much higher than what Kirk and I had found in the woods of the Arnot Forest. Fortunately, no Africanized honey bees or *Varroa destructor* mites were found.

A still higher density of wild colonies was found in a remarkable study conducted by a team of biologists led by M. Alice Pinto at Texas A&M University in 1991–2001. This group worked in the Welder Wildlife Refuge, a 31.2-square-kilometer (12.2-square-mile) nature preserve in southern Texas. Their aim was to track the "Africanization" of a population of wild honey bees living in the southern United States, and they did so by sampling the colonies living in this wildlife refuge before, during, and after the arrival of Africanized honey bees from Mexico. Africanized honey bees are derived from a founder population of an African subspecies, *A. m. scutellata*, that was introduced to Brazil from South Africa in 1956. The purpose of this introduction was to crossbreed a tropical-evolved African subspecies with several temperate-evolved European subspecies already in Brazil to create a honey bee well suited to tropical conditions. However, several colonies of *A. m. scutellata* escaped from the quarantine apiary, thrived in the Brazilian climate, and spawned strong populations of wild colonies of this subspecies throughout the American tropics.

The vegetation in the Welder Wildlife Refuge is a mix of open grassland, chaparral brushland, scattered mesquite trees (*Prosopis* spp.), and groves of live oaks (*Quercus virginiana*) (Fig. 2.7). Several times a year, for 11 years straight, a team of biologists from Texas A&M University searched a 6.25-square-kilometer (2.4-square-mile) study area within the refuge for wild colonies of honey bees, and they collected samples of worker bees from each colony they found. Nesting cavities were abundant in the woodland areas; nearly all (85%) of the colonies were found in cavities in oak trees. When the mitochondrial DNA of these bees was analyzed to determine their maternal ancestry, it became clear that for the first three years

FIG. 2.7. Researcher in a grove of live oaks (*Quercus virginiana*) within the Welder Wildlife Refuge. He has just sampled workers from the colony of Africanized honey bees nesting inside the tree directly behind him. The nest's entrance is visible just above the top of the insect net.

of the study (1991–1993) the queen bees living in the study area had been mainly descendants of several European subspecies: 68 percent *A. m. ligustica* and *A. m. carnica* (both from southern Europe), 26 percent *A. m. mellifera* (from northern Europe), and 6 percent *A. m. lamarckii* (from northern Africa). Over the next several years, however, the queen bees living here became primarily descendants of the southern African subspecies, *A. m. scutellata*. And what did the surveys of the colonies living in the study area reveal about the colony density during the first four years, when the population was dominated by European honey bee colonies? They showed that the density of the wild colonies in this mixture of grassland, brushland, and woodland habitats was remarkably high: 9–10 colonies per square kilometer (ca. 24 colonies per square mile)!

In Europe, where *Apis mellifera* is a native species, three teams of researchers have investigated the abundance of wild colonies. One team in Poland, led by Andrzej Oleksa at the Kazimierz Wielki University in Bydgoszcz, studied a population of wild colonies living in the lowlands of northern Poland, just south of the Baltic Sea. The landscape here is dominated (68%) by agricultural areas—cultivated fields, meadows, and orchards—and the rest is covered mostly (27%) by forests. The population of honey bees living in this region of Poland still consists primarily of *A. m. mellifera*, the native dark honey bee of northern Europe.

Andrzej Oleksa and his colleagues focused on assessing the occurrence of wild colonies in rural avenues—linear stands of old trees along countryside roads, an example of which is shown in Figure 2.8. These researchers inspected 15,115 large trees in 201 avenues that were carefully chosen to provide uniform coverage of their 15,000-square-kilometer (6,000-square-mile) study area. In total, they searched along 142 kilometers (88 miles) of avenue fragments and found 45 colonies of honey bees, which indicated a density of 0.32 colonies per kilometer (0.51 colonies per mile) of avenue. Knowing the density of avenues on the landscape, they estimated the overall density of wild colonies living in avenue trees at 0.10 colonies per square kilometer (0.26 colonies per square mile). These researchers point out that their estimate is surely an underestimate of the total abundance of wild colonies, because it did not take account of wild colonies present in the woodlands, which cover 27 percent of their study area. They also note that some colonies nesting high in the avenue trees could have been overlooked. Nevertheless, their estimate of wild-colony density is valuable, for it reveals that rural avenues are serving as a refuge for wild colonies. Moreover, it shows that wild colonies of honey bees still exist in a place where the natural environment has been largely replaced with agriculture and where beekeeping is extremely popular. Beekeepers maintain some 4.4 colonies per square kilometer (11.4 per square mile) in the region of Poland where this study was conducted.

Just to the west of Poland, in Germany, Robin Moritz and colleagues at the University of Halle have investigated the abundance of honey bee colo-

FIG. 2.8. Rural avenue in northern Poland lined with a mixture of European ash (*Fraxinus excelsior*) and Norway maple (*Acer platanoides*) trees.

nies living in natural areas. This group worked in three widely spaced sites spread north to south within Germany. Two sites were in national parks: the 318-square-kilometer (123-square-mile) Müritz National Park, in the Müritz lake region north of Berlin, and the 25-square-kilometer (10-square-mile) Harz National Park, in the forested Harz Mountains in central Germany. The third site was a rural spot in Bavaria (southern Germany) west of Munich. These researchers did not adopt the direct, brute-force approach of exhaustively searching for the colonies living in their three study sites. They took instead an indirect approach based on genetic analyses. Specifically, they had approximately 10 virgin queen honey bees conduct their mating flights within each study site, and then they analyzed the genes of each queen's worker offspring to determine how many colonies had produced the drones that mated with each queen. The beauty of this approach is that it made use of the astonishing promiscuity of queen honey bees to get a strong sampling of the drones in the study area; each queen mates with 10–15 males, so the worker offspring of the 10 queens provided a sampling of 100–150 drones from the colonies living in the area where these queens mated. The genetic analysis revealed that there were 24–32 colonies producing drones in the three locations. If we assume that both queens and drones fly 900 meters (0.56 miles), on average, to reach a mating site, then the area over which these 24–32 colonies were dispersed was that of a circle with a radius of 1.8 kilometers (1.1 miles) and an area of 10.2 square kilometers (3.9 square miles). Therefore, the estimated average density of colonies for the three study locations was 2.4 to 3.2 colonies per square kilometer (6.2 to 8.2 colonies per square mile). It must be noted, however, that in Germany beekeepers sometimes can (and do) place their colonies inside the country's national parks, so the estimates of colony density reported in this study represent the combined densities of managed and wild colonies.

Recently, another team of biologists in Germany, Patrick L. Kohl and Benjamin Rutschmann, graduate students at the University of Würzburg, have made surveys of wild colonies in largely undisturbed European beech (*Fagus sylvatica*) forests in central and southwestern Germany: the 160-square-kilometer (62-square-mile) Hainrich forest in Thuringia, and several forest clusters in the 850-square-kilometer (328-square-mile) biosphere reserve in the Swabian Alb mountain range in Baden-Württemberg. The Hainrich forest site is off-limits to beekeepers seeking locations for their hives. In this forest, the researchers used the same method that Kirk Visscher and I used in the Arnot Forest (beelining), but in the forest clusters in the Swabian Alb region they inspected 98 known nest cavities of the black woodpecker (*Dryocopus martius*). This is the largest woodpecker in northern Europe, and it excavates nest cavities that are spacious enough (20-plus liters/5.3-plus gallons) to be suitable nesting sites for honey bees. In the Hainrich forest, the team found or inferred the locations of nine wild colonies, from which it calculated an estimate of the wild-colony density here: 0.13 colonies per square kilometer (0.34 colonies per square mile). In the forests in the Swabian Alb Biosphere Reserve, they inspected 98 beech trees with old woodpecker nest cavities and found seven occupied by honey bees. Knowing the density of the woodpecker-cavity trees in the region, they calculated an estimate of the wild-colony density in the Swabian Alb forest patches of at least 0.11 colonies per square kilometer (0.28 colonies per square mile). Of course, both figures are minimum estimates of the actual colony density, because the investigators could not be certain they had found all the colonies living within their search areas.

Biologists in Australia, where European honey bees have lived as an introduced species since 1822, have also used national parks to measure the density of honey bee colonies living in undisturbed habitats. This team, led by Benjamin Oldroyd of the University of Sydney, worked in the 755-square-kilometer (292-square-mile) Barrington Tops National Park, the 83-square-kilometer (32-square-mile) Weddin Mountains National Park, and the immense, 3,570-square-kilometer (1,378-square-mile) Wyperfeld National Park. All three parks are in southeastern Australia, but their vegetation types range from subtropical rain forest to semiarid *Eucalyptus* woodland. Like the German investigators, the Australian biologists used an indirect, genetic approach that produced estimates of the density of the wild colonies at their three study sites. Here again, the goal was to determine how many colonies produced the drones present in a given location. The Australian researchers, however, estimated this based on the genetics of drones captured in drone traps suspended from helium-filled weather balloons in drone congregation areas (see chapter 7), rather than the genetics of workers sired by drones residing within a study area. They captured drones in two drone congregation areas in each national park, and from the genetics of these bees they calculated estimates of colony density that ranged from 0.4 to 1.5 colonies per square kilometer (1.0 to 3.9 colonies per square mile). Beekeeping is banned within the large national parks used by these Australian investigators, so their results are not influenced by the presence of managed colonies.

Fig. 2.9. *Left:* Eastern hemlock (*Tsuga canadensis*) bee tree in the Shindagin Hollow State Forest. *Right:* Close-up of the nest entrance. Photo taken in late November, when the colony was quiet inside its nest.

TABLE 2.1. Estimates of the density of wild colonies of European honey bees

Study location	Colonies/sq. km	Colonies/sq. mile
Arnot Forest, NY (USA)	1.0+	2.5+
Oswego, NY (USA)	2.7	6.9
Welder Wildlife Refuge, TX (USA)	9.3–10.1	23.8–25.8
Shindagin Hollow State Forest (USA)	1.0	2.5
Agricultural land (Poland)	0.1+	0.26+
Two national parks and a forest (Germany)	2.4 - 3.2	6.2–8.2
Hainrich Forest and Swabian Alb (Germany)	0.1+	0.3+
Three national parks (Australia)	0.4–1.5	1.0–3.9

The most recent study of the density of wild colonies in undisturbed forest habitat comes from a survey conducted in 2017 by Robin Radcliffe, who thoroughly searched—by means of bee hunting—a 5-square-kilometer (1.9-square-mile) portion of the Shindagin Hollow State Forest in New York State. This is a 21-square-kilometer (8-square-mile) area of old-growth forest southeast of Ithaca and about 30 kilometers (20 miles) east of the Arnot Forest (Fig. 2.9). Robin located five colonies, which yields an estimate of one colony per square kilometer (2.5 colonies per square mile), virtually identical to what Kirk Visscher and I found in the Arnot Forest in 1978 and what I have found there in later surveys conducted in 2002 (see below, this chapter) and in 2011 (to be discussed in chapter 10).

In review: What have we learned about the abundance of wild colonies of European honey bees? We see in Table 2.1, which summarizes the findings of the studies just described, that the density of colonies living in natural areas varies greatly, but that generally it is estimated to be somewhere in the range of one to three colonies per square kilometer (2.6–7.8 colonies per square mile). Evidently, the density of wild colonies is rather low, which means that on average the honey bees' nests are widely spaced. In the Arnot Forest, for example, the average distance between each colony and its nearest known neighbor is 0.87 kilometers (0.54 miles). Clearly, wild colonies living in natural settings are generally spaced far more widely than managed colonies living in apiaries. We will see in chapter 10 that this difference in colony spacing can have enormous consequences for the health of the bees.

HAVE THE WILD COLONIES BEEN KILLED OFF BY *VARROA DESTRUCTOR*?

Varroa destructor is a small but dangerous mite that parasitizes honey bees. A full-grown female of this species is no bigger than the head of a pin, so it is just a small fry compared to a worker bee (Fig. 2.10). Nevertheless, a heavy infestation of these tiny, reddish-brown, disk-shaped creatures can kill a colony of honey bees. What makes these mites so deadly is that they feed on the fat body (energy storage) tissue in immature bees (larvae and pupae), and this feeding spreads viruses that can weaken the developing bees and, still worse, can induce crippling physical defects such as shrunken abdomens and shriveled wings. The latter is symptomatic of high levels of the deformed wing virus, probably the most damaging of all the pathogens for which these mites are vectors. The deadliness of *V. destructor* is amplified by the rapid development of individuals; the mites grow from egg to adult in less than a week. This means that even if the population of *Varroa* mites infesting a colony starts out small in spring, it can, through the magic of exponential growth, explode by late summer. And when a colony has a high mite load, the worker bees it produces start their lives with such high virus titers that they are too sick to work. Eventually, the colony's population collapses for lack of sufficient gains of vigorous young bees to offset the losses of older bees dying from predation, accidents they suffer while working outside the hive, and the accelerated senescence that comes from being parasitized by *Varroa* mites.

Varroa mites originally lived only in the mainland regions of eastern Asia. There they still have a stable host-parasite relationship with their original host, the eastern hive honey bee, *Apis cerana*, whose range is all of Asia east of the vast deserts in Iran and south of the towering mountains of central Asia. Unfortunately, these mites made a successful host shift from *Apis cerana* to the western hive honey bee, *Apis mellifera*, and since

FIG. 2.10. Adult female of the mite *Varroa destructor* on a worker honey bee.

doing so they have spread, and today their distribution is nearly worldwide, like that of their new host, *A. mellifera*. This host shift occurred sometime in the early 1900s, after beekeepers had moved colonies of *A. mellifera* from western Russia and Ukraine to Primorsky, the easternmost region of Russia. This region is far outside the native range of *A. mellifera* but is within that of *A. cerana*. Once the range of *A. mellifera* overlapped that of *A. cerana*, the mite began infesting colonies of *A. mellifera*. This perhaps began when colonies of *Apis mellifera* robbed honey from colonies of *Apis cerana*, but it may have been unwittingly fostered when beekeepers tried to strengthen their colonies of *A. mellifera* by giving them brood (larvae and pupae) from colonies of *A. cerana*. Russian beekeepers then spread *V. destructor* to Europe in the 1950s or 1960s when they shipped queens of *A. mellifera*, carrying the mites, from the Primorsky region to western parts of the Soviet Union. *Varroa* mites were first reported in *A. mellifera* colonies living in Europe in

1967 in Bulgaria, in 1971 in Germany, and in 1975 in Romania. These mites also spread to northern Africa when hundreds of honey bee colonies were sent from Romania and Bulgaria to Tunisia and Libya in 1975 and 1976, as parts of foreign aid programs. Evidently, *V. destructor* reached South America when Japanese beekeepers moved colonies of *A. mellifera* carrying *V. destructor* (acquired in Japan) into Paraguay in 1971. Somehow these mites also reached Brazil in 1972.

Varroa destructor entered North America relatively recently and via two routes. The first was probably into Florida, where it is likely that in the mid-1980s either a beekeeper smuggled in some Brazilian queen bees that were carrying the mites or a swarm of Africanized honey bees (infested with *Varroa* mites) came off a cargo ship. Records from the Bureau of Plant and Apiary Inspection for the state of Florida show that between 1983 and 1989 swarms of Africanized honey bees were found on eight ships—usually in shipping containers—that had arrived from Central or South America. These records indicate that in at least one ship, the "bees were varroa infested." The second pathway was in Texas, where in the early 1990s swarms of Africanized honey bees infested with *Varroa* mites flew north across the U.S.–Mexico border.

My first encounter with a *Varroa* mite came in June 1994 at my laboratory in Ithaca. I remember the moment vividly: I was labeling young worker bees with paint marks in preparation for an experiment when, to my amazement, I spied a *Varroa* mite scuttling across the thorax of the bee I was about to label with a dot of yellow paint. Sighting this mite was dismaying, for I knew that *V. destructor* is a deadly parasite, but I hoped that it marked just the arrival of these mites at my laboratory. In late August, however, I observed a chilling scene that made it clear the mites had arrived well before 1994 and that my laboratory's colonies were now heavily infested: dozens and dozens of worker bees, many with shriveled wings, were crawling feebly through the grass in front of my hives. Clearly, many of the worker bees in my colonies were ill, but still I hoped for the best. This seemed like a realistic hope at the time, for when I inspected the 19 colonies in my laboratory's apiary a month later, in September 1994, I found them all well populated with worker bees and possessing large honey stores, which suggested that all were in good shape for surviving the coming winter. Several months later, though, I learned how badly *Varroa* mites can rob honey bees of their health. By April 1995, only two of the 19 colonies remained alive. The experience of 89 percent colony mortality over the winter of 1994–1995 was a chilling lesson on the fearful virulence of the *Varroa* mite.

There seemed only one way to help the bees: treat them with miticides. I started doing so in the summer of 1995, using fluvalinate (Apistan), and now, some 20 years later, my students and I treat many of the colonies that we keep *in our apiaries* with mite-killing chemicals, usually formic acid or thymol-based medications. We do so because for many of our experiments we need colonies that are not stressed by heavy infestations of *Varroa destructor*.

Back in the mid-1990s, when my students and I were learning how to control the *Varroa* mites infesting the managed colonies living in our apiaries, I was also fretting about what was happening to the wild colonies living in the woods around Ithaca. Certainly, nobody was treating them with the life-saving miticides, so I wondered: Were they dying out? Or were they already all gone? That they had been wiped out seemed possible, maybe even likely. And even if a few were still alive, I feared that any such survivor colonies might be merely the issue of a beekeeper's miticide-treated colony that had cast a swarm. If so, then those still alive were probably just short-lived colonies doomed by the mites spreading the deadly viruses.

My concern for the wild colonies was deepened by three things that suggested that this population of colonies had indeed been killed off. First, starting in the mid-1990s, I had difficulty finding honey bees on the dandelion flowers that carpet the lawns and fields around Ithaca in late April and early May. Not good news. Second, also in the mid-1990s, I noticed that I was receiving very few phone calls asking me to collect a swarm that had settled on a tree or building around the Cornell campus, in contrast to the 1980s, when I would get several of these swarm calls each summer. Definitely bad news. And third, in 1995, a paper by two highly reputable honey bee researchers at the University of California at Davis, Bernhard Kraus and Robert E. Page Jr., reported that *Varroa destructor* "has had a devastating effect on the demography of feral bee colonies throughout California." They also suggested that "there is no widespread, general preadaptation to *Varroa* mites in honey bees in California." Truly dreadful news.

My concerns rose still higher when, in June 1997, I read an article in one of the beekeeping magazines, the *American Bee Journal*, written by Dr. Gerald Loper, a staff scientist at the USDA's Honey Bee Research Laboratory in Tucson, Arizona. In the article, Loper reported his findings from a long-term study he was conducting on a population of wild honey bee colonies living in the mountains of the Sonoran Desert north of Tucson. Starting in 1987, he had located 247 nesting sites (mostly rock crevices) that were or had been occupied by wild colonies at one time or another. He knew from genetic analyses that all these colonies were European honey bees; indeed, 68 percent of the colonies had the mitochondrial DNA haplotype of the dark European honey bee, *Apis mellifera mellifera*. Each year, he inspected the sites in early March to assess the colonies' survival over winter, and he checked them again in June to assess the results of their swarming. Where possible, he also collected samples of worker bees from these colonies—by breathing into their nest entrances and then netting 50–150 bees—to inspect them for tracheal mites (*Acarapis woodi*) and *Varroa* mites.

The findings that Gerald Loper reported painted a somber picture: this population of wild European honey bee colonies was decimated by the arrival of both types of mites, but especially *Varroa destructor* (Fig. 2.11). In 1992 and 1993, before *Varroa* arrived in the study area, 120–160 colonies had been living in the 247 nesting sites, but over the period 1994–1996, during which time nearly every colony became infested with *Varroa*, the number of occupied sites plummeted, leaving only 12 colonies still alive in March 1996. This population of wild colonies would perhaps have died out if Africanized honey bees had not begun to infiltrate it, starting in 1995. By the late 1990s, it was starting to grow back, and today it is again thriving. Reading this paper left me with mixed feelings. I was thoroughly

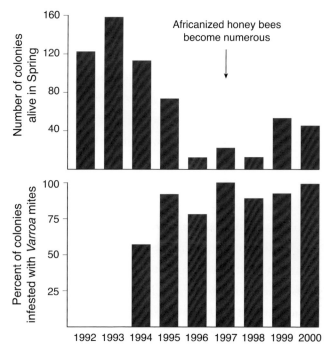

FIG. 2.11. Results of surveys of a population of wild colonies living in the mountains north of Tucson, Arizona. Tracheal mites spread through the area in 1991–1993, and *Varroa* mites arrived in 1993. The population's genetics switched from mainly European to mainly Africanized bees in 1997–1998.

impressed by Loper's long-term study of this population of wild honey bees living in the mountains of southern Arizona, but I was also deeply dismayed by its grim report of a population of wild colonies of European honey bees collapsing soon after the arrival of *Varroa destructor*.

Given Gerald Loper's findings from Arizona, together with what I was seeing firsthand around Ithaca, I believed in the early 2000s that the wild colonies of honey bees probably had vanished from the forests south of Ithaca. And as someone who cannot live without wild things, I mourned their passing. At the same time, however, the inner voice of curiosity kept posing a question: Is it *really the case* that the wild colonies are all gone? My curiosity feeds me many questions, far more than I can tend to, so most of them get left aside, but I could not ignore this one about the wild honey bees, for nearly at my doorstep lay a unique resource from which I could seek a solid answer: the honey bees of Arnot Forest. I realized that this research forest was the one place in all eastern North America for which we had solid, baseline information about the abundance of wild colonies before *Varroa destructor* had set foot in this continent, thanks to the census of the wild colonies living in this forest that Kirk Visscher and I had made back in 1978. If I repeated this work, would I find some survivor wild colonies, or would I confirm their supposed extinction? I knew that I needed to find out, and in 2002 I returned to the Arnot Forest, taking steps to conduct this second census in a manner as close as possible to the first one. One was to make my census in the *same season* as before: from mid-August to late September. Another was to conduct the census in the *same way* as before: using the methods described by George H. Edgell in his charming book, *The Bee Hunter*.

I began the second census on the afternoon of 20 August 2002, in a field of goldenrod (*Solidago canadensis*) just inside the northeastern entrance to the Arnot Forest. This was the same season when I had begun my three weeks of bee hunting for the first census, back on 26 August 1978. The day was sunny and hot, and although no rain had fallen for several weeks, many of the goldenrod plants had unfurled their bright yellow inflorescences, so things looked perfect for me to find bees. But would I find any? I expected the answer would be no. As I climbed from my truck, I figured that I would probably spend the afternoon tromping around looking for foraging honey bees, not find any, and return home with the knowledge that, yes, indeed, the deadly duo of the *Varroa* mite and the deformed wing virus had wiped out the population of wild colonies of the Arnot Forest. For the first 10 minutes, this expectation seemed correct. I found no honey bees, but I did encounter numerous bumble bees (*Bombus* spp.), whose presence told me that, despite the drought, the goldenrod flowers were offering nectar and pollen that was attractive to bees. Then I spotted a honey bee on a shining goldenrod inflorescence (Fig. 2.12)! A few seconds later, she was buzzing furiously inside my bee box. A couple more minutes of searching revealed another honey bee on other goldenrod flowers nearby, and soon I had a

Fig. 2.12. Worker bee collecting nectar and pollen on a goldenrod (*Solidago* sp.) inflorescence.

second prisoner in my bee box. Within an hour, I had spotted, captured, fed, and released six worker honey bees.

My success in finding these bees showed me that there were still honey bees foraging in these woods. But where were they coming from? A bee tree in the Arnot Forest or a beekeeper's hive outside the forest? By the end of the afternoon I knew the answer, for by then I had two rip-roaring lines of bees leaving my syrup-filled comb, one leading north and one leading south. Both lines pointed to locations deep in the Arnot Forest, not places where beekeepers would put hives.

For the next six weeks, I devoted every available hour of every fair-weather day to bee hunting in the Arnot Forest. Classes at Cornell started in late August, and my class on animal behavior met midday on Mondays, Wednesdays, and Fridays, so most weekdays I could hunt for just a few hours in the afternoon. Also, the nights soon started to grow chilly, so some days the foraging bees only began appearing on the flowers late in the morning. In my favor, however, the drought persisted, sunny weather prevailed, and the bees, apparently unable to find flowers brimming with rich nectar, mobbed my feeder comb whenever I baited it with my anise-scented sugar syrup.

All told, I hunted in the forest for 117 hours spread over 27 days, and during this time I started beelines from 12 clearings spread over the western half of the forest (Fig. 2.13). As was the case back in 1978, I did not make a complete, forest-wide survey of the wild colonies living in the Arnot Forest. Nevertheless, I did find eight wild colonies! Each one had taken up residence in a sturdy, live tree: two sugar maples (*Acer saccharum*), two white ash (*Fraxinus americana*), one eastern hemlock (*Tsuga canadensis*), one white pine (*Pinus strobus*), one quaking aspen (*Populus tremuloides*), and one red oak (*Quercus rubra*). To have found eight trees occupied by honey bees delighted me, for it showed that there were about as many wild colonies living in the Arnot Forest in 2002 as there had been back in 1978, when I had found nine wild colonies.

How could this be, given that *Varroa* had been in New York State for most of the previous decade? One possibility is that the honey bees in the Arnot Forest lived in such isolation that they had not been exposed to *Varroa*. The fact that few of my beelines pointed out of the forest (just those pointing west from sites 3, 5, and 9, as shown in Fig. 2.13) showed that there were few, if any, managed colonies living near the Arnot Forest. It was possible, therefore, that the colonies living in this forest had simply not yet been exposed to *Varroa*.

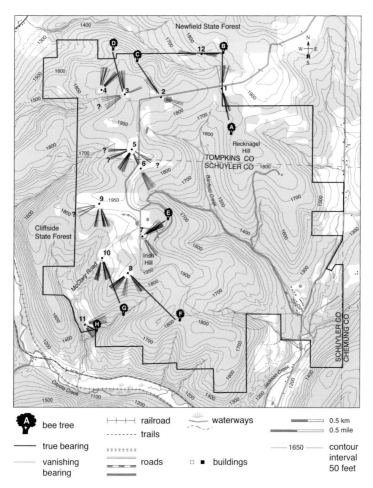

FIG. 2.13. Map of the Arnot Forest showing the locations where beelines were started (1–12) and where eight bee trees were discovered (A–H) during bee hunting in August and September 2002.

ARE THE ARNOT FOREST COLONIES INFESTED WITH *VARROA DESTRUCTOR*?

To see whether the wild colonies in the Arnot Forest were or were not infested with *Varroa* mites, I needed to induce several wild colonies to take up residence in standard, movable-frame hives, for this would enable me to measure their mite loads. The easiest way to acquire wild colonies occupying hives located in the Arnot Forest was to capture swarms in bait hives set out in the forest. So, in early May 2003, before the start of the swarm season, I set out five bait hives near the sites 1, 2, 5, 7, and 10 shown in Figure 2.13. Each one was an old Langstroth hive in which I had installed eight frames of worker comb and two frames of drone comb. This arrangement provided the ratio of worker comb to drone comb (4:1) that I had found in natural nests (discussed in chapter 5). I reduced the entrance of each hive with a block of wood so that it was a rather small, 16-square-centimeter (2.5-square-inch) opening, which is what the bees desire. Finally, I mounted each bait hive on a platform in a tree about 4 meters (ca. 12 feet) off the ground, with its entrance facing south (Fig. 2.14). My goal was to offer the bees nesting cavities whose properties (cavity volume, entrance area, entrance height off ground, etc.) would match the nest-site preferences of European honey bees and so would provide dream homes for honey bee swarms in the Arnot Forest.

So that I could easily measure the mite loads of any colonies that might inhabit my bait hives, I equipped each one with a *Varroa* screen, which is simply a screen through which mites, but not bees, can fall. I sandwiched this screen between the wooden box holding the combs (the hive body) and the wooden bottom board (the hive floor). When I wanted to measure a colony's mite load, all I had to do was insert a sticky board—a sheet of cardboard whose upper surface was coated with vegetable oil—beneath the *Varroa* screen and then count the mites trapped on it after 48 hours.

This plan worked well. Three of my five bait hives were occupied during July 2003, and in August 2003 I started getting monthly mite-drop counts from these three wild colonies. The results of these assays, shown in Table 2.2, were crystal clear: all three colonies were infested with *Varroa destructor*. Eventually, it also became clear that all three colonies were surviving just fine despite the mites, for each colony's mite population was rather stable during the late summer and fall of 2003, dropped markedly over the winter of 2003–2004, and increased only slowly and gradually over the summer of 2004. Also, when I inspected these colonies in late

FIG. 2.14. One of the bait hives installed in trees in the Arnot Forest to attract swarms and so acquire some wild colonies living in movable-frame hives. Installation shown is typical: Langstroth hive mounted about 5 meters (16 feet) off the ground, facing south, and with an entrance opening of 16 square centimeters (2.5 square inches).

TABLE 2.2. Monthly assays of *Varroa* mite populations in wild colonies living in hives in the Arnot Forest. Each assay is the number of mites that dropped onto a sticky board over a 48-hour period at the start of the month.

Date	Colony 1	Colony 2	Colony 3
August 2003	30	14	21
September 2003	16	21	39
October 2003	36	3	22
May 2004	2	2	1
June 2004	3	11	2
July 2004	2	10	4
August 2004	3	5	7
September 2004	16	15	13
October 2004	42	40	22

summer in 2003 and in 2004—on 4 September 2003 and on 29 August 2004—I found that all three were in excellent condition; each had a strong population of bees, several frames of brood, and several frames of honey. None contained bees with deformed wings or showed signs of any other disease.

In mid-October 2004, I moved colony 2 from the forest to my laboratory at Cornell so that I could rear queens from it the following spring. I left colonies 1 and 3 in the forest, intending to continue monitoring their *Varroa* mite loads indefinitely. Colony 2 survived the winter of 2004–2005 at my laboratory in excellent health. Sadly, however, the two colonies that I left behind in the forest were discovered and destroyed by a black bear (*Ursus americanus*) sometime between my last colony check in mid-October 2004 and my next colony check in mid-April 2005. I knew that it was a bear that had killed both colonies because I found calling cards left by a bear at both sites: claw marks in the bark of the supporting trees, hive boxes lying overturned beneath the trees, and frames of combs scattered about where the bear had feasted on brood and honey.

We will see in chapter 10 that bears have difficulty finding wild colonies that live inconspicuously in natural cavities high in trees. It is clear, though,

Fig. 2.15. A bear-proof bait hive suspended from a tree limb in the Arnot Forest. Installed by David T. Peck (shown), a PhD student studying the bees' mechanisms of behavioral resistance to the mite *Varroa destructor*.

that at least one bear in the Arnot Forest had learned that a squat wooden box mounted in a tree sometimes contains honey bees and that when it does, it offers a sumptuous feast of honey and bee brood. Now when my students and I trap swarms in the Arnot Forest, we suspend our bait hives from tree limbs (Fig. 2.15), making sure that each bait hive hangs far beyond the reach of any black bear.

By the end of the summer of 2004, it was clear that the Arnot Forest was well populated with wild colonies of honey bees, that these colonies were infested with *Varroa* mites, and that *somehow* these colonies were not dying from their mite infestations. It remained a profound mystery, though, how these wild colonies were surviving without ever being treated with miticides. What was most puzzling was that the populations of *Varroa* mites in these wild colonies had not grown to dangerously high levels in August and September as they would do in late summer in my managed colonies if the colonies were not treated with a potent miticide, such as formic acid, oxalic acid, or some blend of essential oils. How exactly were these wild colonies keeping their mite populations under control? And, more broadly, how does the general biology of these wild colonies—nest structure, seasonal rhythms of growth and reproduction, food collection, defense mechanisms, life history, and still more—differ from what we see in colonies managed by beekeepers?

In chapters 5 to 10, we will review what my colleagues and I, and dozens of other biologists, have learned so far about the answers to these questions. What comes next, in chapters 3 and 4, however, is a review of the history of the relationship between honey bees and humans. It is a history that extends back into the mists of prehistory—even to times before our ancestors were humans. We will see how, over the last 10,000 or so years, *Apis mellifera* became a semidomesticated species whose members now often live as managed colonies located in agro-ecosystems, suburban landscapes, and other artificial environments. These managed colonies are the ones that we humans interact with most frequently and most easily. It is not surprising, therefore, that until recently we have known rather little about the natural lives of our most important pollinator.

3

LEAVING THE WILD

> Each time I cut a dripping square of wild
> honeycomb and eat it, wax and all,
> I marvel at its perfection, which no processing
> could possibly improve.
> —Euell Gibbons, *Stalking the Wild Asparagus*, 1962

The honey bee, *Apis mellifera*, has been extolled as "humanity's greatest friend among the insects," for its contributions to our agriculture and to our understanding of animal behavior. This bee should also be celebrated as humanity's *oldest* friend among the insects. After all, the pleasure we experience when eating honey was surely shared by our early ancestors, probably even those who were not humans. Most of the history of our association with this familiar bee can never be known, but we do know that the ancestry of honey bees—all the bees in the genus *Apis*—extends back to Oligocene times, some 30 million years before the present. The evidence of the honey bee's ancient origin comes from fossils unearthed in the 1800s from the fine-grained lignite of Rott, Germany, which provides paleontologists with some of the most exquisitely detailed of all insect fossils. These include a lovely fossil honey bee flawlessly preserved in profile (Fig. 3.1). This specimen and one more found in these paper-thin coal shales are described as members of the species *Apis henshawi*. Both fossils are worker bees, as indicated by the pollen press seen clearly on the hind leg of one of the two specimens. That they are workers tells us that they are social bees and so were once members of a colony with a queen, workers, and drones. All social bees living today store honey in their nests as strategic energy reserves, so it is highly likely that the nests of these fossil honey bees also contained honeycombs. If so, then their nests were bonanza food sources for our distant primate ancestors, who surely found honey marvelously delicious, just as we and all our great ape relatives—chimpanzees, bonobos, gorillas, and orangutans—do today.

That honey bees have lived for tens of millions of years in Europe, and probably also the adjacent continents Africa and Asia, means that honey bees have *always* been part of our natural world. The fossil record of the genus *Homo* shows that modern humans (*Homo sapiens*) arose about 300,000 years ago in Africa and that we then spread across Asia and Europe, all places where honey bees had already been living in the wild for many millions of years. Indeed, there is a strong possibility that the earliest modern humans encountered the same species of honey bee that we find today in Africa, western Asia, and Europe, *Apis mellifera*. The oldest known fossils of this species have been found in copal (fossilized tree resins) from Africa, and although the ages of these copal fossils are not known precisely, some of them may be more than a million years old.

For most of our history, our ancestors lived in hunter-gatherer societies. Recent studies by anthropologists of existing hunter-gatherers indicate that hunting for honey bee colonies—to feast on their nutritious brood and delicious honey—has long been an important foraging activity for our species. Honey is, after all, not just an amazingly delicious food but also an exceptionally energy-rich food, packing more than 13,000 kilojoules per kilogram (1,450 calories per pound). In the Hadza of northern Tanzania, a hunter-gatherer society with a rich tradition of honey hunting, both men and women rank honey as their favorite food. Honey is also a critically important food for these people because it provides a seasonal complement to game meat as a source of calories. Plentiful rains from November to April produce a time of tall grass and flowering trees, and during these six months the Hadza have their poorest success in hunting for meat

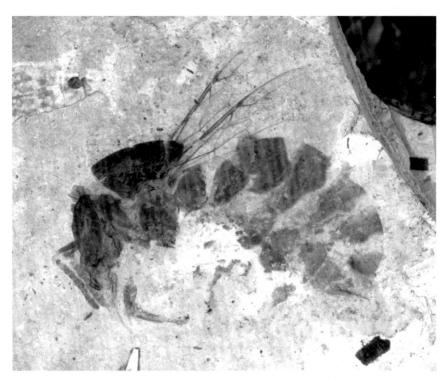

Fig. 3.1. Specimen of the fossil honey bee *Apis henshawi* Cockerell.

but their richest success in hunting for honey. During the rainy/honey season, a Hadza man will spend about five hours a day foraging for honey and will, on average, bring home approximately 1.5 kilograms (3 pounds) of honey. Similarly, in another society of African hunter-gatherers, the Efe of the Ituri Forest in the Democratic Republic of the Congo, the rainy/honey season also runs from November to April, and throughout these six months these people subsist largely on bee brood (eggs, larvae, and pupae) and honey. Efe men and women work together honey hunting, and each person, on average, collects more than 3 kilograms (6.6 pounds) of brood and honey in a day. Some 80 percent of the Efe's caloric intake during their rainy season comes from honey alone. Given these and many other examples of honey hunting as a traditional form of human foraging, there can be little doubt that honey hunting is as old as humanity itself.

Perhaps, though, the most compelling evidence that honey was a highly desirable and important food for our distant ancestors comes from rock art paintings of humans and other animals found on the walls of caves and rock shelters in southern France, eastern Spain, and southern Africa. Those found in the mountains in eastern Spain date from around 8,000 years ago. A beautiful painting discovered in 1917 in the Cueva de la Araña (cave of the spider) in the province of Valencia (Fig. 3.2, left) is the first direct record of ancient honey hunting to be found. It depicts a man who has climbed huge aerial roots or giant vines (or perhaps ropes) that hang from a steep cliff and has plunged his right arm into the nest of a honey bee colony occupying a crevice high on the cliff's face. With his left hand, he holds a bag for the honeycombs that he will steal from the bees. Meanwhile, several giant honey bees swirl around him. There is another human figure climbing up, carrying another collecting bag, but he stands far below and is probably safe from the bees' counterattack. A more complex depiction of ancient honey hunting was discovered in 1976 in the Cingle de la Ermita del Barranc Fondo (belt of the hermit of the canyon bottom), a deeply eroded riverbed in Spain's province of Castellón (Fig. 3.2, right). Here we see a tall ladder running up a cliffside to a nest of honey bees. There are twelve people standing at the base of the ladder, perhaps waiting for a share of the honey, and five others climbing up a sophisticated ladder built of two side ropes connected by rigid rungs. The two highest climbers and the one fourth from the top have a doubled-up posture, evidently grasping the ladder's rungs securely with both hands and feet, but the individual third from the top has begun falling and has arms and legs flailing in the air! Likewise, the fifth honey hunter, at the bottom, is either falling or jumping off the ladder. Both paintings portray the great risk of honey *hunting* and (by implication) the powerful attraction of honey *eating*, for these Mesolithic people.

FROM HONEY HUNTING TO HIVE BEEKEEPING

Until several thousand years ago, every honey bee colony lived in the wild, and probably only a tiny percentage of these colonies were ever plundered

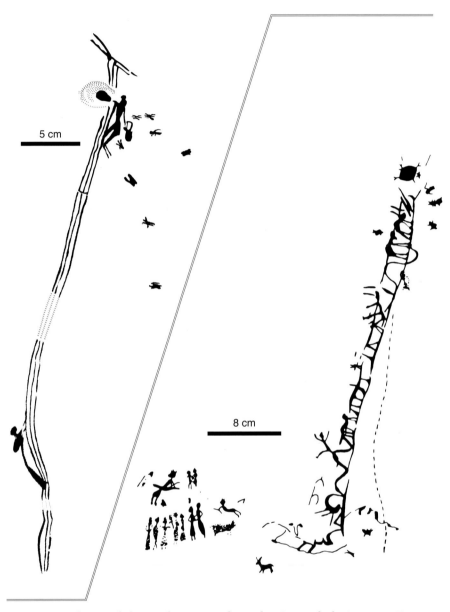

FIG. 3.2. *Left:* Mesolithic rock painting from the Cueva de la Araña, in Bicorp, Valencia, Spain, showing honey gathering from a wild colony of honey bees. *Right:* Another Mesolithic rock painting, from Barranc Fondo, Castellón, Spain, showing honey gathering from a wild colony.

by honey hunters. It seems likely, therefore, that early humans had only a small impact on honey bees; perhaps their principal effect was to bolster natural selection favoring colonies that chose cryptic nest sites and defended themselves fearsomely. Only when honey hunting began to be superseded by hive beekeeping—that is, when people began keeping colonies in man-made structures—did the impact of humans on honey bees begin its rise to the sky-high level that exists today in many parts of the world. The origin of hive beekeeping probably occurred shortly after, or along with, the invention of agriculture in the Fertile Crescent of the Middle East some 10,000 years ago. This was when some of our ancestors became small-scale farmers and started to manipulate the lives of plants and animals, including honey bees, to make them more productive for human purposes.

The earliest known evidence of hive beekeeping is the stone bas-relief carving shown in Figure 3.3, which dates to 2400 BCE, or nearly 4,500 years ago, when honey and dates were the chief sweetening materials in Egyptian cookery and beekeeping was an important Egyptian industry. This sculpture is now displayed in the Neues Museum in Berlin, but it was originally part of the pharaoh Nyuserre's temple to the sun god Re at Abū Jirāb, a site about 16 kilometers (10 miles) south of Cairo. On the left side of the panel, we see a beekeeper kneeling by a stack of nine horizontal hives, whose tapered shape suggests they were made of fired pottery. The three hieroglyphs above this beekeeper are the letters for the Egyptian word *nft* (to create a draft), so evidently the man is using the time-honored method of using smoke—the smoker (missing) is between him and the hives—to pacify bees and drive them off their honeycombs. In the center and on the right, we see other men handling honey in a production line that ends with one individual, perhaps an official, affixing a seal on a vessel to safeguard its precious contents.

Further direct evidence of hive beekeeping in antiquity was discovered in 2007 by archaeologists who found 30 intact hives, along with the remains of another 100–200 hives, while excavating the ruins of the Iron Age city of Tel Rehov, located in the Jordan Valley in northern Israel. Radio-

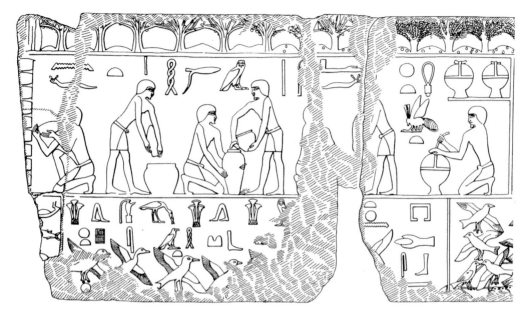

Fig. 3.3. Oldest evidence of beekeeping, from the sun temple of the pharaoh Nyuserre, which was constructed nearly 4,500 years ago. On the far left, a kneeling man puffs smoke toward a stack of nine horizontal hives. In the middle, two standing men pour honey from smaller pots into larger vessels, the taller vessel being steadied by a kneeling man. On the right, a kneeling man ties a seal on a container filled with honey; on a shelf above him are two similar containers that also have been sealed shut.

carbon dating of spilled grain found near the hives indicates that this apiary dates to 970–840 BCE, hence to nearly 3,000 years ago. Each hive is an unfired clay cylinder whose length (ca. 80 centimeters/ 32 inches), outside diameter (ca. 40 centimeters/ 16 inches), and entrance opening (diameter 3–4 centimeters/ 1.3–1.6 inches) matches those of the traditional hives used in the Middle East today. What is perhaps most remarkable about this find is that these ancient cylindrical hives, the oldest yet found, are stacked horizontally and parallel—like logs in a woodpile—to form three rows about 1 meter (ca. 3 feet) apart, each one three tiers high. This shows that this nearly 3,000-year-old apiary was organized in the same way as those of traditional beekeepers in the Middle East today (Fig. 3.4).

Eva Crane, in her monumental work *The World History of Beekeeping and Honey Hunting* (1999), describes the methods of the ancient Middle Eastern beekeepers, assuming that their ways of working with the bees match those of traditional beekeepers in Egypt today:

1. The beekeeper usually worked at the back of the stack of hives, to avoid being stung by the guard bees poised at each hive's entrance.
2. After opening one of the hives from the back, he smoked the bees to calm them and to drive the queen toward the front of the hive. He then removed combs containing honey that he had sliced free with a sharp, flat-bladed tool shaped like a large spatula fixed to a long, wooden handle.
3. He left behind in the hive any combs that he had cut and that contained brood, propping them in position on the hive floor with sticks slightly longer than the tube's diameter so the combs ran perpendicular to the long axis of the hive.
4. In the swarming season, he opened the hives from the front to inspect the colonies' brood combs, and he either cut out unwanted queen cells to inhibit swarming or he removed combs with queen cells and worker bees and installed them in an empty hive to make a split (a new colony with a queen).

We can see, therefore, that already several thousand years ago, beekeepers in Egypt and the lands east of the Mediterranean Sea were managing colonies in ways that were beneficial for people but were not altogether benevolent for the bees: packing colonies in crowded apiaries, stealing their honey, and manipulating their reproduction. When hive beekeeping spread west and north from the Middle East to regions around the Mediterranean Sea, the relationship between beekeepers and their bees evidently remained much like what existed in biblical times. The most detailed writer on beekeeping in the Roman world was Lucius Junius Moderatus Columella, a farm owner who lived in Rome in the first century CE. Most of book 9 in his 12-volume work *De re rustica* (On agriculture) is devoted to beekeeping. He provides sound advice for getting started as a beekeeper:

Fig. 3.4. An apiary of stacked, cylindrical hives made of mud in central Egypt. The spaces between the hives have been filled with mud to prevent swarms from moving into them. Each hive has a small, circular opening on the front end, together with one to four white markings that tell the beekeeper when the colony was established.

cluster hives in a walled apiary beside the beekeeper's dwelling so they will "be under the master's eye," and arrange them in at most "three rows of hives one above another." This arrangement suggests that the Roman beekeeper typically had horizontal hives (like those found at Tel Rehov), from which honeycombs were harvested from one end. Although Columella does not specify a certain shape or size of hive, he does advise to use well-insulated hives made of the thick bark of cork trees—that is, the cork oak (*Quercus suber*)—not those made of earthenware, because colonies housed in ceramic hives "are burnt by the heat of summer and frozen by the cold of winter." Also, he provides detailed guidance on colony management, such as opening hives in the spring to remove all the filth that has collected over winter, uniting weak colonies, capturing and hiving swarms, cutting out queen cells to control swarming, transferring brood combs, killing drones ("bees born of larger size than the rest"), and moving hives between spring and summer locations (to "give the bees a more liberal diet from the late-flowering blossoms of thyme, marjoram, and savory"). He also advises a late-summer honey harvest by cutting out honeycombs from one end of each hive, using sharp knives, but leaving at least one-third of the honey. Then, "when the winter is already causing apprehension," he recommends chinking holes in the hives with "a mixture of clay and ox-dung" and, finally, "heaping stalks and leaves on top of them" to fortify them against cold weather. Clearly, beekeepers in the time of Columella loved their bees and wanted to care for them, but they also wanted their bees to produce large quantities of honey and beeswax for harvest.

After the Roman period, traditional beekeeping north of the Mediterranean region developed along two distinct trajectories. One was the practice of tree beekeeping (in German, *Zeidlerei*), which consisted of harvesting honey from colonies living in tree cavities (either natural or man-made) that were accessed by tightly fitting doors. This manner of beekeeping was conducted throughout the enormous belt of deciduous forest that then stretched some 3,000 kilometers (ca. 1,800 miles) across northeastern Europe, from the Baltic coast to the Ural Mountains. Unbroken by mountains or sea, this expanse of sparsely populated forests was highly favorable to the settlement of wild colonies of honey bees. It contained ancient willow, basswood, hazel, and oak trees, which provided snug nesting sites, along with herbaceous plants such as raspberries and brambles in the forest openings, which supplied the bees with nectar and pollen. The work of tree beekeeping was conducted by small teams of men. Each one covered a large woodland area—a bee forest—that might contain 100 or more trees containing an accessible nest cavity, of which perhaps only 10 or so were occupied by honey bees at any one time. Evidently, the low density of colonies living in these bee forests was caused by the limited supply of bee forage, not a lack of availability of suitable nest cavities.

Some of the large trees that these tree beekeepers monitored enclosed natural cavities that had been fitted with a door, but most contained artificial nest cavities that the beekeepers had prepared with care. To make one, the beekeeper climbed 5–20 meters (15–60 feet) up the side of a large tree using a leather climbing strap, trimming off limbs as he worked his way up (Fig. 3.5). Then, perched high in the tree and usually facing its south side, he chopped a vertical slot about 10 centimeters (4 inches) wide and about 1 meter (ca. 3 feet) tall in the tree's trunk. This created the access opening for this nest cavity. Next, he used a long-handled chisel to excavate a 40–60 liter (ca. 10–16 gallon) chamber inside the access opening. He also chiseled out a smaller opening for the nest entrance; usually it faced south and was positioned about halfway up the cavity. This smaller opening was roughly 5 centimeters (2 inches) wide and 10 centimeters (4 inches) tall, and into it, to reduce the entranceway, the beekeeper pounded a carved wooden plug with two vertical slots, one on each side and each about 1 centimeter (0.4 inch) wide. Finally, he fitted a rabbeted door tightly into the access opening to secure the nest cavity from bears, hornets, woodpeckers, and other enemies of the bees. While all this work was being done high in the tree, another beekeeper was carving the landowner's mark—a goat's horn, a bow, or just grooves in various arrangements—into the bark at the base of the tree.

In early summer, the tree beekeepers inspected the nesting sites they had prepared to see which ones were occupied. Then, in late summer or autumn, they revisited the occupied ones and harvested a portion of the honeycombs they contained. There are no records of tree beekeepers installing swarms in empty tree cavities, and it seems unlikely that they could do this, so they must have relied almost exclusively on attracting wild swarms to their artificial nest cavities.

Tree beekeeping was an important activity in medieval times within eastern Germany, Poland, the Baltic region, and Russia. Indeed, it played a major role in the economies of these heavily forested lands, because honey and beeswax, and in some places furs, were the only natural sources of wealth. Honey was always important for making mead, and by 700 CE, beeswax was much needed by Christian churches, abbeys, and convents for making candles. A few centuries later, by around 1000 CE, the Russian

FIG. 3.5. Bashkir tree beekeeper who has climbed high up in a tree in which a hollow was prepared to house a wild colony of honey bees. His tools include climbing ropes, ax for prying out the rabbeted doors (stacked above him), bee veil, smoker, and wooden pail with lid. Having opened the tree cavity, he is extracting honey combs and placing them in the pail. The nest entrance is just to the right of his hands. Photo taken in the South Ural region in the republic of Bashkortostan, Russian Federation.

princes, boyars, and monasteries owned many bee forests, and a special class of peasant, the *bortnik*, served as the tree beekeepers. (In Russian, the earliest word for "hive" was *bort*, a hollow tree trunk; hence *bortnik* was the tree beekeeper.) *Bortniki* usually worked in pairs or larger groups to look after the wild colonies and harvest their honey and wax. The closeness of the Russian peasants to honey bees is shown by their term of affection for a worker bee—*Bozhiya ptashka*, God's little bird—and by their view that killing a bee was a sin. It is not surprising, therefore, that in 1854, Peter Prokopovich, the father of modern beekeeping in Russia, reported that a *bortnik* would collect from each bee tree no more than one wooden pail of honeycomb (about 6 kilograms/13 pounds). Evidently, the tree beekeepers in Russia tried to conduct sustainable harvests of the honey from the colonies living in their bee trees.

Honey bee colonies were disturbed only lightly by the practices of tree beekeeping. Tree beekeepers merely provided wild colonies with suitable nest sites and collected a modest fraction of each colony's honey stores toward the end of summer. Eventually, however, tree beekeeping gave way to hive beekeeping—at first, using upright log hives made of tree trunks—as the vast forests of northeastern Europe were cleared for agriculture, large trees became valued as sources of timbers and boards, and prohibitions arose against gouging out hollows in valuable trees. Tree beekeeping is, however, still practiced by Bashkir beekeepers in the South Ural region of Russia, specifically within the Bashkir Ural, a 450-square-kilometer (175-square-mile) region of forested mountains that includes the Bashkiria National Park in Bashkortostan. Within this protected wilderness area are some 1,200 trees with man-made nesting cavities for the bees, of which approximately 300 are occupied each summer.

The second trajectory along which traditional beekeeping developed north of the Mediterranean region emerged in northwestern Europe, in the lands that now include western Germany, the Netherlands, Britain, Ireland, and France. Large trees were not always plentiful in these places, and the most widely used traditional hive was a large, inverted basket called a skep. (The word *skep* comes from *skeppa*, an Old Norse word for a big basket, so evidently it entered the English language after Vikings began raiding and settling in England, starting around 800 CE). Skeps were built at first of woven plant stems (wicker) coated with clay and cow dung for insulation and waterproofing, but later they were made also of coiled straw. Both wicker and coiled-straw skeps were inverted, and the open mouth was set on a flat stone or wooden stand (Fig. 3.6). Most skeps were kept under cover, often on a shelf built against a house wall with some form of protection overhead but sometimes in a freestanding lean-to shelter with roof, back, and sides. Sometimes skeps were housed in recesses built in stone walls (bee boles), especially those associated with monasteries or other church property.

Beekeeping with skeps was often called swarm beekeeping because it depended on the production and hiving of swarms in early summer; the beekeepers then left the colonies alone to store honey from flows later in the summer and finally killed a certain portion of all the colonies to harvest their honey while leaving the rest alone to overwinter. Constant watch was required during the swarming season to capture the swarms, for the beekeeper needed to have several times as many colonies at the end of summer as he had had at its start. Often a swarm was caught by holding an inverted skep beneath a branch where bees had settled and then shaking the branch so the bees dropped into their new home. Sometimes a swarm was caught by placing a swarm catcher—a tube of netting held in shape with several hoops—over the skep's entrance when a swarm began to issue. Some beekeepers knew that if they heard a sound like that of a bugle horn coming from a skep that had cast a prime swarm, then they should expect an afterswarm to issue soon. Beekeeping with skeps required abundant swarming, and beekeepers encouraged this by making their skeps small so that their colonies became crowded in late spring and early summer. The skep sizes recommended in English beekeeping books from the 1500s to 1800s range from 9 to 36 liters (2.4–9.5 gallons), and were typically about 20 liters (5.3 gallons), so they were much smaller than modern hives. For comparison, a single 10-frame Langstroth hive body has a volume of 42 liters (11.1 gallons).

FIG. 3.6. Woodcut showing two wicker skeps on a wooden stand and a beekeeper wearing a hood with a woven screen insert.

Various methods were used for killing colonies living in skeps so their honey and wax could be harvested. These include standing the skep over a pit containing burning sulfur, or placing the skep in a closed sack that was immersed in water. Sometimes, however, skep beekeepers did not kill their colonies to harvest their honey but instead cleared the honeycombs of bees by driving them from one skep into another. This process involved smoking the bees to induce them to fill up on honey, then turning the occupied skep upside down and attaching an empty skep over the occupied one, and finally hitting the sides of the bottom (occupied) skep for several minutes to stimulate the bees to run upward into the empty skep. Unfortunately, a colony driven in late summer into a hive without combs often did not survive winter unless it was given some honey-filled combs. A second, nonlethal practice for harvesting honey from colonies in skeps was to slice the bottoms off some of the honeycombs after clearing them of bees by using smoke. This method matches that used for harvesting honey in tree beekeeping.

Skep beekeeping had several advantages over the methods of hive beekeeping that preceded it, including the Egyptian beekeeping method with clay pipes and the Roman beekeeping with hollow logs. Beekeeping with skeps rather than pipes or logs meant that colonies could be moved to places associated with strong nectar flows (times of intense nectar collection), such as the Lüneburger Heath south of Hamburg, Germany, where there is a huge ling heather (*Calluna vulgaris*) bloom in August and September. Beekeeping with skeps also made it easy for a beekeeper to inspect some of the combs in a colony's nest, so he or she could assess in a general way the colony's strength, honey stores, and preparations for swarming (presence of queen cells). But beekeepers using skeps still could not make thorough inspections of their colonies' combs, so they could not always determine whether a colony was queenright (had a laying queen), had filled its combs with honey, was preparing to swarm, or was diseased. Also, because skep beekeepers relied heavily on killing colonies to harvest their honey, they needed to encourage swarming, which required housing their colonies in small hives. This kept their colonies small, so their honey production per colony was low. Moreover, by the early 1800s, there was growing opposition to the merciless killing of colonies housed in skeps to take their honey. Beekeepers needed a better hive.

FROM FIXED-COMB TO MOVABLE-COMB HIVES

In 1848, Lorenzo Lorraine Langstroth, a 38-year-old Congregational pastor, resigned his ministry in Greenfield, Massachusetts, due to ill health and moved to Philadelphia, Pennsylvania, where he opened a school for young women and started a business as a commercial beekeeper. In the latter endeavor, he concentrated on glass-jar beekeeping: producing comb honey inside glass tumblers and small bell jars. The mid-1800s was a time when buyers wanted their honey in the comb, to be sure of its purity, so glass jars filled with honeycombs sealed by the bees were a premium product. Langstroth guided his bees to build their honeycombs inside his jars by placing empty jars upside down over holes cut in a board (a honey board) set atop a hive and then covering the jars with a wooden box (a super) to put them in darkness. The bees behaved as they would in nature and filled the dark voids above their dark, brood-filled combs with lovely white combs stuffed with honey.

Langstroth's work as a commercial beekeeper exposed him to the wonders of the behavior and social life of the bees. It also motivated him to design better hives than the ones he was using, which were squat wooden boxes, 15 centimeters (6 inches) deep and about 45 centimeters (18 inches) square. Each hive contained 12 wooden bars that were arranged in parallel and spaced at intervals of about 3.5 centimeters (1.4 inches) from center to center and that were set into rabbets in the front and rear walls of the hive. Each bar was the top support for a separate comb. Langstroth liked these hives because each one had a large top surface on which he could set many jars to be filled with comb honey. But he disliked how the bees attached their brood combs to the walls inside his hives, which meant that when he needed to remove brood combs to inspect them, he faced the bothersome task of slicing the combs free from the hive's walls.

Langstroth knew that for beekeeping to be more practical and more humane, beekeepers needed a better hive. Ideally, this would be a hive in which the bees could live and work naturally and in which the beekeeper could access the bees easily. This would enable the beekeeper to make inspections, give assistance, and remove honeycombs, all without undue damage to the combs, injury of bees, or waste of honey.

He began designing a better hive by addressing the difficulty he had of simply getting the covers off his hives. Each hive's cover lay directly on the comb bars, so of course Langstroth's bees glued his hive covers to the comb bars using the antimicrobial tree resins (propolis) they had collected to coat the inner surfaces of their nests. In 1851, he solved this problem by cutting more deeply—by a scant 9 millimeters (0.35 inch)—the rabbets in the front and rear walls of his hives on which the comb bars rested. This lowered the tops of the comb bars so they lay about 9 millimeters below the top edge of the hive and thus of the underside of the cover. Langstroth was pleased to find that his bees now left open the narrow space above the comb bars. Wonderful! This is how he discovered the bee space, a structural rule followed by the bees, in which corridors 7–9 millimeters (0.28–0.35 inch) high are left open for passage. In natural nests, corridors of this height are found along the edges of combs, where they function as passageways between the two sides of each comb (Fig. 3.7).

It is curious that initially Langstroth saw his discovery of the bee space only as a solution to the problem of the bees attaching his hive covers (or the honey boards supporting the jars for comb honey) to the comb bars in his hives. Throughout the summer of 1851, he enjoyed the newfound ease of opening his hives, but he still struggled with the messy task of slicing combs free from his hives' walls whenever he wanted to pull out combs for inspection. It was not until that fall, on October 30, that he realized that the bee space would solve this problem too. He described his insight as follows:

> Uprights might be fastened to the bars, so as to give the same bee space between [them and] the front and rear walls of the hive, and so change the slats [or bars] into movable frames. . . . In a moment, the suspended movable frames, kept at suitable distances from each other and the case containing them, came into being. Seeing by intuition, as it were, the end from the beginning, I could scarcely refrain from shouting out my "Eureka!" in the open streets.

Langstroth's journal entry for 30 October 1851 contains sketches of his new plan for movable frames, which includes fastening pieces of "clean worker comb" to the bar, or drawing "a thin line of wax across the center of the bar" to guide the bees to build their comb in the plane of the frame. He also realized the value of adding a bottom piece to his "compound bar" so that a bee-space corridor would exist between the frames and the cover,

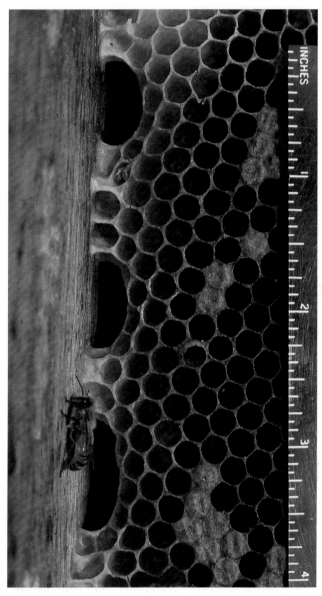

Fig. 3.7. Three passageways on the vertical edge of a comb built inside a tree hollow.

the frames and the walls, *and* the frames and the bottom board. This way, each movable frame would be surrounded by a bee space except at the two points of suspension (Fig. 3.8).

By the time Langstroth had his insight to build hives with suspended movable frames, each separated by a bee space from its neighbors and from the ceiling, walls, and floor of the hive, it was too late in the year for him to test his ideas in his apiary. Nevertheless, his journal entries for November 1851 show that he was confident that his concept of combs in movable frames would be of the utmost importance for the future of beekeeping, because it "will give the apiarian perfect control over his bees."

In 1853, Langstroth published his book titled *Langstroth on the Hive and the Honey-Bee: A Bee Keeper's Manual*. In it, he described his invention of a "movable-comb bee-hive" and his use of it in a program of practical and profitable beekeeping. By making it easy for beekeepers to open their hives and then inspect and manipulate the bees and their combs, Langstroth made it possible for beekeepers to perform such operations as dividing strong colonies to make artificial swarms, providing weak colonies with honey or brood, finding and replacing queens, examining colonies for pests and diseases, helping cleanse colonies of pathogens and parasites, and removing honey. Moreover, beekeepers could do all these things whenever they wanted and with minimal damage to the bees' home and without any injury whatsoever to the bees themselves.

By giving beekeepers virtual command over the lives of the bees dwelling in their hives, Langstroth made it possible for beekeepers to produce much more surplus honey per colony than before, especially in locations with climates and nectar and pollen sources favorable to beekeeping. Beekeepers quickly adopted Langstroth's hive design. The widespread embrace of his movable-frame hive was helped by the fact that the timing of his invention coincided with the period in which powered machinery made woodworking faster and cheaper. This meant that, despite their precise construction, movable-frame hives were affordable in many parts of North America and Europe. In the late 1800s, land transport also became more mechanized with the development of railroads, which enlarged the

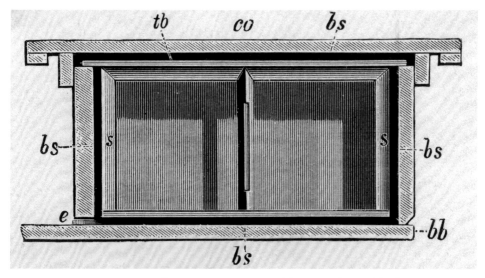

FIG. 3.8. Cross section of Lorenzo L. Langstroth's original movable-frame hive, from Cheshire (1888). co = cover; tb = top bar; bs = bee space; s = side of frame; e = entrance; bb = bottom board. This is the first drawing of a Langstroth hive in which the bee space was so labeled.

markets for honey and made commercial beekeeping more profitable. This rise in profitability stimulated further advances in beekeeping technology that boosted still further the productivity of the colonies managed by beekeepers. These included wooden section boxes for comb honey production, centrifugal honey extractors for flinging the bees' honey out of their combs, wire-reinforced beeswax comb foundation for strengthening the combs built within the new wooden frames, queen excluders for segregating the brood chamber and honey combs in a hive, chemical repellents and mechanical "bee escapes" for removing bees from honey combs before a honey harvest, and still more.

The proliferation of inventions for boosting a colony's honey productivity was accompanied by fundamental changes in colony management for the same purpose. The principal goal of commercial beekeepers living in North America and Europe became, and remains to this day, to make their colonies grow large in the spring and early summer, but prevent them from swarming, and then harness the energy of each colony's massive workforce to accumulate honey stores far greater than the bees will ever need. The beekeeper then harvests and sells the surplus. This manner of beekeeping is based on housing colonies in spacious hives comprising five or more hive boxes to create a 200-plus liter (ca. 53-plus gallon) nesting space with a capacity to hold 100 or more kilograms (220-plus pounds) of honey. We shall see that this amount of living space is three to fives times greater than the amount colonies seek when they choose their homesites by themselves (see chapter 5). Another way to increase a colony's size, and thus boost its capacity for honey production, is to install more than one queen in a hive. The queens sharing a hive are prevented from killing each other by being separated by queen excluders—metal sheets or wooden boards with slots through which workers, but not queens, can squeeze. This manipulation creates a colony with a supersize brood nest and an enormous population of forager bees. In regions where nectar flows are intense, the operation of multi-queen colonies has rewarded beekeepers with colossal crops of honey, sometimes more than 500 kilograms (1,100 pounds) of honey per hive.

When we look back across the 4,500-year history of beekeeping, in which honey bees were first housed in clay pipes and hollow logs, then in straw skeps and simple wooden boxes, and most recently in the sophisticated movable-frame hives of today, we see clearly that Lorenzo L. Langstroth and the other inventors of modern beekeeping have given beekeepers better hives. Unfortunately, as we shall see in the coming chapters, modern beekeeping has not given the bees better lives.

4

ARE HONEY BEES DOMESTICATED?

> The honeybee [is] capable of being tamed or
> domesticated to a most surprising degree.
> — *Lorenzo L. Langstroth, Langstroth on the
> Hive and the Honey-Bee, 1853*

Domestication is the process of human selection and breeding of wild species to obtain cultivated variants that thrive in man-made environments and that produce things useful to humans, such as food, clothing, assistance in hunting, pulling power, and companionship. It is how we have teamed up with other species to improve our lives. The practice of human-directed selection started at least 15,000 years ago in Eurasia, when dogs were domesticated from wolves to serve as hunting partners. About 10,000 years ago, its scope broadened greatly when our ancestors living in the Middle East began to shift their means of subsistence from food collection (hunting and gathering) to food production (herding and farming). The transition to agriculture involved the intensive domestication of crops, livestock, microbes (e.g., brewer's yeast), and pets. As a rule, the process of domestication produces organisms with traits that enable them to thrive in environments managed by humans but cause them to struggle in the wild. A familiar example of this is the way that corn (or maize, *Zea mays*) plants no longer have an effective mechanism for seed dispersal because the seeds stay tightly clustered on the cob.

In this chapter we will address the question: Are honey bees truly domesticated? This is an important question, because to answer it we must examine the special relationship that exists between honey bees and human beings. To begin, let us note that while it is true that *Apis mellifera* is often included in lists of the 18 or so animal species that are domesticated, and while it is also true that beekeepers own and control (somewhat) their bees, the human-animal relationship for honey bees is fundamentally different from that for cattle, chickens, horses, and other farm animals. In all these species, selection is steered almost entirely by human hands for life in man-made environments and with human assistance. In honey bees, however, selection is still steered mainly by natural selection for life in natural environments and without human assistance. We will see the evidence of this again and again throughout this book: the honey bee remains superbly adapted for living on its own in the wild.

THE PATH TOWARD DOMESTICATION

The bas-relief sculptures depicting beekeeping in the Egyptian temple to the sun god Re at Abū Jirāb (Fig. 3.3) show us that honey bees were already living under the care of people some 4,000 years ago, but these sculptures do not reveal when the first steps were taken toward domesticating the honey bee. They do, however, provide a clue: the sophistication of the beekeeping activities depicted in these sculptures—the skilled use of smoke and the careful sealing of storage vessels—indicates that the origins of beekeeping must predate these sculptures. The latest evidence on this matter comes from archaeologists who have reported compelling evidence of widespread exploitation of *Apis mellifera* in the earliest farming communities in the Middle East. Specifically, they have found the chemical fingerprint of beeswax on many fragments of pottery vessels collected from the sites of prehistoric farming communities in Anatolia (a region within eastern Turkey) that date to 9,000 years before present. Honey bees were probably important to these early farmers both for their honey—a rare sweetener for them—and for their beeswax, which probably had technological, cosmetic, and medicinal applications. Given that the close association between human beings and honey bees dates back to the onset of agriculture, it is likely that honey bees, along with sheep and goats, were among the first creatures to start moving down a path toward domestication when agriculture emerged and spread out of Anatolia and the Fertile Crescent about 10,000 years ago.

What might have motivated ancient farmers to start keeping colonies of honey bees near their homes and under their care? Perhaps the first beekeepers were individuals who especially enjoyed the delectable honey they knew was hidden away in the nests of these mysterious little creatures. There can be no doubt that whenever one of our Neolithic ancestors bit into a chunk of honeycomb, tasted its dizzying sweetness, and smelled the appealing aroma of the honey oozing forth, he or she had an intensely pleasurable experience. Perhaps this was even a joyous experience, since golden honey was the only strong sweet known to these people. I suspect that when Moses, some 3,500 years ago, reported God's promise to the wandering Israelites to deliver them to "a land flowing with milk and honey," his words had a significance that we cannot appreciate fully today, given how much cane sugar, high-fructose corn syrup, honey, maple syrup, and other sweeteners we now consume.

To understand fully the origins of beekeeping, though, we must consider not just the strong *motives* possessed by the earliest beekeepers but also the *opportunities* for domestication provided by the bees—of which there are two. Each is a behavioral trait that predisposed honey bees to start living around humans that had settled into an agricultural way of life. The first is the bees' predilection for nesting in cavities the size of a water pot or large basket (20–40 liters/5.3–10.6 gallons). It may be, therefore, that the first dwelling places of the honey bee located near human homes were empty pots and overturned baskets that had been left lying outdoors and were occupied by wild swarms. This scenario seems especially likely in the grassland regions of the Fertile Crescent, where bee forage must have been abundant but natural nesting cavities were probably scarce. If this hypothesis is correct, then it was the honey bees themselves, not human beings, that took the first step toward having bee colonies reside in man-made structures (hives) arranged in clusters (apiaries) near human dwellings.

The second, and probably more important, behavioral trait of honey bees that predisposed them to domestication is described by Lorenzo L. Langstroth in the second chapter of his 1853 manual for beekeepers, *Langstroth on the Hive and the Honey-Bee*. Its title is intriguing: "The honeybee capable of being tamed or domesticated to a most surprising degree." Here Langstroth explains to prospective beekeepers that even though honey bees can be as fiercely defensive of their nests as hornets, they are decidedly different from hornets in that *they are not always highly defensive*. He goes on to explain that worker honey bees are amazingly reluctant to sting once they have filled their crops (honey stomachs) with honey (Fig. 4.1), and that it is this striking feature of their behavior that makes possible the taming of these otherwise fearsome stinging insects.

There are two distinct contexts in which it is adaptive for worker bees to stuff themselves with honey and become averse to stinging. One is when they are in a swarm. Swarming bees tank up with honey—indeed, they nearly double their body weight in doing so—before they leave their old home in order to be fully energized for the flight to their new dwelling place and for the work of fitting it out with beeswax combs. But why are these honey-laden bees so reluctant to sting? The answer is simple: the act of stinging is fatal for a worker honey bee, and a swarm needs as many worker bees as possible once it has moved into its new nest site. As we shall see in chapter 7, the greater the number of bees in a swarm, the higher the probability the colony will survive its perilous first winter in its new home.

The second circumstance in which it is highly adaptive for worker bees to engorge on honey and then refrain from stinging is when their home is threatened by fire, a danger they sense by smelling smoke. A field study recently conducted by Geoff Tribe, Karin Sternberg, and Jenny Cullinan has revealed how colonies of the Cape honey bee (*Apis mellifera capensis*) in South Africa benefit from imbibing honey and becoming passive when they smell smoke. Seven days after a wildfire incinerated a 988-hectare (2,441-acre) swath of the Cape Point Nature Reserve, these investigators in-

Fig. 4.1. Worker bees filling up on honey.

spected 17 nesting sites within the charred landscape that they knew had been occupied by wild colonies before the fire. Each colony occupied a rock-walled cavity located either beneath a boulder or in a cleft within a rocky outcrop (Fig. 4.2). The research team discovered that all 17 colonies were still alive, even though several had suffered partial destruction of their nests: some melting of the propolis "firewall" at the nest entrance and (less often) of the beeswax combs deeper in the nest cavity. Evidently, the bees had filled up with honey upon smelling the smoke, had retreated as deeply as possible into their fireproof nest cavities, had survived the wildfire, and were sustaining themselves on the honey they had cached in their

Fig. 4.2. Wild honey bees in South Africa flying from their undamaged nest shortly after wildfire has swept past their home in the rocks. The heat of the fire has triggered the scorched, brushy plants (*Leucadendron xanthoconus*) around the nest to open their orange-brown seed heads.

bodies. A week or so later, plants known as fire-ephemerals would sprout and start to bloom, so soon these bees would be able to resume foraging.

This investigation of wild honey bee colonies surviving a wildfire shows us how the bees' engorgement response to smoke is adaptive for the bees living in a fire-prone region of South Africa. What it reveals, however, is a bit different from the standard explanation for why honey bees fill up on honey and become quiet when they smell smoke: to prepare *for abandoning the nest to escape the fire*. I think the standard explanation is probably incorrect, for I suspect it is unlikely that a colony threatened by fire can successfully evacuate its nest site and fly off through flames and smoke, especially since its queen is apt to be gravid and therefore a perilously clumsy flier.

One wonders, when did humans discover the magical potency of a few puffs of smoke to pacify thousands of irascible bees? The beekeeper blowing smoke toward a hive in Figure 3.3 shows us that 4,000-plus years ago Egyptian beekeepers already knew this trick for calming their colonies. It is possible, though, that the power of smoke to disarm a honey bee colony had been stumbled upon much earlier, indeed long before the origins of Egyptian beekeeping, back when humans were still just bee hunters, not yet beekeepers. Archaeological evidence indicates that the controlled use of fire was universal among humans some 120,000 years ago.

ARTIFICIAL SELECTION WITH HONEY BEES IS BARELY 100 YEARS OLD

We humans boost the productivity of our domesticated animals, cultivated crops, and useful microbes in two general ways: by changing their genes and by manipulating their environments. I am reminded of this fact whenever I drive through the rich farmlands north of Ithaca, a place that truly is a land "flowing with milk and honey." Dairy farms and bee yards are common here, and seeing them turns my thoughts to what dairy farmers and beekeepers now do to make their livestock as profitable as possible. Over the last 50 years, dairy farmers have boosted their production of milk per cow both by changing the genetics of their cows—black-and-white Holsteins have replaced the once familiar Dutch Belteds and brown Jerseys—and by transforming their living conditions. For example, dairy cows no longer spend summer days grazing in grassy fields. Most now live year-round in individual stalls or group spaces (freestalls) inside immense, open-sided shelters, where they are fed protein-rich corn and alfalfa, and often antibiotics and hormones, all to pump up their milk production. Sex is also a thing of the past for these cows. Calves are separated from their mothers within days of birth, and when a mother cow's milk production slackens, she is impregnated artificially using semen from a bull selected for his record of siring cows that are first-rate milkers. Once "mated," the cow is on course for another round of work as a unit of production on the factory farm.

Honey bees share with dairy cows the fate of being economically important animals that are thoroughly manipulated by humans to boost their productivity. But unlike Holstein cows, which require daily care from humans to thrive, honey bees remain capable of living on their own. Why is this? Specifically, why is it that we humans have not altered the genetics of honey bees through breeding to the point where they, like dairy cows, need steady help from us to survive? The answer is not that honey bees lack the critical ingredient of all breeding programs: differences among individuals in traits that are heritable (have a genetic basis) and have a high economic value. In bee breeding, the colony is the individual, and colonies vary in many ways that reflect their genetic differences and are economically important. These include honey production, pollen collection, gentleness, proclivity to swarm, propolis collection, wintering ability, and disease resistance.

What is it, then, that has prevented beekeepers, until recently, from breeding their colonies to have high honey production, low defensiveness, strong disease resistance, or some other desired trait? The answer is mainly one thing: beekeepers lack tight control over the reproduction of their colonies. An animal breeder shapes the future generations of his stock by controlling their reproduction, letting only those individuals with desirable traits produce offspring. Until the late 1800s, however, beekeepers could not control which of their colonies had the greatest reproductive success, that is, the greatest success in producing the queens of the future colonies and the drones that would inseminate these queens. Beekeepers had to leave these matters up to the bees, and therefore up to natural selection. What beekeepers needed were ways to favor the reproduction of certain queens and certain drones, namely those from their best colonies, so that they could promote the genetic success of their best bees.

This situation began to change in the mid-1800s following Langstroth's invention of the movable-frame hive (Fig. 3.8). Hives of his design made it possible for beekeepers to examine their colonies without seriously disrupting them and to take out swarm cells—queen cells in colonies preparing to swarm—from their best colonies to produce high-quality queens for requeening poor colonies and starting new colonies to enlarge their apiaries. However, it was not until the invention of efficient methods of artificial queen rearing by Gilbert M. Doolittle, which he reported widely in his 1889 book *Scientific Queen-Rearing*, that it became possible for beekeepers to begin breeding strongly from their best colonies. Usually the virgin queens reared from these superior colonies mated freely, in which case the breeding was accomplished without artificial selection among drones. But sometimes the selected virgin queens were taken to remote places (e.g., islands or high mountain valleys) stocked with selected colonies producing drones, in which case the breeding also included some artificial selection of the drones. Fully controlled bee breeding based upon strong artificial selection of both queens and drones only began to become possible in the 1920s with the invention by Lloyd R. Watson (for his PhD thesis at Cornell University) of the tools and techniques for the artificial insemination of queen bees (Fig. 4.3). Watson was skilled in designing, making, and using micromanipulators, and this led him to refer to artificial insemination of queen honey bees as "instrumental insemination," the term that is generally used today by bee breeders. It was not until the 1940s, however, once Harry H. Laidlaw had refined the procedures and syringes for injecting semen deep into a queen bee's oviducts, so that the spermatozoa can migrate easily into her spermatheca, that the artificial insemination of queen bees became reliable. At last, beekeepers had complete control of who inseminated their selected queens and so could control fully the genetics of their new colonies.

An early example of well-controlled bee breeding comes from a program that bred for resistance to American foulbrood (AFB), a disease of developing bees caused by the bacterium *Paenibacillus larvae*. Because AFB spreads easily between colonies (mainly through robbing), it is the most virulent of the brood diseases of honey bees. The program of breeding for AFB resistance began in 1934 when O. Wallace Park and F. B. Paddock, entomologists at Iowa State College, and Frank C. Pellet, an editor of the *American Bee Journal*, began a search for colonies that beekeepers judged to have some resistance to AFB. In 1935, they assembled, in a testing yard in

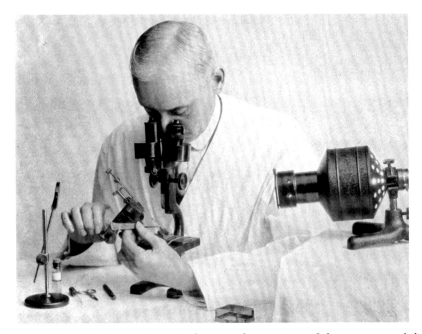

FIG. 4.3. Lloyd R. Watson in 1928 showing the position of the operator while conducting an instrumental insemination of a queen bee with the original model of the apparatus. Both elbows are on the desk, and the left hand is steadied against the stage of the microscope.

Iowa, 25 such colonies drawn from various parts of the United States. Each colony was tested for resistance by the insertion into one of its brood combs of a rectangle of comb containing approximately 200 cells, of which 75–100 contained AFB scales—the dried remains of larvae killed by AFB. The colonies responded by either removing the introduced comb, cleaning out the infected cells in it, or doing nothing. Most of the colonies contained AFB-killed brood at the end of the summer, but seven (28%) showed no signs of the disease and were considered resistant. The next step in this breeding program came in 1936, when these investigators established a semi-isolated apiary in the middle of a 100-square-kilometer (ca. 40-square-mile) citrus orchard in Texas, in which queens and drones were reared from the resistant colonies and allowed to mate. Twenty-seven colonies headed by these queens were then given a test for AFB resistance

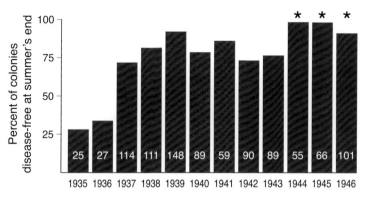

FIG. 4.4. Striking progress in breeding for resistance to American foulbrood. The percentage of colonies that remained disease-free after being inoculated with AFB spores increased over 12 years of selective breeding. The numbers within the bars indicate the number of colonies used each year. The asterisks indicate years in which the queens' matings were controlled by instrumental insemination.

using the same inoculation-comb procedure used the previous year. Nine colonies (33%) showed no sign of AFB when carefully inspected at the end of the summer. Over the next 10 years, this process of rearing and mating (in semi-isolation) queens and drones from the most resistant colonies was repeated, and impressive progress was made in raising the percentage of resistant colonies (Fig. 4.4). This was especially so in the years starting in 1944, when the queens were instrumentally inseminated to prevent outcrossing due to incomplete isolation at the mating yard; the proportion of AFB-resistant colonies climbed to nearly 100 percent.

The striking results from this program of artificial selection for AFB resistance by O. Wallace Park and colleagues in Iowa, along with later studies on the genetics of this resistance by Walter C. Rothenbuhler in Ohio, are impressive examples of what can be done in bee breeding. This initial work on breeding for resistance to brood disease has been developed further by selection programs for resistance to *Ascosphaera apis*, the fungus that causes chalkbrood, and to tracheal mites (*Acarapis woodi*) and *Varroa destructor* mites. All these programs have focused on breeding for hygienic behavior—the removal and disposal of diseased brood (larvae and pupae)—because multiple studies have shown that better hygienic behavior endows colonies with greater resistance to chalkbrood and mites, just as it does to American foulbrood. Hygienic behavior is also an attractive target for selective breeding because it is a colony-level trait that is easily measured. A comb containing brood is removed from a hive, a small area of this brood comb is frozen with liquid nitrogen, the comb is returned to the hive, and then the removal of the freeze-killed brood is measured after fixed time intervals. Colonies that remove it within 24 hours are considered hygienic; those that take longer are considered nonhygienic. Many commercial queen producers in the United States score their breeding stock using this freeze-killed brood assay, because there is evidence that hygienic behavior is an important mechanism of resistance to *Varroa destructor*, just as it is to American foulbrood and chalkbrood. Colonies with hygienic workers do remove more *Varroa*-infested pupae than do those without hygienic workers.

Another successful program of artificial selection with *Apis mellifera* was conducted in the 1960s by William P. Nye and Otto Mackensen, working for the U.S. Department of Agriculture. These researchers produced inbred lines of honey bees that ranked high and low as collectors of pollen from alfalfa, a plant that honey bees visit mainly for nectar and for which they are usually rather ineffective pollinators. In a test performed in a location in northern Utah with diverse sources of pollen, using bees in the fifth generation of selection, they found that the percentages of pollen collectors returning with loads of alfalfa pollen were 54 percent and 2 percent, for the high and low lines, respectively. Several commercial seed companies in the United States extended this research. In these studies, multiple lines were selected and then crossed (to avoid inbreeding), followed by further selection and more crossing. After three years of breeding, colonies in the selected lines collected 68 percent of their pollen from alfalfa when placed in test areas containing fields planted with alfalfa, safflower, cotton, melons, and sugar beets, while the control colonies in the same location (the check stock) gathered only 18 percent from the alfalfa. The alfalfa bee story is another convincing example of what can be done in the breeding of honey bees. In the end, though, these selected lines of honey bees were not widely adopted in the commercial production of alfalfa seed, probably because they were much less effective than certain species of solitary bees that are regular, and efficient, visitors of small legumes, including alfalfa.

NO DISTINCT BREEDS OF HONEY BEES

Despite the decisive successes in breeding honey bees for hygienic behavior to strengthen their resistance to American foulbrood, chalkbrood, and tracheal and *Varroa* mites, and in breeding honey bees for alfalfa pollination behavior, there is no evidence that artificial selection has altered in any general way the behavior of honey bees. Why is it that the breeding of honey bees has had few, if any, strong and lasting effects? Certainly, it is not for lack of variation among colonies in traits relevant to beekeeping. Beekeepers know that some colonies glue everything together in their hives with propolis, while others hardly use it all; that in some colonies the bees run about when the hive is opened, while in others they stay calm (Fig. 4.5); and that some colonies sting fiercely when disturbed, while others are much less defensive. Moreover, when we look across the honey bees living in their homelands in Europe, we see variations in color, morphology, and behavior associated with geography—the differences among Italian bees, Carniolan bees, Caucasian bees, Irish black bees, and so forth—but we do not see distinct breeds of honey bees.

How different this is from the other animals that are hugely important to humans. Consider dogs. Like modern honey bees, modern dogs are the descendants of ancient European animals, specifically gray wolves (*Canis lupus*), but unlike modern honey bees, modern dogs (*Canis familiaris*) have been profoundly shaped by their domestication. This is clear from the diversity in form and behavior among their breeds, for example: German shepherds, beagles, dachshunds, Labrador retrievers, Chihuahuas, Irish wolfhounds, Scottish terriers, and dozens more. The same is true for

FIG. 4.5. Worker honey bees (and several drones) standing calmly, and still distributed evenly, on a comb that has been pulled from a hive.

modern cattle, which are the domesticated descendants of the stately, reddish-brown, long-horned wild oxen (aurochsen, *Bos primigenius*) that we know from Ice Age cave paintings. From this ancient species, which is now extinct but was still alive in Roman times and was described by Julius Caesar as "little smaller than an elephant," we have bred modern cattle to give us meat, hides, and milk, as well as muscle strength and stamina for pulling plows and wagons. And as with dogs, the effectiveness of artificial selection in cattle cannot be doubted, given the diversity in form and function among the breeds of modern cattle (*Bos taurus*): Holstein-Friesian, Belted Galloway, Texas Longhorn, Brown Swiss, Black Angus, Scottish Highland, and hundreds more.

Why is it that bee breeders, unlike dog breeders and cattle breeders, have changed *Apis mellifera* so little over the past 10,000 or so years? Part of the answer is that we have had the full tool kit for breeding honey bees—artificial queen rearing and instrumental insemination—for less than 100 years. What is probably a much bigger part of the answer, however, is that the tools now available for artificial selection with honey bees have not been used widely and persistently. Indeed, I suspect that in most places where *Apis mellifera* lives in Europe and North America, the effects of artificial selection are minimal because most queen bees mate outside of human control: high in the air and with whatever drones they encounter. Many, perhaps most, of these drones will come from colonies living in trees, buildings, and the hives of beekeepers who do not manipulate the genetics of their colonies. This situation means that any changes in the genetics of honey bees created by the work of bee breeders will, over time, be erased. It also means that, in many (perhaps most) places, the genetics of honey bees is shaped far more by natural selection for traits that boost the genetic success of colonies living on their own than by artificial selection for traits that can boost the profits from colonies owned by beekeepers. This explains why it is that honey bees do not need our manufactured hives, and instead are still perfectly at home in a hollow tree. Indeed, most beekeepers have watched with dismay how readily their bees will abandon the former for the latter during the swarming season. The step back to living in the wild is but a short one for the bees housed in our hives.

APIS MELLIFERA, A SEMIDOMESTICATED SPECIES

While we humans have successfully manipulated the genes of maize, dogs, and cattle, we have not made fundamental changes in the genes of honey bees. We have, however, made great use of our second basic way of boosting the gains from our plant and animal partners: manipulating their environments. The first stage of doing so with honey bees occurred thousands of years ago when we induced colonies to make their homes in our hives. This gave us control over where honey bee colonies live, and it enabled us to reach into their homes and take from them what we wanted: honeycombs and beeswax. These days, the technology of apiculture has advanced to where beekeepers have many sophisticated means for controlling the living conditions of their colonies to raise their production of the goods (honey, pollen, beeswax, royal jelly, and occasionally venom) and the services (pollination) that we seek from honey bees.

The most basic of all the environmental manipulations performed by beekeepers to boost the profits from their colonies is to move them to where they will make a valuable honey crop or where a fruit or vegetable grower needs their pollination services. Many beekeepers in Scotland, for example, move colonies to heather moorlands in August to get crops of the richly aromatic, reddish-orange honey made from the nectar of the ling heather (*Calluna vulgaris*). This is the Rolls-Royce of honeys, so it is profitable to transport hives up to the purple-heather-covered hills in autumn (Fig. 4.6). Similarly, some beekeepers in New York State will move colonies to large fields planted with buckwheat in hopes of getting crops of the dark and peculiar-flavored but prized honey this plant yields. I know one beginner beekeeper who, unfamiliar with the odor of the buckwheat nectar that his bees were collecting, feared that something had died near his hive.

The most impressive, indeed mind-boggling, example of beekeepers moving their colonies for pollination service is the trucking of some 1.5 million colonies—more than half of all the colonies in the United States—to the almond groves in the Central Valley of California. This involves intense management of the bees. Even before they are loaded onto an armada of tractor trailers for their cross-country trips to California from places as distant as Florida and Maine, many colonies are fed sugar syrup and pollen patties to stimulate their brood production so they will meet the colony-size requirement of their pollination contracts. Once the colonies reach California, in late January or early February, they are dispersed among the 325,000 hectares (800,000 acres) of almond orchards that stretch from Sacramento to Bakersfield. When the almond bloom is over, in early March, some beekeepers will truck their colonies north to the apple and cherry orchards in Washington State to fulfill more pollination

FIG. 4.6. Two hives of honey bees that have been taken to a heather moorland in the Scottish Highlands. The purple hillside in the background, near Nairn, is covered with ling heather (*Calluna vulgaris*) in bloom.

contracts, while others will head east to make crops of honey from the vast fields of alfalfa, sunflowers, and clover in North and South Dakota.

A beekeeper can increase the income earned with his or her colonies by manipulating not just *where* they live but also *how* they live. For instance, in a natural nest, a colony's beeswax combs are firmly attached to the

Fig. 4.7. Wooden frames holding combs filled with honey in a honey super that is sitting atop a hive.

ceiling and walls of the cavity, but in a modern hive the bees are induced to build their combs in rectangular frames made of wood or plastic that hang like folders in a filing cabinet inside the stack of rectangular wooden boxes that form a hive. This arrangement makes it easy for the beekeeper to remove honey-filled combs by pulling the frames holding honeycombs straight up and out the top of a hive, rather as a file clerk extracts a folder (Fig. 4.7). It also makes it easy for the beekeeper to inspect (or leave undisturbed) the brood-filled combs—those in which the queen has laid eggs and young bees are developing—which rest in frames hanging in the lower boxes of a hive. Another way that a beekeeper strongly manipulates the activities of his or her bees is by inserting into each frame a foundation, a sheet of beeswax (or plastic) that has the hexagonal pattern of honey bee comb embossed on each side. When the bees come upon a sheet of foundation, they take it to be the start of a comb and they complete it, or as beekeepers say, they "draw it out." In doing so, the bees build the new comb following the hexagonal cell pattern on the foundation. This usually guides the bees to build combs whose cells are of the smaller size used for rearing worker bees, not the larger size needed for rearing drones. In this way the bees are inhibited from rearing drones, and this boosts their colony's honey production. It also, however, reduces its reproductive success.

Besides the invention of tools like movable frames and comb foundation, the craft of modern beekeeping rests on sophisticated methods of colony management. Perhaps the most important one is the control of swarming. Beekeepers achieve this primarily by housing their colonies in large hives so the colonies do not become overcrowded. Some beekeepers will also insert frames with empty comb between the frames with brood-filled comb, to prevent congestion in the central brood-nest part of each colony's nest. A few will even cut out queen cells—the special cells in which new queens are being reared—to disrupt this essential step in a colony's preparations to swarm. By preventing swarming, the beekeeper forces the colony to invest less in its reproduction (fewer swarms) and more in its growth and survival (more combs, more bees, and more honey) than it would if it were left alone. Table 4.1 lists the principal tools and techniques used in modern beekeeping to boost colony productivity.

Because beekeepers have not made fundamental changes in the genetics of their livestock, in the way farmers and herders have done with their cattle, sheep, chickens, and other farm animals, it is a mistake to include the honey bee in the list of animals that are truly domesticated. At the same time, however, because beekeepers strongly manage the environments in which their honey bee colonies live, both on the landscape scale (by trucking colonies about) and on the beehive scale (by managing the bees' living quarters), it is a fact that the honey bee is not a totally wild species. I suggest, therefore, that we consider *Apis mellifera* a semidomesticated species, one whose genetics we have changed very little but whose environment we can, and often do, change very much. I also suggest that we recognize that even though we have pressed millions of colonies of this industrious

TABLE 4.1. Key tools and techniques used by modern beekeepers to boost colony productivity

Tool	Effect
Movable-frame hive	Easy manipulation of colony and its nest's contents
Comb foundation	Strong control of location and type of comb in hive
Queen excluder	Clear separation of brood from honey stores in hive
Section boxes for comb honey	Honey combs are built in ready-for-sale containers
Queen cages	Easy transport and shipping of queen bees
Centrifugal honey extractor	Efficient spinning of honey from combs
Honey uncapping knives and machines	Efficient uncapping of honey combs
Smoker	Powerful calming of bees
Artificial queen cells	Mass production of queen bees
Bee escapes/boards	Easy removal of bees from honey combs
Chemical repellents/fume boards	Easy removal of bees from honey combs
Feeder frames and pails	Easy feeding of bees to stimulate brood production
Pollen traps	Easy collection of pollen for stimulative feeding
Pollen substitutes	Replacement of pollen for stimulative feeding
Medications	Reduced levels of disease
Bee suits and gloves	Colonies can be worked rapidly

Technique	
Swarm control	Strong colonies for pollinating and honey making
Housing colonies in large hives	Boost colony's investment in honey production
Commercial queen rearing	Mass production of queen for establishing colonies
Stimulative feeding	Accelerated brood production and colony growth
Moving colonies to work sites	More pollination work and better honey production

bee to live under our care, near our homes, and into our service, there remain countless colonies of this marvelous bee living without our care, far from our homes, and apart from our aims. These wild colonies show us that honey bees have not yielded their nature to us, for whenever they live on their own, they still follow a way of life set millions of years ago.

5

THE NEST

> He must be a dull man who can examine the
> exquisite structure of a comb,
> so beautifully adapted to its end, without
> enthusiastic admiration.
> —*Charles Darwin, On the Origin of Species, 1859*

The aim of this book is to review what we know about how honey bee colonies live in the wild, so naturally it focuses mainly on the bees themselves. We must, however, also give close attention to the nests that the bees build. This is because it is the qualities of a colony's inert nest, as much as the abilities of its lively bees, that determines how long a colony survives, how much it reproduces, and thus how well it achieves genetic success. We should expect, therefore, that when a colony builds a nest, its workers follow genetically based rules of behavior that guide them to find a good nesting site, secrete beeswax in a timely manner, skillfully build the hexagonal-celled beeswax combs, and laboriously coat their nest's walls with antiseptic tree resins. We will see in this chapter that the nest of a honey bee colony contributes hugely to its survival and reproduction.

By looking at the nest of a wild colony as a survival tool that extends beyond the bees' own bodies, we will become aware that beekeepers risk disrupting the adaptive biology of their bees by housing them in movable-frame hives that are crowded together in apiaries. We shall see that when colonies live under a beekeeper's supervision in wooden boxes, rather than on their own in hollow trees, they are forced to cope with homes that are often jumbo-sized, poorly insulated, and closely spaced. Recognizing these alterations of the bees' natural housing conditions, and their effects on colony health and biological fitness, raises many questions about optimal hive design and best management practices in beekeeping. It also highlights some fundamental conflicts of interest between the bees and their keepers.

NATURAL NESTS IN TREES

Most of what we know about the nests of honey bee colonies living in the wild comes from a study that I made in the mid-1970s with my first scientific mentor, Professor Roger A. Morse, then the professor of apiculture at Cornell University. Back then, I was in my early 20s and was starting to investigate what a swarm of honey bees seeks when it chooses its future dwelling place. My first goal was to describe the natural nests of honey bees, because I figured that doing so would provide clues about what constitutes a dream homesite for a honey bee colony. Then, once I knew the typical properties of their natural nesting sites—cavity volume, entrance size, entrance height, and so forth—I would conduct experiments to determine whether what I had found for each property of the bees' natural nests (e.g., the typical size of the entrance opening) was an expression of the bees' nest-site preferences or merely an indication of what was available to the bees.

Professor Morse and I decided that we would try to describe at least 20 natural nests. I knew already, from roaming in the woods around my parents' home outside Ithaca, the whereabouts of three bee trees. We located several dozen more by running an advertisement in the local newspaper, the *Ithaca Journal*. It read "BEE TREES wanted. Will pay $15 or 15 lb. of honey for a tree housing a live colony of honey bees." This worked surprisingly well, and in less than 10 days we had a list of 36 bee trees, all of them within 15 miles of Ithaca. Great! The next step was to work with a technician in the Entomology Department at Cornell, Herb Nelson, who had worked as a logger in the Maine woods and could help me fell these bee trees safely. Most were massive, old individuals. We began by conducting reconnaissance at each bee tree, to measure things like the height and direction of the nest entrance, but mainly to see if we could get Roger Morse's pickup truck close to the tree. Truck access to a bee tree's location was essential, because to dissect a nest properly, we would need to load the section of tree trunk housing the nest onto the truck for transport back to the Dyce Laboratory for Honey Bee Studies at Cornell, where I could slowly and carefully and comfortably dissect the nest. Twenty-one of the 36 bee trees allowed us a close enough approach with the truck, so we managed to collect 21 natural nests. To do so, Herb and I felled each tree, sawed out the portion of its trunk that contained the bees' nest, wrestled it into the truck's bed, and hauled it back to the lab. There I would carefully split open the log to expose the nest (Fig. 5.1) and start my dissection. The other 15 bee trees were too deep in the woods to allow us to collect their nests, but at 12 of them we managed to measure several key features—height, area, and compass direction—of the entryways to the honey bee homes that they harbored.

The entrances of the 33 bee-tree nests that we studied showed several striking features. First, most (79%) consisted of just one opening (e.g., see Figs. 1.1, 2.9, and 5.1); the others had two to five. Second, these entrance openings were usually knotholes (56%), but sometimes they were fissures in the tree's trunk (32%) or gaps among its roots (12%). Third, they tended to face in a southerly direction (23 nests) rather than a northerly one (10 nests). Fourth, most were in the bottom third of the nest cavity (58%) rather than the middle (18%) or top third (24%). And fifth, the average size of the nests' entrances was rather small, only 29 square centimeters (4.5 square inches), and the most common size was just 10–20 square centimeters (1.5–3.0 square inches) (see Fig. 5.2, top). In comparison, the standard entrance opening of a Langstroth hive is several times larger, approximately 75 square centimeters (almost 12 square inches). We began to wonder, do wild colonies prefer rather small, easily guarded entrance openings?

FIG. 5.1. The first of the 21 bee-tree nests that were dissected in 1975. *Left:* The intact tree, with the knothole that served as the colony's nest entrance visible in the left fork. *Right:* The section of the tree housing the nest has been brought to the laboratory and carefully split open to reveal the nest inside. The combs containing honey (in cells covered with yellow cappings) are in the top, and those containing brood—eggs, white larvae, and pupae (in cells with brown cappings)—are below. The nest entrance is on the left side, about two-thirds of the way up the cavity.

The results regarding entrance height for the bee-tree nests proved especially illuminating. This is because what we learned about this nest-site property from the 33 bee trees sampled in the mid-1970s was contradicted—and, as it turns out, corrected—by some later work. The bottom portion of Figure 5.2 shows that the 49 entrance openings of the 33 bee-tree nests that we examined in the mid-1970s were mostly low on the tree. About half were less than 1 meter (ca. 3 feet) off the ground, though there were also others more than 5 meters (16.4 feet) up. Given these results, I thought, OK, honey bees living wild in tree cavities generally have low nest entrances, just as they do when they live in beekeepers' manufactured hives. This seemed to be a reasonable conclusion at the time, but since then I have learned that it was dead wrong.

What corrected my thinking about the entrance heights of colonies living in the wild is what I discovered bit by bit—between 1978 and 2011—when I made three surveys of the wild colonies living in the Arnot Forest. Whenever I found a bee tree in one of these surveys, I did my best to locate the entrance of the bees' home and to measure its height, and most times I succeeded. I measured entrance heights by either climbing the bee tree and extending a tape measure or using a forester's clinometer. There were only two colonies possessed of entrance openings so well hidden in the bee tree's crown that I could not spot them, even with powerful binoculars. Figure 5.2 shows that the 21 entrance openings of the nests that I found in these Arnot Forest surveys were all well above ground level. The lowest was 4 meters (13 feet) up, and 90 percent were more than 5 meters (16.4 feet) high. Moreover, only one of these Arnot Forest nests had more than one entrance opening, and it had just two. Incidentally, this pattern of wild colonies usually having nest entrances high above ground level was recently confirmed by a friend, Robin Radcliffe, when he conducted a survey of the wild colonies living in a 5-square-kilometer (ca. 2-square-mile) portion of the Shindagin Hollow State Forest that is adjacent to his farm (see chapter 2). Robin beelined his way to five colonies, one in the wall of a hunter's cabin just outside the forest, and four in old-growth eastern hemlock trees inside the forest.

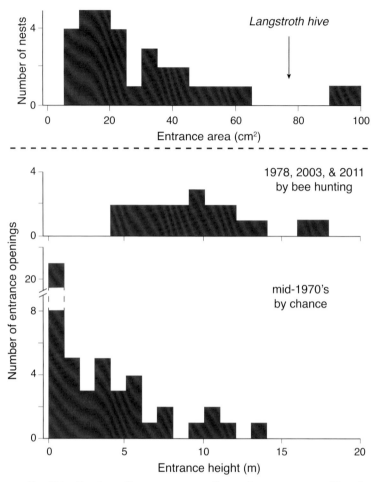

FIG. 5.2. *Top:* Distribution of entrance areas for 32 bee-tree nests. Not shown is the 204-square-centimeter (31.6-square-inch) value for one nest. *Bottom:* Distributions of entrance opening heights for 33 bee-tree nests found by chance (in the mid-1970s, $n = 49$ openings) and for 20 bee-tree nests found by bee hunting (in later years, $n = 21$ openings).

The average height of the entrances of Robin's four bee-tree colonies is 9.5 meters (31.2 feet).

Why is there such a striking difference between the two sets of entrance-height results shown in Figure 5.2? Looking back, I now recognize that my mid-1970s sample of bee-tree nests was unintentionally biased in favor of nests that had low entrances. Such nests, relative to ones whose entrances are up in the leafy crowns of trees, are much more likely to be found by chance, which is how the nests studied in the mid-1970s were discovered by the farmers, villagers, and forest owners who responded to our advertisement in the *Ithaca Journal*. In contrast, when I located bee trees in the Arnot Forest in 1978, 2002, and 2011, I avoided this sampling bias. I discovered these nests not by noticing them inadvertently but by intentionally tracking them down using the methods of a bee hunter—that is, by letting foragers from these nests lead me to their homes wherever they might be. Success in finding wild colonies by bee hunting is biased very little, if at all, by nest-entrance height. The stark difference between the two distributions of entrance height presented in Figure 5.2 taught me an important general lesson about studying the bees: *always be on guard against unintentional sampling biases*. More specifically, the data presented in Figure 5.2 show us that wild colonies living out in the woods around Ithaca usually reside in tree-cavity homes whose entryways are high up and therefore well hidden from ground-dwelling animals, most importantly (as we shall see in chapter 10) from black bears.

When Roger Morse and I shifted our attention from the exteriors of our bee-tree nests to their interiors, we were not surprised to see that the tree cavities occupied by the bees were generally tall and cylindrical: on average, 156 centimeters (62 inches) tall and 23 centimeters (9 inches) in diameter. This is consistent with the shape of tree trunks. But we were surprised to find that most of the wild colonies were living in tree cavities much smaller than beekeepers' hives. The average volume of the bees' nest cavities (not including one very large outlier) was just 47 liters (12.4 gallons), which is only slightly larger than the 42 liters (11.1 gallons) of one deep box (or hive body) in a Langstroth hive (Fig. 5.3). This means that, on average, a wild colony's nest cavity has only one-quarter to one-half of the living space of a typical beekeeper's hive. At this point we wondered, do wild colonies prefer rather small and snug nesting sites?

Certainly, the bees living in these smallish tree cavities were making good use of their living space, for each colony had nearly filled its nest

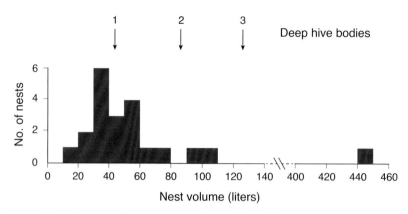

FIG. 5.3. Distribution of nest-cavity volumes for 21 bee-tree nests.

cavity with multiple sheets of beeswax comb (just eight combs, on average). Each comb formed a wall-to-wall curtain spanning the relatively slender cavity, but the bees had built small passageways through the combs where they attached to a cavity's walls and ceiling (Fig. 3.7). These openings served, no doubt, to allow the bees to crawl easily from one comb to the next. Most combs hung free along their bottom edges, leaving several centimeters (an inch or more) of open space between the bottoms of the combs and the nest cavity's floor. In recently occupied cavities, which contained yellow or light-brown combs, the floor was usually covered with a layer of soft, dark, rotten wood several centimeters thick. But in ones that were filled with dark combs, and probably had been occupied for years, the cavity's floor was coated with a dry, hard layer of tree resins (propolis) several millimeters (up to ca. 0.10 inch) thick, which made it shiny and waterproof. Likewise, the ceilings and walls of these long-occupied nest cavities had hard coatings of tree resins about 0.5 millimeter (0.02 inch) thick (Fig. 5.4). Evidently, when swarms had moved into these cavities, the workers had promptly chewed off the soft, rotted wood from their inner surfaces to expose firm wood for attaching their combs, and then had gradually coated the wall and ceiling surfaces between the combs with tree resins. We presumed that the bees had done all this work to seal off the innumerable cracks and crevices where molds and bacteria would

Fig. 5.4. Close-up of the propolis layer that a honey bee colony has built by coating the inner surfaces of its nest cavity with resins. The propolis has been chipped off in the upper-right region to show the substrate of decayed wood with its countless cracks and crevices, an ideal environment for bacterial growth.

otherwise thrive. Usually, the bees had extended their resin work to the surfaces just outside the entrance opening, filling the fissures in the bark here with propolis. This smoothed the bark surfaces around this busy spot. I suspect that this improves the flow of bee traffic at this unavoidable point of congestion; perhaps it also helps the bees disinfect their tarsi (feet) before they enter their home.

One more noteworthy feature of the walls of the bees' nest cavities was their thickness and thus their insulation. On average, the length of the passageway through the tree's trunk that the bees had to walk when entering or leaving their nest was slightly more than 15 centimeters (6 inches), and in one case it was 74 centimeters (29 inches)!

Eight of the 21 wild colonies whose nests we analyzed had filled their nest cavity with combs. On average, the total area of the combs in each nest was 1.17 square meters (12.6 square feet). This is the amount of comb that is held in 13 of the full-size (deep) frames that fit in a Langstroth hive. Most of each colony's comb was, as expected, the smaller-celled worker comb, but there was also a sizable portion of the larger-cell drone comb, 17 percent, on average (range, 10%–24%). I regret to report that we did not measure the sizes of the worker cells in these bee-tree nests. But working from photographs of the combs in three of these wild colonies' nests (e.g., Fig. 3.7), I have determined the mean wall-to-wall dimension of worker cells in the brood combs of three of the wild colony's nests: 5.12, 5.19, and 5.25 millimeters (0.201, 0.204, and 0.206 inch). So, on average, their worker cell size was 5.19 millimeters. For comparison, in my managed colonies, which have built their combs on standard beeswax comb foundation purchased from various manufacturers, the worker cells have a somewhat larger mean wall-to-wall dimension: 5.38 millimeters (0.212 inch). Once I did a study that looked at small-cell comb as a means of controlling infestations of *Varroa* mites (discussed in chapter 10), and in the pilot-work stage of this study I induced colonies to build combs on small-cell foundation. The cells of the worker comb built by these colonies had an average wall-to-wall dimension of just 4.82 millimeters (0.190 inch). We can conclude, therefore, that the wild-colony nests studied in the mid-1970s contained worker comb whose cell size was between that of the standard-cell worker comb and small-cell worker comb found in beekeepers' hives but much closer to the former than the latter.

The bees in these wild colonies had organized the use of their combs in the way that is familiar to all beekeepers, storing honey in the upper region of the nest, rearing brood below, and creating a band of cells holding pollen in between, though often with additional cells of pollen scattered within the colony's brood-nest region. Most of the nests were collected and dissected in late summer—late July and August—and except for one colony that was queenless, all were thriving. The average sizes of the worker bee and drone populations of the queenright colonies were 17,800 and 1,004, respectively. It was clear that the colonies' production of more workers and drones would soon be limited by the filling of their combs with honey. On average, the colonies had 15.1 kilograms (33.2 pounds) of honey stores and had already filled roughly 50 percent of the cells in their nests with honey: 56 percent of the worker cells and 48 percent of the drone cells. They had brood in 25 percent of their worker-comb cells and 26 percent of their drone-comb cells, so 19 percent of their worker cells and 26 percent of their drone cells were vacant. Overall, it appeared that these colonies were making good progress toward having the 25-plus kilograms (55 pounds) of honey stores and the large population of young bees that each would need to survive the winter. We examined all the combs closely for signs of brood diseases—American foulbrood, European foulbrood, sacbrood, and chalkbrood—but found none. Our study was conducted some 10 years before the first detections of tracheal mites (*Acarapis woodi*, in Florida, in 1984) and *Varroa* mites (*Varroa destructor*, in Florida, in 1987) in North America, so naturally we did not find either of these two parasites.

NEST-SITE SELECTION

The tree cavity or rock crevice that houses a wild colony's nest is the center of the universe for its inhabitants. It is the spot where these bees have built their nest, the place they will defend with their lives, and the only site on earth to which they return from miles around bearing loads of nectar and pollen. Both the nesting site and the beeswax combs inside are parts of the colony's set of survival tools that extend beyond the bodies of its members. It is obvious to anyone who has peered inside a wild colony's nest and admired its combs (Fig. 5.5) that these labyrinthine structures are products of the bees living there. After all, the beeswax used to build each

FIG. 5.5. Combs in the nest of a honey bee colony living in a hollow tree.

comb is a secretion of the bees' bodies, and the marvelous hexagonal-cell structure of each comb is a product of the bees' behavior. What is less obvious, though, is that the hollow tree or rock pile that shelters this intricate nest is also part of the colony's extended tool kit for survival. As we shall see, although honey bees do not *build* their nesting sites, they do *carefully choose* them, so the cavity that a colony occupies is also a product of its members' behavior.

The honey bee's process of choosing a dwelling place unfolds during colony reproduction (swarming), which occurs mainly in late spring and early summer (May–July) in the Ithaca area. The first step in this house-hunting process begins even before a swarm has left the parent nest. A few hundred of a colony's oldest bees, its foragers, cease collecting food and turn instead to scouting for new living quarters. This requires a radical switch in behavior. These bees no longer visit brightly lit, sweet-scented sources of nectar and pollen; instead they investigate dark places—knotholes, cracks in tree limbs, gaps among roots, and crevices in rocks—always seeking a snug cavity suitable for housing a honey bee colony.

Upon discovering a potential homesite, a scout spends nearly an hour examining it closely. Her inspection consists of a few dozen trips inside the cavity, each one lasting about one minute, alternating with trips outside. While outside, the scout scurries over the nest structure around the entrance opening and performs slow, hovering flights all around the nest site, apparently conducting a detailed visual inspection of the structure and surrounding objects. While inside, the bee scrambles over the interior surfaces, at first not venturing far inside the cavity, but with increasing experience pressing deeper and deeper into the remote corners of the hollow. When her examination is complete, the scout will have walked some 50 meters (about 150 feet) or more inside the cavity and so will have crossed all its inner surfaces. Experiments that I conducted with a cylindrical nest box whose walls could be rotated freely while its bright entrance opening remained stationary—so, in effect, the scout stepped onto a treadmill when she went inside the nest box—have shown that a scout bee judges a potential nest cavity's volume by sensing the amount of walking required to circumnavigate it. Exactly how a scout bee judges a cavity's roominess using the information gained from the walking (and occasional flying) movements that she makes inside the cavity remains a mystery.

The long duration of the nest-site inspections made by scout bees suggested that they assess multiple properties of a site to judge its suitability. Moreover, the regularities in certain nest-site properties that Roger Morse and I had found—in entrance area, entrance height, cavity volume, and so forth—supported the idea that bees have strong preferences in their housing. It was also possible, however, that the consistencies we found merely reflected what was generally available in tree cavities. At this point, I turned to surveying the scientific and beekeeping literature for information about the nest-site preferences of honey bees, but I found almost nothing, just one article in a French beekeeping magazine on how to build attractive bait hives for catching wild swarms. This situation surprised me, because I knew that beekeepers had worked for centuries to design the perfect hive, and I figured they might have looked to the natural living quarters of honey bee colonies for guidance, but evidently they had not. At the same time, finding this gap in our knowledge delighted me, for I realized then that my curiosity about the bees' natural homes had drawn me to a region of uncharted territory in the biology of *Apis mellifera*.

The method I developed for asking the bees about their nest-site preferences was simple: set out nest boxes that differed in certain properties and see which ones were occupied preferentially by wild swarms. More specifically, I set out nest boxes in groups of two, three, or four, with the boxes in each group identical except for one property, such as entrance area or cavity volume. The boxes within each group were spaced about 10 meters (33 feet) apart on similar-size trees (or a pair of power-line poles) where they were matched in their visibility, wind exposure, and location (Fig. 5.6). Each group of boxes served to test one (potential) nest-site preference, and it did so by giving swarms a choice between one box whose properties all matched those of a *typical* nest site in nature (e.g., average entrance size, average cavity volume, etc.) and another box (or boxes) identical to the first box except in one property (e.g., entrance

FIG. 5.6. Two nest boxes mounted on power-line poles. The two boxes offer identical nesting sites (same cavity volume and shape, same entrance height and direction, etc.), except that the one on the right has a smaller entrance opening (12.5 square centimeters/2 square inches) than the one on the left (75 square centimeters/12 square inches).

size), the value of which was *atypical*. In this way, wild swarms were tested for a preference in the one variable in which the boxes differed. For example, to test whether the distribution of entrance areas shown in Figure 5.2 reflects a preference for small entrance openings, I set up pairs of cubical nest boxes that were identical except that one box had an entrance area of 12.5 square centimeters (ca. 2 square inches), which was typical, and the other box had a large entrance area of 75 square centimeters (ca. 12 square inches, the size found in a Langstroth hive), which was atypical.

Planning this investigation was easy, but executing it was hard. Altogether, I built 252 nest boxes in the winter of 1975–1976, and I deployed them in small groups over the countryside around Ithaca in the summers of 1976 and 1977. Luckily, wild swarms were plentiful, so the experimen-

TABLE 5.1. Nest-site properties for which honey bees do or do not show preferences, based on nest-box occupations by swarms. $A > B$, denotes A preferred to B; $A = B$ denotes no preference between A and B.

Property	Preference	Function(s)
Entrance size	$12.5 \text{ cm}^2 > 75 \text{ cm}^2$	Colony defense and thermoregulation
Entrance direction	south > north facing	Colony thermoregulation
Entrance height	5 m > 1 m	Colony defense
Entrance location	bottom > top of cavity	Colony thermoregulation
Entrance shape	None: circle = slit	(Both shapes work well)
Cavity volume	10 < 40 > 100 liters	Storage space for honey; colony thermoregulation
Cavity shape	None: cubical = tall	(Both shapes work well)
Cavity dryness	None: wet = dry	(Bees can waterproof a leaky cavity)
Cavity draftiness	None: drafty = tight	(Bees can caulk cracks and holes)
Combs in cavity	with > without	Economy of comb construction

tal plan worked. My nest boxes attracted 124 swarms, enough to reveal many of the bees' secrets about what they seek in a homesite.

Table 5.1 summarizes the results of this study. We see that the bees in these wild swarms revealed preferences in four aspects of the entrance openings of their nest cavities. This was not surprising, given that this passageway is the interface between the colony and the rest of the world. All of a colony's food, water, resin, and fresh air comes in through this opening; all its waste and debris goes out of it; and all attacks by predators will focus on this point of greatest vulnerability. We also see that the wild swarms expressed preferences about just two features of the cavity itself: its volume and whether it was furnished with beeswax combs (from a previous colony).

FUNCTIONS OF THE BEES' HOUSING PREFERENCES
Entrance Size

I set up 14 test sites where swarms had a choice between two nest boxes that differed only in entrance size, and at six of them a wild swarm occupied one of the boxes, always the one with the smaller, 12.5-square-centimeter

FIG. 5.7. The right half of a hive entrance (2 × 14 centimeters/0.8 × 5.5 inches) that was reduced in late summer when the bees built a propolis wall over most of the opening, leaving just the two small passageways shown. The one on the left is barely wide enough for three bees to crawl through simultaneously.

(2-square-inch) entrance opening. This was unsurprising, because a small entrance helps a colony defend itself against animals that want to steal its honey. Most beekeepers around Ithaca, for example, know to reduce the entrance openings of their hives in the autumn, especially once the frosts have destroyed the flowers, and the yellow jacket wasps (*Vespula* spp.), now starving, try desperately to get to the bees' stores of honey. A small entrance probably also helps a colony stay warm in winter by minimizing draftiness in its nest cavity. This may also explain why some colonies, in late autumn, will reduce their nest's entrance by closing off most of the opening with a propolis wall pierced by just a few openings, each one just large enough to allow passage of a bee or two (Fig. 5.7).

Entrance Direction

Test sites where the paired nest boxes faced southeast, south, or southwest had markedly higher probabilities of occupation than those where the boxes faced northwest, north, or northeast. Evidently, the bees prefer their nest entrances to have a southerly exposure. A study conducted by

Tibor I. Szabo in Alberta, Canada, found that a hive with a south-facing entrance, relative to one with a north-facing entrance, had a lower probability of becoming plugged by ice and snow in winter, and that having the hive entrance open all winter improved nest ventilation and colony health. A south-facing entrance probably also helps the bees by providing a solar-heated site from which they can take off to conduct cleansing flights—to eliminate accumulated body wastes—on mild days in midwinter.

Entrance Height

I established eight test sites with paired nest boxes that were either high and low, and I caught six swarms at them, all in the high box of the pair. This indicated a clear preference for nest entrances high off the ground, consistent with the distribution of entrance heights for the nests of wild colonies located by bee hunting (Fig. 5.2). In chapter 10, I will describe a natural experiment in the Arnot Forest that shows one way that a wild colony benefits from having its nest entrance high up: it lowers the risk of detection by bears. It may also reduce the likelihood of nest damage in winter by woodland rodents such as deer mice (*Peromyscus maniculatus*).

Entrance Location

This variable was tested by setting out 12 pairs of nest boxes. Both boxes in each pair provided a cavity that was 100 centimeters (ca. 40 inches) tall and 20 centimeters (ca. 8 inches) wide and deep. But one box had its entrance opening at floor level, while the other had it flush with the ceiling. Ten of these 12 pairs of nest boxes attracted a swarm: eight moved into the box with a bottom entrance, and two into the box with a top entrance. In chapter 9, we will look at a study by Derek M. Mitchell, an engineer and physicist, that sheds light on the benefits to the bees of *not* having an opening near the top of their nest cavity.

Cavity Volume

This is the nest-site variable that I investigated most closely in my studies of nest-site selection by honey bees living in the wild around Ithaca, and it is the one that has most strongly attracted further studies by other biologists in North America. I began by setting up 14 test arrays, each of which consisted of four cubical nest boxes of different sizes: 10, 40, 70, and 100 liters (ca. 2.6, 10.6, 18.5, and 26.4 gallons). Over the next two months, I found that no swarms had occupied the 10-liter boxes, but that 11 swarms had moved into the larger boxes. This told me that 10 liters is too small for the bees. To see whether there is a size that is too large for them, I created 10 more test arrays that consisted of paired nest boxes of just two sizes, 40 and 100 liters. Now I found a striking pattern: seven swarms occupied the 40-liter boxes, and none moved into the 100-liter ones. Given these results, I concluded that swarms in my study area strictly avoid 10-liter cavities (too small) and strongly prefer a 40-liter cavity to a 100-liter one (too large). I did not explore further the upper limit of acceptably sized nest cavities for wild colonies in the Ithaca area, but I did investigate further the bees' lower limit for an acceptably sized nest cavity. In tests where wild swarms could choose between 10- and 25-liter (2.6- and 6.6-gallon) nest boxes, or between 17.5- and 25-liter (4.6- and 6.6-gallon) ones, I found they readily occupied the 25-liter nest boxes but never the 10-liter ones and almost never the 17.5-liter ones. This makes sense, because a 10-liter, and even a 17.5-liter nesting cavity—which has only about 40 percent of the volume of a deep Langstroth hive body—is too small to hold the ca. 20 kilograms (44 pounds) of honey that a colony living in the Ithaca area will consume over winter.

I reported my findings in 1977, and soon afterward two investigators at the University of Illinois, Elbert R. Jaycox and Stephen G. Parise, reported their studies that showed that artificial swarms of yellow-bodied Italian honey bees (*A. m. ligustica*) avoided 5.2-liter (1.4-gallon) cavities, but accepted 13.3- and 24.4-liter (3.5- and 6.4-gallon) ones, and that swarms of black-bodied Carniolan honey bees (*A. m. carnica*) rejected 5.2- and 13.3-liter cavities, but accepted ones with volumes of 24.4, 43.5, and 85.1 liters (6.4, 11.5, and 22.5 gallons). Their findings indicate that honey bees of different geographic races have different lower limits for an acceptably sized nest cavity, and that bees of races native to colder regions (e.g., the

Carniolan bees, native to the Carnic Alps, between Austria and Italy) require bigger cavities, probably to hold larger stores of honey.

The idea that the bees' nest-site preferences are adaptively tuned to their native climate is reinforced by a large-scale study conducted in 1981 by Thomas E. Rinderer and colleagues at the USDA's Honey Bee Breeding, Genetics, and Physiology Research Laboratory in Baton Rouge, Louisiana. They studied nest-site selection as a function of cavity size by wild swarms of European honey bees around Baton Rouge. To do so, they erected 10 wooden platforms 3 meters (ca. 10 feet) off the ground, and on each platform they set six nest boxes: two each of 5, 10, and 20 liters (1.3, 2.6, and 5.3 gallons); or two each of 20, 40, and 80 liters (5.3, 10.6, and 21.1 gallons); or two each of 40, 80, and 120 liters (10.6, 21.1, and 31.7 gallons). This team found that swarms occupied mainly the 20- or 40-liter boxes, and rejected the very small (5-liter) and very large (80- and 120-liter) boxes.

One more study, conducted by Justin O. Schmidt in the late 1980s and early 1990s, compared the nest-site preferences of European and Africanized honey bees with respect to cavity volume. This investigation focused on determining the minimal size of an acceptable homesite. It found that when wild swarms of European honey bees living in Arizona, and also wild swarms of Africanized bees living in Costa Rica, were given a choice between artificial nest cavities (pulpwood pots) with volumes of 13.5 or 31.0 liters (3.6 or 8.2 gallons), the swarms of both types of honey bee would occupy cavities providing only 13.5 liters of nesting space. It also found, however, that the European bees (but not the Africanized bees) occupied preferentially the pulpwood pots that provided a 31.0-liter nest cavity.

The acceptance of 13.5-liter cavities by some of the swarms of European honey bees living in Arizona surprised me at first, but then I realized that this result suggests that these bees have become locally adapted—that is, have experienced changes in the genes controlling their nest-site preferences—to living in this place without harsh winters. Although wild colonies of European honey bees living in central New York cannot survive winters living in 13.5-liter nesting cavities, evidently wild colonies of these bees can do so in southern Arizona, and this may have relaxed selection pressure on these bees to avoid occupying such small living quarters.

Combs in Cavity

This variable was tested by setting out 12 pairs of 40-liter (10.6-gallon) nest boxes. One box in each pair contained a 300-square-centimeter (46-square-inch) slab of dark, old beeswax comb propped against one wall, while the other box contained no comb. Four of these 12 pairs of nest boxes attracted a swarm: three moved into the box with the comb, one into the box without the comb. This suggested that the bees do prefer a cavity that comes furnished with some comb. Also, when I inspected the box with comb at the site where the swarm had occupied the box without comb, I found that it contained a large colony of yellow jacket wasps (*Vespula germanica*), one powerful enough to repel me! This meant that the box with comb at this site had not been available to the honey bees. All in all, I had three clear choices of the comb-containing box over the empty box. Not a definitive result. Nevertheless, I believe that it is likely that bees strongly prefer sites that are already equipped with combs.

Tibor I. Szabo, working in Alberta, Canada, reported that swarms installed in hives containing a full set of combs produced nearly twice as much honey across a summer compared to swarms placed in empty hives: 81 and 43 kilograms (178 and 95 pounds), respectively. There is also the fact that tree beekeepers in medieval Russia valued much more highly tree cavities that had been occupied by bees compared to those that had not.

Nest-site Properties Not Important to Bees

I thought that bees might prefer a nest entrance that is tall and thin over one that has the same area but is circular, because the former might be more easily defended. The wild swarms, however, showed no preference in this nest-site property. They also showed no preference for a box whose floor was covered with 2 liters (0.5 gallons) of *dry* sawdust relative to one in which the floor covering was *soggy* sawdust. Furthermore, the wild swarms showed no preference for a box with solid walls relative to one whose front and side walls were each perforated with 25 holes, each one 6.35 millimeters (0.25 inch) in diameter.

The bees' lack of choosiness about cavity dryness and draftiness puzzled me until I took down the occupied boxes and opened them to transfer the swarm bees inside to standard hives. I was surprised to find that the bees living in the boxes with the sawdust-covered floors had tidily removed this loose stuff, both the dry and wet forms. I also was surprised to find that the bees that had occupied the drafty boxes had plugged all the holes in their walls with propolis. Suddenly, it all made sense: honey bees are not choosy about nest-site details that they can fix after moving in. Of course, they cannot modify things like the entrance height or cavity size, so they must get these things right by carefully choosing their homesites.

What I found most surprising, however, was the absence of a preference for a tall box (interior dimensions 100 centimeters tall by 20 centimeters wide and deep/39.4 inches tall by 7.9 inches wide and deep) relative to a cubical box with the same volume. I knew that out in the woods, which is their natural habitat, the bees live in tree hollows that are usually tall and rather narrow. I also knew that they arrange the contents of their nests with a conspicuous vertical segregation of the honey (stored above) and the brood (reared below). Therefore, I strongly expected them to prefer cavities in which their combs would have a large vertical dimension, but when I studied the variable of cavity shape by putting out 12 pairs of nest boxes, with one tall box and one cubical box in each pair, I found no sign of a preference. Nine of the 12 pairs of nest boxes attracted a swarm: three moved into the tall box, six into the cubical box. Evidently, the bees do not have a strong preference for tall nest cavities.

COMB BUILDING

Finding a snug nest cavity is just the first of several hurdles that a wild colony of honey bees must clear if it is to survive its first year. Its next challenge is building the beeswax combs whose cells will provide the fledgling colony with cradles for its brood and cupboards for its food (Fig.

Fig. 5.8. Comb in the brood nest showing cells containing eggs, larvae, and pollen.

5.8). This construction work begins without delay, because without combs the new colony can neither begin rearing the worker bees it needs to bolster its strength nor start stockpiling the honey it needs to survive the coming winter. It is not surprising, therefore, that many of the bees in a swarm have scales of fresh beeswax protruding from the ventral surface of their abdomens (Fig. 5.9), an unmistakable sign that these bees are primed for comb building. Another indicator of the pressing need for plentiful wax in fledgling colonies is the remarkably broad age range of the swarm bees with fully functional wax glands. It is not just the middle-aged worker bees—the ones that are usually the primary wax producers in a colony—but also the older worker bees. Normally, these elderly bees are engaged in foraging and have nothing to do with making wax and so possess degenerate wax glands, but when they are members of a swarm they rejuvenate their wax glands and so help fill their colony's critical need for beeswax. Indeed, the thickness of the wax-gland epithelium—a measure of a bee's

FIG. 5.9. White beeswax scales protruding from the four pairs of wax glands on the underside of a worker bee's abdomen.

rate of wax production—is the same for the middle-aged bees and the elderly bees in a swarm.

The energetic cost of synthesizing the wax needed to build a complete nest is huge. We have seen already that a typical nest of a wild colony consists of several curtains of beeswax comb whose total area is about 1.2 square meters (ca. 12.9 square feet). Each of these combs is double-sided, so the total surface area is approximately 2.4 square meters (ca. 25.8 square feet). Approximately 80 percent of the surface area of these combs consists of the smaller worker cells (about 82,000 total) and the other 20 percent consists of the larger drone cells (about 13,000 total). To build this immense structure, a colony's worker bees must produce some 1.2 kilograms (2.6 pounds) of beeswax, which represents the lifetime wax production of approximately 60,000 worker bees, or about five times the population of the swarm that founds a colony. On average, a worker bee in a swarm contains about 35 milligrams (0.001 ounce) of a 65 percent sugar solution, so collectively the 12,000 bees in a typical swarm carry an energy supply of some 275 grams (0.6 pounds) of sugar. Given that the weight-to-weight efficiency of beeswax synthesis from sugar is at most about 0.20, and that one gram of wax yields about 20 square centimeters (ca. 3 square inches) of comb surface, the complete conversion of an average swarm's sugar supply into comb will produce about 1,100 square centimeters (1.2 square feet) of comb surface—less than 5 percent of the comb in a completed nest. The other 95 percent of the nest's construction must be funded by the income of the colony's nectar collectors over the following months and years.

The cost of nest building can also be reckoned in comparison to the energy consumed by a colony over winter. Knowing that honey is an approximately 80 percent sugar solution, and knowing that bees have approximately 20 percent efficiency in converting energy into wax, we can calculate that producing the 1.2 kilograms (2.6 pounds) of wax in a typical nest requires expending the energy contained in some 7.5 kilograms (16.5 pounds) of honey. This much honey (as we will see in the next chapter) is about one-third of the honey consumed by a colony over a winter to fuel its heat production. Clearly, the cost of nest construction is a major item in a colony's energy budget during its first year, which means that colonies can benefit greatly from energy conservation during nest building. In fact, a colony's very survival may depend on such frugality. We will see in chapter 7 that only about 20 percent of the incipient (founder) colonies living in the woods around Ithaca survive to the following spring. Most perish in winter when they exhaust their honey stores.

Nest building starts as soon as the swarm bees have moved into their chosen tree hollow. The work begins with some of the bees chewing off the loose, crumbly punkwood from the cavity's ceiling and upper walls. Soon other bees begin coating the cleaned surfaces with tree resins (propolis). This site-preparation work is partly a long-term investment in nest hygiene (see below), but initially it functions mainly to create smooth, hard, varnished surfaces to which the soon-to-be-built combs will be attached. Meanwhile, most of the bees in the swarm assemble themselves into a mass of linked chains of bees suspended from the ceiling and walls of the cavity. For the next several days, nearly all the bees in the colony—except some foragers and water collectors—will hang essentially motionless in the dark interior of their new home, all the while secreting tiny scales of wax from glands on the underside of their abdomens (Fig. 5.9).

The comb construction commences when individuals with well-developed wax scales disconnect themselves from their sisters, climb upward through the hanging braids of bees, and deposit their wax on the cavity's ceiling or walls. To remove a wax scale from one of the wax-gland pockets on the underside of her abdomen, a bee presses a hind leg firmly against the ventral surface of her abdomen and then slides it rearward until one or more of the larger spines in the leg's pollen comb skewers a scale and dislodges it from its wax-gland pocket. Next, the hind leg bearing the scale is drawn toward the head, where the wax scale can be grasped by the forelegs and chewed by the mandibles. The scale is chewed to mix it with a mandibular gland secretion that makes it more plastic, then it is deposited on the surface where the comb building is getting started or is already underway. Initially, these wax deposits produce just small piles of wax, but eventually the piles merge into a ridge of wax several millimeters (ca. 0.25 inch) long. At this point, the sculpting of cells begins. First, a cavity the width of a worker cell is excavated in one side of the wax ridge, and the excess wax is deposited along the sides of the hole. This work is repeated on the other side of the ridge of wax, but here two cells are dug, such that the center of the first cell on one side is between the two cells on the other side. Next, the raised edges of these cells are sculpted into thin lines to form the bases of each cell's six walls, with the adjoining walls laid out at an angle of 120 degrees. This gives each cell a hexagonal cross section from the start. As additional bits of wax are deposited, the bases of additional cells begin to take shape at the appropriate distances from the preexisting cells, and the walls of the first cells are raised by adding rough particles of wax to the top of each wall and then shaving each one down on both sides to form a thin, smooth, plane of wax in the middle (Fig. 5.10, middle). The cutaway wax is then piled up, together with some fresh wax, on the top of the wall, and the process is repeated. So, bit by bit, a wonderfully thin wall of wax grows steadily outward, always crowned by a broad coping.

Throughout this process of comb construction there runs a theme of economy in the use of the energetically expensive beeswax. The most conspicuous expression of this frugality is the cell shape in the combs of honey bees: a right hexagonal prism capped on the inner end by a trihedral pyramid. Because the cells of honey bee nests were originally circular in

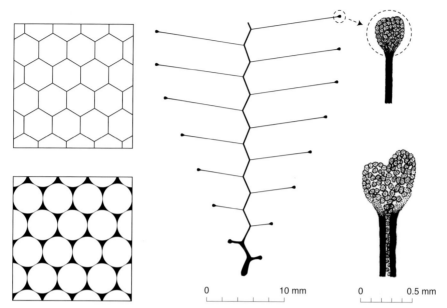

FIG. 5.10. *Left:* Illustration of the economy achieved by building hexagonal rather than circular cells in wax. *Middle:* Cross section through a comb showing the way bees precisely plane down cell walls during construction to minimize cost of comb building. *Right:* Cross sections of the outer margins of cell walls built by normal bees (top) and by bees from which the six distal segments of their antennae were amputated (bottom). Normally, loose bits of wax are packed together at the tip of the cell wall while the rest of the wall is a single, thin layer of wax. Bees with antennal operations built costly, triple-layered cell walls.

cross section, as they still are in the nests of all other bees, one can think of the comb in a honey bee nest as an assemblage of identical cylinders compressed into hexagonal prisms. The perimeter of a hexagon of a given area is 5 percent longer than the perimeter of a circle with the same area, but because each hexagonal cell in a comb shares its walls with other cells, whereas circular cells do not, building the cell walls in a comb with hexagonal cells requires only about 52 percent of the wax needed to build those in a comb with circular cells. For example, the average wall-to-wall distance of the worker cells in the nests of the wild colonies that I dissected was 5.20 millimeters (0.20 inches). A hexagon of this size has an area of

23.40 square millimeters (0.04 square inches) and a perimeter of 18.01 millimeters (0.71 inch). A circle with the same area has a perimeter of only 17.15 millimeters (0.67 inch). But because each cell wall in a hexagonal-cell comb is shared by two cells, the effective perimeter per hexagonal cell is just 18.01 millimeters divided by two, which equals 9.00 millimeters (0.35 inch), and 9.00 millimeters divided by 17.15 millimeters equals 0.52, hence the figure of 52 percent mentioned above. A comb of circular cells also requires wax to fill the spaces between the cell walls (Fig. 5.10, left), so the volume of wax required to build a hexagonal-cell comb with a given number of cells is, all things considered, less than half that required to build a circular-cell comb with the same number of cells.

Several other features of the comb construction process, beside the hexagonal cell design, also contribute to economy in wax use by honey bee colonies. One is the skill of worker bees in shaving down the wax partitions between cells, which leaves the walls of the cells only 0.073 millimeter (0.003 inch) thick. An experimental investigation of the sensory abilities of worker honey bees has revealed that a worker's antennae play a critical role in the process of sensing cell-wall thickness. When the six distal segments of both antennae of several hundred bees were amputated, and their comb building was studied, it was found that their building practices were grossly disrupted. Some cells were built with holes gnawed in their walls, while other cells were built with walls twice as thick as normal (Fig. 5.10, right). Because the temperature and composition of the bees' building material (beeswax) are constant, and because the shape of the cells in their combs is uniform, it is likely that a worker bee can judge the thickness of a cell wall by pressing on it with her mandibles and noting the elastic resiliency of the substrate with her antennae.

Bees achieve further economy in wax production by constantly recycling old wax. Whenever a bee emerges as an adult, the fragments of the cap of her brood cell are carefully bitten off, by the emerging bee or nearby nurse bees, and then are stuck to the cell's rim for reuse later. Likewise, queen cells—the special, large, peanut-shaped cells in which queen bees are reared—are built from bits of wax cut from adjoining worker cells, and once vacated they are torn down to free wax for other purposes.

Perhaps the most important way that a honey bee colony achieves economy in wax use is by carefully timing its comb building, limiting it to periods when the benefits of comb construction outweigh the costs of wax production. Consider first the situation of a swarm that has just moved into an empty tree hollow. Here the benefits of comb building are huge, for comb is vital to the colony's future. A fledgling colony cannot begin to rear brood or store food until it has built some combs, so it makes sense that it invests in a burst of comb building. This phenomenon was described recently by one of my PhD students, Michael L. Smith, who monitored the lives of several honey bee colonies from the moment they occupied their nest cavities to when they died. Each colony started out as an artificial swarm of average size—some 12,000 worker bees plus a queen—and was installed in a large, glass-walled observation hive that provided a 38-liter (10-gallon) nest cavity. Each swarm was fed sugar syrup *ad libitum* before it was installed in its hive, so its members were stuffed with energy-rich food when they moved into their new homesite, as is usually the case for natural swarms. For the next 20 months, the three colonies were left undisturbed, except that once a week, at night, the insulation boards covering the glass walls of each hive were removed to census the colony's worker population and drone population; trace the area of its worker comb and drone comb; and record which comb cells contained brood, pollen, honey, or nothing at all.

Figure 5.11 shows how each colony quickly built, over the first three weeks, 2,000–4,000 square centimeters (ca. 300–600 square inches) of comb surface. This much comb comprises only 8,500 to 17,000 worker cells—less than 20 percent of the number found in a completed nest—but it was enough for each colony to begin rearing worker bees within a few days of moving into its new home. Doing so was critically important, because the development period of a worker honey bee is 21 days, which means that for the first three weeks after moving into its new home, a colony's population shrinks steadily, as old workers are dying off and

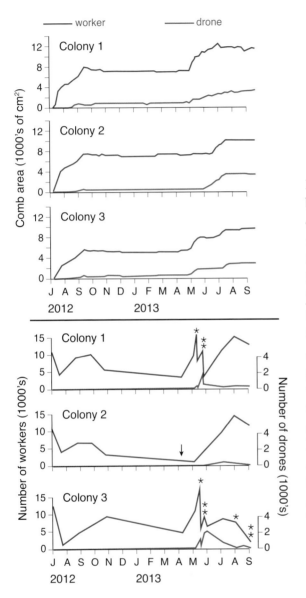

FIG. 5.11. *Top:* Histories of comb building in three unmanaged colonies, each of which began as a 12,000-bee swarm in July 2012. Magenta line, worker comb; blue line, drone comb. Each colony had just two pulses of comb building in the first year: one when it moved into its empty hive and one in late August and early September, when goldenrod (*Solidago* spp.) plants were producing copious nectar. Comb areas shown represent both sides of each comb. *Bottom:* Population dynamics in the same three colonies. Magenta line, workers; blue line, drones. The casting of a swarm (or an afterswarm) is marked by an asterisk (or two). Arrow in colony 2 plot marks when colony was given a queen to replace its original queen, which had died.

young ones are not yet emerging to replenish the workforce. Indeed, Figure 5.11 shows that by the end of the first three weeks, each colony's worker population had fallen to just 20–50 percent of its original level, and that even by summer's end none of the colonies had regained its original workforce size.

After its initial burst of comb building to initiate brood rearing and food storage, a colony faces a dilemma with respect to further comb building. On the one hand, building more comb can help a colony to efficiently exploit unpredictable nectar flows by giving the colony spare storage space, which helps keep the nectar-receiver bees from becoming mired in long searches for empty cells in which to deposit the fresh nectar. On the other hand, building more comb can shrink a colony's honey stores through the costly production of wax. Back in the 1990s, another one of my PhD students, Stephen C. Pratt, decided to investigate how the bees deal with this dilemma. Using a technique called stochastic dynamic programming, he started his work by modeling the situation that a colony faces in deciding when, and how much, to invest in comb building. His goal was to determine the optimal timing of additional comb construction during a colony's first summer based on the availability of nectar to be collected in the field, the amount of nectar already stored in the comb (as honey), and the amount of comb already present in the nest. Stephen's modeling work revealed that an established colony will achieve nearly optimal timing of comb building by limiting the construction of additional comb to times when two requirements are met: 1) the colony has filled its comb above a threshold level, and 2) the colony is busy collecting more nectar.

Stephen tested this prediction by performing experiments in which he monitored the comb building by colonies living in three-frame observation hives in which one frame was filled with brood, one frame was partially filled with honey, and one frame was empty (to provide space for comb building). He moved each study colony (one at a time) to a location without natural nectar sources, so that he could control its rate of nectar intake by adjusting the availability of "nectar" (sugar water) from a feeder. Figure 5.12 shows the results of one experiment that tested the prediction that a colony that is collecting nectar starts to build additional comb only after a threshold level of comb fullness has been reached. In phase 1, the colony had a high rate of nectar intake, but its comb fullness was kept at a low level. In phase 2, the high rate of nectar intake continued, and now the colony could fill its storage comb with honey. Finally, in phase 3, the colony

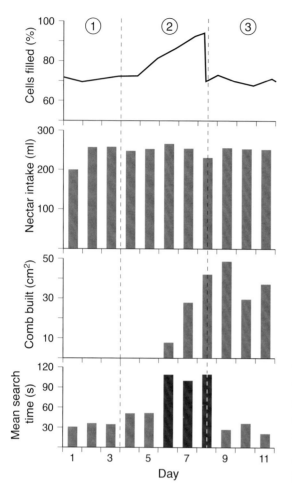

FIG. 5.12. Results of an experiment that tested the role of comb fullness in a colony's decision to start building additional comb. *Phase 1:* Colony had a high rate of "nectar" intake (collection of a 65% sucrose solution from a feeder), but its honey-storage comb was kept at a low level of fullness. *Phase 2:* The "nectar" intake continued, but now the bees were permitted to fill their storage comb with honey. *Phase 3:* Colony was returned to the conditions of phase 1. The search time variable plotted at the bottom is the time taken by a returning nectar forager to find a food-storer bee willing to receive her nectar load. The three mean search times indicated by dark bars are significantly higher than the others.

was returned to the conditions in phase 1. As predicted, the colony built no comb in phase 1 but then started to do so in phase 2 once its comb fullness had reached a threshold level (about 80%). In phase 3, however, the colony did not cease construction even though the fullness of its storage comb had been reduced (by Stephen) to a low level. This experiment shows that colonies will start building comb before they have filled all the existing storage comb in their nest. I suspect that the bees follow this strategy—which entails some risk of premature comb construction—to be sure that they have adequate storage space in the event of a large honey flow. This experiment also shows that once a colony has begun building comb, it will continue doing this so long as it enjoys a strong influx of nectar and thus a good supply of the fuel for comb building. Stephen also performed experiments in which colonies were deprived of a high rate of nectar intake, but their storage comb had a high level of fullness (85% or more). He found that these colonies did *not* build comb. Taken together, these two experiments show that Stephen's modeling analysis was correct; a colony invests in additional comb building only when it has reached a threshold level of comb fullness *and* it is experiencing a strong influx of nectar.

How do worker bees monitor both the comb fullness and the nectar intake rate of their colony so that they know when to start comb building at the right time? One possibility is that when a colony's storage combs get close to being full, the nectar receivers—the bees that unload the returning nectar foragers—experience increasing difficulty in finding cells for storing the fresh nectar, and they respond to this coincidence of high nectar influx and increasing comb fullness by starting to secrete wax and build comb. In support of this hypothesis is our knowledge that the nectar receivers and the comb builders are both middle-aged bees, about 10–20 days old, which means that nectar receivers are the right age to become comb builders. Further support of this hypothesis comes from the experiment presented in Figure 5.12. In performing this experiment, Stephen noticed that the average search time of a nectar *forager*—the time she spent searching in the hive for a bee who would receive her nectar load—rose markedly when the colony's comb construction started up. This conspicuous rise in the search times of the nectar foragers may have been caused by many of the colony's middle-aged bees switching from nectar receiving to comb building. However, when Stephen applied paint marks to 30–40 percent of the nectar receivers in a colony, and then induced this colony to start building comb, he found that the marked bees made up less than 5 percent of the comb builders. This surprising result shows that future work on the control of comb building needs to focus on the behavior of *unemployed comb builders*, to find out how these bees acquire information about their colony's comb fullness and nectar intake.

CONTROL OF COMB TYPE

Besides deciding *when* to build comb, colonies must decide *what type* to build, worker comb or drone comb. Both types of comb are used for honey storage, but only the large cells in drone comb can be used for rearing the large, heavy-bodied drones. A colony living in the wild devotes a surprisingly high percentage of the total comb area in its nest to drone comb: 17 percent, on average (Fig. 5.13). It does so because having plenty of drone comb enables a colony to invest heavily in producing drones, which are critical to its reproductive (genetic) success. Figure 5.11, from Michael L. Smith's study, shows, however, that during the first several weeks after moving into their new homesites, all three colonies in this study built only worker cells and reared only worker bees. This makes sense, for it is worker bees that will build the combs and collect the honey and pollen that are essential for the survival of a fledgling colony.

Eventually, however, a healthy colony will grow strong enough to invest in building both drone comb and worker comb. How does it regulate the proportion of each comb type in its nest? In principle, each comb-builder bee—or perhaps the queen—must be informed about the current relative amount of each kind of comb in the nest, but how is this possible given the large size of the whole nest relative to that of a worker bee? The queen is uniquely suited to evaluate this feature because she spends much of her time—when she is not resting or being groomed—walking around on the comb and measuring the sizes of empty cells to decide which kind of egg to lay: fertilized eggs in worker cells and unfertilized ones in drone cells. If she keeps a running count of the two cell types, then she might be able to perceive any imbalance between the two types of comb and might communicate this to the workers so they can correct it by adjusting the type of new comb they build.

Stephen C. Pratt investigated the information pathways that the bees use to guide their decisions about which type of comb to build. He did so by setting up queenright, full-size colonies housed in two 10-frame hive bodies and then adjusting the number and type of combs in each hive and the

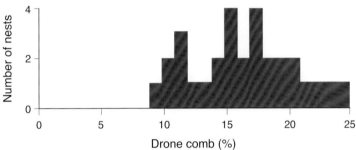

FIG. 5.13. *Top:* The cell diameters in natural honey comb have a bimodal distribution: the smaller size on the left (approximately 5.2 millimeters/0.20 inches wall to wall) is used to rear worker brood, and the larger size on the right (approximately 6.5 millimeters/0.26 inches wall to wall) is used to rear drone brood. *Bottom:* The percentage of drone comb by area in the nests of 8 wild colonies collected in upstate New York, and 22 simulated wild colonies. Values are clustered around the mean value of 17 percent.

access by the workers to the combs in their hive, to create different experimental treatments (Fig. 5.14). Stephen's experiments revealed three things. First, the bees in a colony require contact with the drone comb for its presence to inhibit their building of additional drone comb. Second, contact with the drone comb by the queen plays no part in this inhibition. Third, the inhibition of building more drone comb can come from the drone

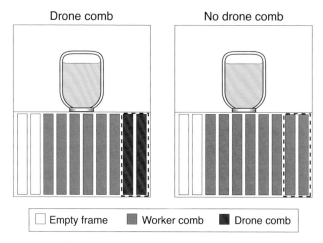

Fig. 5.14. Design of the experiment testing the need of the bees to have direct contact with drone comb for proper regulation of drone-comb building. Bees (workers and queen) were separated from two drone combs by a wire-mesh screen in the Drone Comb treatment. As a control, two worker combs were similarly screened off in the No Drone Comb treatment. In both treatments, the hive had two empty frames where the bees built comb. A feeder jar filled with sugar syrup (to encourage comb building) sits atop each hive.

comb itself, though the inhibition is even stronger if the drone comb is filled with brood. These findings show that the workers are not responding to a volatile chemical signal built into the drone comb, since they require direct contact to respond to its presence. These findings also show that the queen is not acting as a central regulator of drone-comb building, even though she is well positioned to collect information on drone comb need. So how do workers decide which type of comb to build? It may be that each comb-builder bee crawls over the existing combs and slowly accumulates information on their cell-size composition by measuring directly some number of cells. And it may be that worker bees use the same mechanism of cell-size measurement as is used by queen bees: inserting the head and forelegs into a cell and sensing how they contact the cell's walls.

To summarize, what we now know about how bees control the type of cells they build is that a comb-builder bee may be informed about the need for each type of cell by noting the relative encounter rates that she has had with *completed* worker and drone cells. We also know that a comb-builder bee may be able to coordinate her work—that is, whether to build worker comb or drone comb—with that of her nest mates by noting the relative encounter rates that she has had with cells *under construction*. Each of these hypotheses, however, awaits a rigorous experimental test.

THE PROPOLIS ENVELOPE

Perhaps the most eye-catching difference between the inside of a hollow-tree home of a wild colony and the inside of a lumber-built hive of a managed colony is the look of their walls. The walls inside a bee tree are coated with tree resins that make them shiny and waterproof (Fig. 5.4), whereas the walls inside a bee hive—even one used for years and years—have no such coating, so they usually look as dull and porous as freshly planed boards. In short, only wild colonies live surrounded by a propolis envelope. Most of this envelope is a thin (less than 1 millimeter/0.04 inch) coating of resin on the nest cavity's walls, but if there are cracks in the walls the bees will fill them with dark seams of resin. Also, if the entrance opening is oversize—hence too drafty or too inviting to invaders—the bees will reduce it by building a sturdy wall of dried resin (Fig. 5.7). Constricting the entrance opening is the most conspicuous use of tree resins by the bees, and it is the origin of the word beekeepers use for the resins bees apply in their hives, *propolis* (*pro*: in front of; *polis*: city or community).

What are the bees' sources of tree resins, and how do the bees collect this sticky stuff? The only tree on which I have observed bees collecting resin (from sticky leaf buds) is an eastern cottonwood tree (*Populus deltoides*) that grows near my laboratory. However, bees have been observed by others collecting resin from the buds and wounds of many other tree species in North America and Europe, especially horse chestnut (*Aesculus hippocastanum*), white poplar (*Populus alba*), quaking aspens (*Populus tremula* and *P. tremuloides*), and various birches (*Betula* spp.). In these species, the leaf buds shine with protective coatings of resin. Bees further collect resin

FIG. 5.15. Honey bee with a load of shiny resin in the corbicula (pollen basket) of her left hind leg.

from conifers—including spruces (*Picea* spp.), pines (*Pinus* spp.), and larches (*Larix* spp.)—that have injury sites oozing the trees' protective resins. A resin-collector bee loads herself with this gluey material by chewing off a bit of resin with the mandibles, then grasping it with the forelegs, passing it to one of her midlegs, and finally shifting it from the midleg to the corbicula (pollen basket) on the same side. This action is repeated until a load of glistening resin fills each corbicula (Fig. 5.15). When a resin collector returns to her nest, she crawls across the combs to one of the places where resin is being used and then she stands there for 5–20 minutes while other bees, resin users, bite off pieces of resin from the two loads in her corbiculae.

The resin collectors and resin users of a honey bee colony are fascinating bees. But who are they? How do they handle resin inside their nests? And how do they control their collection of this building material? Until recently, we had no solid answers to these questions. I was delighted, therefore, when Professor Jun Nakamura, from the Honeybee Science

Research Center at Tamagawa University in Japan, came to my laboratory, in 2002, to try to solve these mysteries. He conducted his investigations using a colony of approximately 3,000 bees that lived in a glass-walled observation hive that was housed in a heated room. To study individual *resin users* and *resin collectors*, he added eight cohorts of zero-day-old bees to the colony, one cohort every three to four days, during the month of May. He quickly learned how rare the resin users can be, for he observed only 10 of his 800 labeled bees engaged in resin-use behaviors: caulking (forcing resin into cracks and crevices) and resin taking (biting resin from the corbiculae of a resin collector). He also learned that all 10 resin users were middle-aged bees. They were doing their resin work when they were 14–24 days old, which was *after* they had quit working as nurse bees (e.g., feeding larvae and eating pollen) and *before* they would begin working as foragers. The resin collectors in Jun's colony were all elderly bees (age range 25–38 days old), as were the nectar, water, and pollen collectors in this colony.

Jun also learned that with resin—unlike with pollen, nectar, or water—there is not a strict partitioning of work between collectors working outside the hive and users working inside the hive. This became clear when he video-recorded 35 resin collectors, starting when each one entered the observation hive. He found that upon entering, each resin collector walked quickly and directly to a place in the top or side of the hive where resin users were getting resin from collectors and using it to fill the cracks between the hive's glass walls and its wooden frame (Fig. 5.16). Five of these resin collectors had carried home a chunk of resin in their mouthparts—along with full loads in their corbiculae—and they walked immediately to a work site and began to do some caulking themselves, using the resin they carried in their mandibles. In contrast, the resin collectors whose loads were entirely in their corbiculae did not engage in caulking. They simply stood around for 5–18 minutes, waiting for bees (resin users) to bite off chunks of their loads and carry them off to sites being caulked.

It is still not known how the resin collectors sensed their colony's need for resin. They could have done this directly (while engaged in using resin,

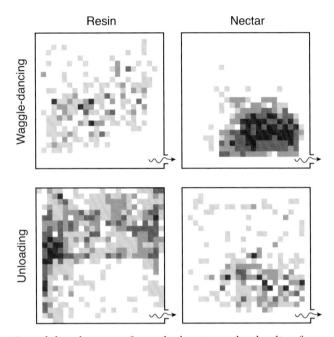

FIG. 5.16. Spatial distributions of waggle dancing and unloading for resin collectors and nectar foragers of a colony living in a two-frame observation hive. Arrow at bottom of each plot denotes the hive entrance.

if they did so) or indirectly (by noting their unloading delays) or both. It seems clear that the resin collectors in the study colony sensed an ongoing need for resin, because of the 102 resin collectors that Jun labeled and watched come and go from his observation hive, 68 (67%) continued collecting resin for the rest of their lives. (The other 34 bees eventually switched to collecting nectar or pollen.) It may be that these resin collectors sensed an ongoing need for resin, because every week or so Jun replaced the dirty glass walls on his observation hive with clean ones, and in doing so he probably stimulated the resin users by reopening the crevices where the glass walls contacted the wooden frame of the hive. We know that the resin collectors are stimulated by finding gaps, crevices, and rough surfaces—all places that are hard to keep clean—in their nest cavities. (Indeed, many of the bee-supply companies in the United States now sell propolis traps: plastic sheets that are perforated with hundreds of narrow slots about the width of a pencil's lead. These sheets are laid across the combs in the topmost hive body, to stimulate the bees to collect and deposit propolis, which will be harvested for use in making antibiotic tinctures and lotions.) Recently, Michael Simone-Finstrom and colleagues at the University of Minnesota have discovered that resin collectors are better than pollen collectors at learning to associate the stimulus of their antennae touching a surface with a 1-millimeter (0.04-inch) wide slot with the reward of a droplet of sugar syrup. The same is true when resin collectors and pollen foragers were tested for their ability to learn to associate touching a rough surface (a small piece of sandpaper) with a sugar-syrup reward. It is not known whether this difference between resin collectors and pollen foragers in learning tactile stimuli is based on their recent experiences or their genetics, but either way, it is likely that this ability helps resin collectors to learn about the presence of gaps, crevices, and rough surfaces in their colony's nest.

Most beekeepers find propolis a messy hindrance to their work, since it can make it hard to crack open a hive and manipulate the frames of comb inside. For the bees, though, propolis must be extremely valuable, for otherwise they would not expend the effort of chewing off bits of sticky plant resins, packing them onto their hind legs, hauling the gluey blobs back to their hive, and then using this sticky stuff to fill the cracks and coat the walls inside their nest cavities. Recent work has revealed that the propolis envelope functions mainly as an antimicrobial surface that lowers a colony's costs of defense against disease. We will examine this subject closely in chapter 10, when we explore the multifaceted matter of colony defense. But before we dive deeply into this and the three other major topics of colony functioning—reproduction, food collection, and thermoregulation—we will look broadly at the life of a wild colony by examining its remarkable annual cycle.

6
ANNUAL CYCLE

> Oh, give us pleasure in the orchard white,
> Like nothing else by day, like ghosts by night;
> And make us happy in the happy bees,
> The swarm dilating round the perfect trees.
> —Robert Frost, "A Prayer in Spring," 1915

One key to understanding the natural lives of honey bee colonies living in cold-climate regions is their unique annual cycle. In winter, when colonies of all the other social bees—bumble bees and social sweat bees—have dwindled away, leaving only a residue of mated queens living alone deep in hibernation, honey bee colonies continue to function as full social groups, each one consisting of some 15,000 worker bees and one queen bee. Moreover, rather than becoming dormant, as is the rule for insects living in cold climates, honey bee colonies remain active and fight the cold. Each one keeps the perimeter temperature of its winter cluster above about 10°C (50°F), even in ambient temperatures of −30°C (−22°F) or colder. To achieve such strong temperature control, honey bees nest inside protective cavities, press tightly together to form a well-insulated cluster, and pool the metabolic heat generated by isometric contractions of their powerful flight muscles. The fuel for this impressive, winter-long heat production is the 20 or so kilograms (some 45 pounds) of honey that each colony has stockpiled in its nest over the previous summer.

Fig. 6.1. Worker honey bee collecting pollen from eastern skunk cabbage (*Symplocarpus foetidus*), which flowers early in the spring. Only the flowers are visible above the muddy soil; the stems remain buried below the surface of the soil, with the leaves emerging later.

The honey bee's annual cycle is special in other ways besides the process of overwintering as a warm and active colony. Shortly after the winter solstice, when the days begin to grow longer but snow still blankets the countryside in Ithaca, each colony raises the core temperature of its winter cluster to about 35°C (95°F) and starts to rear brood. Initially, there are only a hundred or so cells of brood in a colony's nest, but by early spring, when the red maple trees, pussy willow bushes, and skunk cabbage plants (*Symplocarpus foetidus*) have come into bloom and are providing the bees with plentiful nectar and pollen (Fig. 6.1), more than 1,000 cells hold developing bees, and the pace of a colony's growth is quickening daily.

Come late spring, when the bumble bee queens and sweat bee queens are just rearing their first daughter-workers to adulthood, honey bee colonies have already grown to full size—30,000 or so individuals—and are starting to reproduce. Colony reproduction by honey bees involves not only the simple process of rearing males, which fly from their nest to find and mate with virgin queens from neighboring colonies, but also the intricate process of swarming (colony fissioning), in which a labor force of some 10,000 to 15,000 worker bees, together with the colony's mother queen, suddenly departs in a swirling mass to establish a new colony.

The main aim of this chapter is to present an overview of the life of a wild honey bee colony living in the woods around Ithaca. It does so by describing the pattern of events that unfold in these colonies across a year. This view reveals several fundamental themes about how honey bee colonies live in the wild, themes that will unify the subsequent chapters. A second goal of this chapter is to explain why colonies of honey bees living in the cold-temperate regions of Europe and North America lack a period of winter dormancy and therefore possess an annual cycle that is unique among all the insects living in these places. As we shall see, the distinctive annual cycle of *Apis mellifera* as it lives in temperate regions is a blending of its current ecology and its ancient history. Novel adaptations for living in a seasonally cold climate have been superimposed on the physiological and social characteristics that this bee inherited from its tropical ancestors.

ANNUAL CYCLE OF ENERGY INTAKE AND EXPENDITURE

Surviving winter by fighting the cold is energetically expensive. In the heart of winter, a honey bee colony weighs approximately 2 kilograms (4.4 pounds) and consumes energy—mostly for heat production—at a rate of 20–40 watts, roughly the same rate as a small incandescent lamp. Naturally, a colony's long-term survival requires that its energy expenditure over winter be balanced by its energy storage over summer, when energy-rich nectar can be gathered to rebuild its honey stores. A strong flow of energy back and forth between colony and environment is, therefore, a key feature of the natural history of honey bees, and one that provides both biologists and beekeepers with a valuable window on the lives of these bees. By monitoring this energy flow throughout a year, one acquires both a synoptic view of a colony's annual cycle and a detailed picture of one of the major challenges that it faces every year: the need to amass within a short summer season an ample supply of winter heating fuel.

A simple but effective way to monitor the net flow of energy into or out of a honey bee colony is to record changes in its total weight: the weight of the bees, the nest, and the food stored in the nest. Weight is gained when food is brought into a colony but is lost when stored food is consumed, when a colony divides for reproduction, and when a colony's members die. Detailed records of the weight changes of honey bee colonies across a summer or throughout a year have been published for many temperate regions of the world, including the United States, Canada, Germany, and the United Kingdom. Almost without exception, however, these records were collected for apicultural purposes, so they describe patterns of net weight gain and loss for beekeepers' colonies that were managed for honey production and were living in agricultural landscapes. Therefore, the discussion that follows is based primarily on the findings of a study that I conducted with the aim of shedding light on the energy budgets of wild colonies.

The heart of my study was monitoring the weekly weight changes of two unmanaged colonies. Each one occupied a hive that consisted of two deep Langstroth hive bodies (total volume, 84 liters/22.2 gallons), so the nest cavities of my two study colonies were larger than the typical nest cavity of a wild colony (Fig. 5.3). I mounted each colony's hive on platform scales and weighed both hives every Sunday evening from early November 1980 until late June 1983. Except for twice-monthly measurements of brood rearing in late spring, summer, and early autumn, neither colony was disturbed or manipulated once the study began. The least natural aspect of the ecology of these colonies was their location: the Othniel C. Marsh Botanical Garden, which is in a leafy residential neighborhood in the small city of New Haven, Connecticut. (At the time, my wife and I lived in the caretaker's house in this garden, and having the hives nearby made it easy to take readings of their weights every Sunday night.) The forage available to honey bee colonies living in this city is perhaps more plentiful than that available to colonies living deep in forests, so the results of this study probably *under*represent the difficulty of food collection by honey bees living in the wild.

The weight records, presented in Figure 6.2, show that winter was a time of dramatic weight loss for these colonies. On average, the two colonies studied lost 23.6 kilograms (52 pounds) each year between September and April. Except for approximately 1 kilogram (2.2 pounds) attributable to the removal of dead workers from the nest, these weight losses represent consumption of stored food—honey and pollen. I suspect that most of the 20-plus-kilogram (44-plus-pound) cost of overwintering comes from the high cost of rearing brood in winter. It is telling that the two colonies lost weight twice as rapidly in March, when brood rearing was intense (0.84 kilograms/1.85 pounds per week), as in December (0.42 kilograms/0.93 pounds per week), when the colonies were broodless.

Further evidence of the high energetic cost of midwinter brood rearing comes from experiments conducted in the 1930s by Clayton L. Farrar, an entomologist at the University of Wisconsin, who compared the winter weight-loss records of colonies with and without stored pollen, hence with and without winter brood rearing. The weights of colonies with pollen fell 22.7 kilograms (50 pounds), on average, between October and May, whereas those of colonies lacking pollen dropped only 11.8 kilograms (26 pounds) over the same period. Presumably, this 10.9-kilogram (24-pound) difference in average weight loss arose primarily from the higher energetic cost of thermoregulation that a colony incurs when it starts rearing brood in the middle of winter (see chapter 9). Whereas a colony without brood needs only to maintain the surface temperature of its winter cluster at about 10°C (50°F), a colony with brood must maintain its core temperature at approximately 35°C (95°F), the optimal brood-nest temperature for a honey bee colony.

Figure 6.2 also illustrates a second important fact about the annual cycle of honey bee colonies: the brevity of the period each year when colonies experience a net energy gain. On average, the two colonies that I

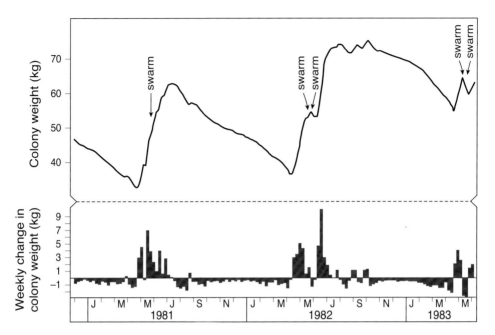

FIG. 6.2. Weekly changes in the weights of two honey bee colonies (hive plus bees and stored food) living in New Haven, Connecticut. The data shown here are representative of the weight-change patterns recorded for both colonies that were studied.

studied gained weight for only 14 weeks each year. What is even more striking, 86 percent of the annual weight gain by these colonies occurred between April 16 and June 30—a period of just 75 days.

The rather grim picture that emerges from this look at the energetics of colonies living in a cold climate is one of an ever-looming energy crisis. A colony consumes 20 or more kilograms (44 or more pounds) of honey each winter, yet it has little time each summer in which to rebuild its food reserves. As we shall see in the next section, this energy problem is especially acute for newly founded colonies, which, unlike colonies that are already established, cannot fall back on food reserves from previous years. Moreover, newly founded colonies must also cover the high costs of building their nests of beeswax combs. It may be that not all honey bee colonies experience such severe energy problems. Colonies living in milder climates, or in habitats richer in forage than those just discussed, may well find other problems—such as predators or shortages of nest sites—far more challenging. Nevertheless, for colonies of *Apis mellifera* living on their own in the northern parts of the species' range in Europe and North America, I think the primary obstacle to their survival is balancing the winter losses and the summer gains in their annual energy budgets.

ANNUAL CYCLES OF COLONY GROWTH AND REPRODUCTION

Correct timing of colony growth and reproduction is essential to honey bee survival in cold climates. As we have seen, a colony of honey bees survives the cold, flowerless days of winter through intensive thermoregulation fueled by a strategic reserve of some 20 kilograms (44 pounds) of honey that it has managed to amass over the previous summer. We will now see that this impressive feat of building a food reserve requires a colony to correctly time its growth and reproduction, for unless it does so, it will not have a sufficiently strong workforce at the right time of year to meet this life-or-death challenge.

The patterns of colony growth and reproduction across a year are easily studied. Colony growth patterns have been described in two ways: by periodically counting the number of brood-filled cells in a colony's nest or by making repeated censuses of a colony's population of adult bees. Describing the annual cycle of colony reproduction is a bit more difficult because it has distinct processes for females (queens) and males (drones). The seasonal pattern of colony reproduction by means of males is easily described by counting the cells containing developing drones. The seasonal pattern of colony reproduction through females could likewise be determined by monitoring the appearance of queen cells, the large and eye-catching cells that cradle developing queens (Fig. 6.3). However, because colonies frequently make false starts in their queen rearing—building queen cells but then destroying them before the developing queens ma-

FIG. 6.3. One of the large, peanut-shaped cells in which queen bees are reared.

ture—a more reliable method is to record the appearance of swarms, which are, after all, the true units of female reproduction for honey bees.

Figure 6.4 shows the growth pattern of the two honey bee colonies that I monitored in New Haven, Connecticut, from 1980 to 1983, based on bimonthly censuses of their brood. Following several months without brood in late autumn and early winter, the colonies begin rearing bees in January or February, evidently in response to increases in day length. At first, fewer than 1,000 cells containing brood are found in a bee hive or a bee tree, but in late March or April this number soars, climbing to a peak of some 30,000 or more developing bees per colony in May or June. Shortly thereafter there appears a gap in brood rearing when, because of the turnover in queens associated with swarming, the colony lacks an egg-laying queen for 10–20 days. Once the new queen eliminates her rivals and completes her mating, brood rearing resumes at nearly full tilt for a few weeks but then starts a gradual decline over the remainder of the summer, finally ceasing altogether in October. This annual pattern holds for honey bee colonies in temperate climates, although the precise timing of the rapid springtime expansion varies markedly with latitude. For example, colonies located in Somerset, Maryland, studied by Willis J. Nolan, entered the intensive growth phase three weeks before the colonies that I studied in New Haven, Connecticut, 250 kilometers (150 miles) to the north. Such geographic differences in annual cycle of brood rearing can be partly under genetic control, reflecting adaptation to the local climate and flora. This has been demonstrated by experiments conducted in France in which colonies were exchanged between Paris (northern France) and the Landes region (southwestern France). Beekeepers in the two locations knew that the principal peaks of brood rearing in their colonies came at different times: early summer in Paris and late summer in Landes. The remarkable finding of this colony-transplant study was that the colonies from each place kept their original (native) annual rhythms of brood rearing in their new locations.

Reproduction by honey bee colonies commences in late spring, shortly after the surge in worker brood production, and is largely completed by

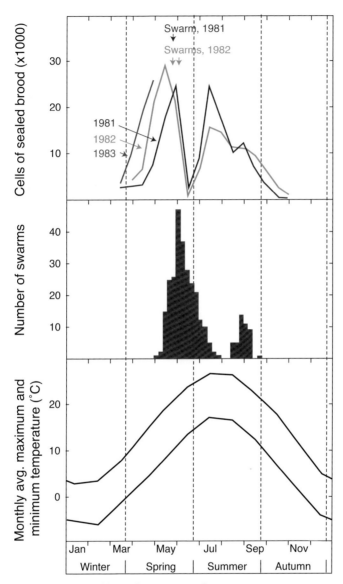

FIG. 6.4. Annual cycles of brood rearing and air temperature (in New Haven, Connecticut) and of swarming (in Ithaca, New York). *Top:* Mean number of cells of sealed brood (pupae) in a colony throughout the year for two years, and during the start of a third year. *Middle:* Seasonal occurrence of 301 swarms collected over the 10-year period of 1971–1981. *Bottom:* Annual pattern of air temperature for New Haven, Connecticut.

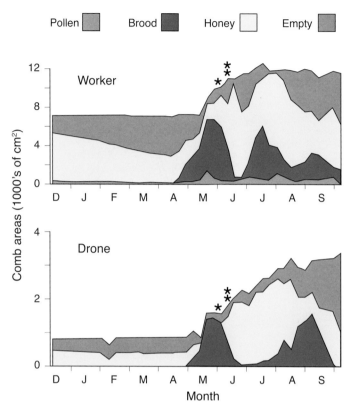

FIG. 6.5. Comb contents of a full-size, unmanaged colony living in a large observation hive in Ithaca, New York, from December 2012 to October 2013. Comb area amounts represent both sides of the colony's single sheet of comb. Asterisks mark the departure dates of the prime swarm and two afterswarms.

the start of summer. For example, in a study conducted at the Rothamsted Experimental Station in southern England, drones were found to constitute 9 percent of a colony's brood in May and June, but only 1 percent in July and August. The peak in drone production in any given locale usually comes a few weeks before the peak in swarm production, for this ensures that colonies will have mature drones ready for mating when the virgin queens produced in the swarming colonies will be conducting their mating flights. Figure 6.5 shows, for example, how a strong colony living in a large observation hive at my laboratory in Ithaca had a burst of drone rearing that began in late April, thus well in advance of this colony's production of a prime swarm and two afterswarms in late May and early June. Figure 6.4 shows that the timing of these three swarms was typical for Ithaca. Of 301 swarms collected in and around Ithaca in the years 1971–1981—in response to swarm calls received at the Dyce Laboratory for Honey Bee Studies at Cornell—84 percent appeared between May 15 and July 15. Reports from throughout North America and Europe confirm this pattern of late spring–early summer swarming, although, as with brood rearing, there exists geographic variation in the peak time of swarming.

One of the most amazing features of honey bee colonies is their ability to begin rearing brood in winter and thereby grow to swarming strength by late spring or early summer. Swarming this early in the year is highly adaptive, because doing so provides the new colonies (founder colonies) with as much time as possible to build their nests, rear more bees, and amass a large store of honey before winter calls a halt to their foraging. Even so, only a small fraction of these founder colonies make it through their first winter alive. On average, for honey bees living wild in the forests around Ithaca, only 23–24 percent of the founder colonies established each summer will be alive the following spring, whereas 78–82 percent of the established colonies—those that have already survived at least one winter—will make it through the cold, snowy months that stretch from November to April.

In the late 1970s, my colleague Kirk Visscher and I were so intrigued by the ability of honey bee colonies to start rearing brood in the middle of our cold, snowy winters that we decided to experimentally test its adaptive significance. We did so by performing two experiments. In the first, we compared colonies whose onsets of brood rearing were normally timed vs. colonies whose onsets of brood rearing were delayed experimentally from midwinter to mid-spring—that is, to April 15, when plentiful forage is normally available. We found a striking difference in the populations of the two types of colonies on May 1: on average, the control colonies contained 10,800 bees, but the treatment colonies contained only 2,600 bees. Furthermore, the control colonies swarmed much earlier than the treatment colonies: mid- to late May vs. late June to early July. This study showed that the amazing ability of honey bee colonies to start rearing brood in the middle of winter greatly helps them build strength for the coming summer.

In our second experiment, we examined the consequences of swarming early vs. late. To do so, we compared the probabilities of winter survival of colonies that we started—using artificial swarms (packages)—on May 20 vs. June 30. We chose these two dates because they are 20 days before and 20 days after the median date of swarming by colonies living in and around Ithaca (see Fig 6.4). We installed each artificial swarm in a one-story Langstroth hive containing 10 frames that held only beeswax foundation, so each colony had to build its combs from scratch, as in nature. In three out of the four years in which we performed this experiment, forage was either extremely sparse or extremely abundant, and in these three years the colonies either all died over the following winter (in the year with meager forage) or they nearly all survived (in the two years with plentiful forage). In the year with only moderate forage, however, nearly all the early swarms survived but all the late swarms starved.

Taken together, these two experiments demonstrated the importance of midwinter brood rearing and early spring swarming to the success of honey bee colonies living in cold temperate regions of the world. In these places, the time window in which colonies can collect a surplus of honey is small, but each colony's need for honey so it can survive the winter is large. Given this predicament, natural selection favors honey bee colonies that begin rearing brood in winter so that they will be strong in early spring. This early growth in colony strength sets the stage for early colony reproduction, which in some years is critical to the survival of the new colonies started up by swarms.

THE UNIQUE ANNUAL CYCLE OF HONEY BEE COLONIES

This chapter began by noting that the annual cycles of honey bee colonies and bumble bee colonies differ markedly. A colony of honey bees is active year-round, maintains a warm microclimate in its nest throughout winter, times its growth and reproduction to peak in early summer, and reproduces by rearing males and casting swarms. A colony of bumble bees, however, becomes active in spring as a solitary queen, then grows by rearing workers, eventually switches to rearing males and queens in summer, and finally disintegrates in autumn. Come winter, the only members of a bumble bee colony still alive are the young, newly mated queens, who, if lucky, will survive the winter and initiate their own colonies in the spring.

At first glance, it might seem that the annual cycles of honey bees and bumble bees differ so markedly because honey bees have achieved a breakthrough in their means of winter survival. On the one hand, a colony of honey bees creates a warm microclimate inside its nest by pooling the metabolic heat produced by its worker bees, which are fueled by the honey stored in the colony's nest. On the other hand, bumble bees solve the problem of winter survival more simply: queen bumble bees add antifreeze materials to their blood and then become dormant in their underground burrows. It is important to note that even though the mechanisms of winter survival of bumble bees are considerably simpler than those of honey bees, they are also much more effective. There is, for example, a species of bumble bee, *Bombus polaris*, that thrives in tundra habitat above the Arctic Circle (latitude 66°34′ N), which is far, far beyond the northern limit of the honey bee, *Apis mellifera*, unless it receives assistance from beekeepers (see Fig. 1.2). Why do honey bees and bumble bees differ so profoundly in their annual cycles and in their mechanisms of winter survival? The answer, I believe, is rooted in the difference between their ancestral environments: warm tropical regions for honey bees and cool temperate regions for bumble bees.

Tropical regions are the ancestral homes for two groups of highly social bees: honey bees and stingless bees. Although colonies of honey bees and stingless bees differ in many ways, they share two fundamental traits that reflect their common heritage as tropical social bees: 1) colony life spans of several years, and 2) colony reproduction by swarming. Probably the reason the colonies of both honey bees and stingless bees have multiyear life spans is that their ancestors lived in the tropics where they had no need for a solitary phase (i.e., a hibernating queen) in a one-year (annual) cycle. It also seems clear that the reason the colonies of both honey bees and stingless bees reproduce by swarming is also because both groups lived originally in the tropics, where incipient colonies would have been highly vulnerable to ant predation if each queen were not accompanied by a throng of worker bees to guard her when she left her natal nest to start a new colony. Robert L. Jeanne, a world authority of the biology of social wasps, has shown that ant predation on unguarded nests of social wasps is much more intense in tropical forests than in temperate woodlands. He did so by monitoring the disappearance of wasp larvae placed in vials accessible only to ants and found that in Costa Rica (latitude 10° N) more than 86 percent of the larvae were removed within 48 hours but in the United States (in New Hampshire, latitude 43° N) less than 50 percent were removed in the same period.

I believe that when *Apis mellifera* expanded north out of the tropics and into temperate regions it was constrained by the complex social organization of its colonies—one highly fecund mother queen supported by thousands of daughter workers—in how it could adapt for living in cold climate places. Certainly this species did not overhaul its social organization to survive winters in bumble bee fashion, as a residue of solitary, dormant queens. Moreover, *Apis mellifera* did not revise its physiology in ways that would enable whole colonies to become dormant. Instead, the honey bee evolved along what was presumably its easiest path to achieve winter survival in temperate regions—by tweaking its existing biology. I suspect that this tweaking included adjusting the nest-site preferences, refining the mechanisms of colony thermoregulation, increasing the size of honey stores, and fine-tuning the annual rhythms of colony growth and reproduction. In summary, I believe that the unique annual cycle of *Apis mellifera* as it lives in temperate regions of the world is best understood as being "built on top" of this bee's original biology as a tropical social insect.

7

COLONY REPRODUCTION

> An individual is fit if its adaptations are such as to make it likely to contribute a more than average number of genes to future generations.
> —George C. Williams, *Adaptation and Natural Selection*, 1966

To understand how reproduction works in honey bees, it helps to note some similarities between bee colonies and apple trees in how they pass on their genes to future generations. First, they both function as simultaneous hermaphrodites, which is to say that they both propagate their genes by producing each summer both female and male units of reproduction: queens or seeds on the one hand, and drones or pollen grains on the other. Second, they both send forth their female units of reproduction tucked inside a large and intricate structure—a swarm of some 10,000 worker bees (Fig. 7.1) or an apple of many thousands of plant cells. Doing so helps ensure the safe dispersal of the queen sheltered within the swarm and the seeds buried inside the apple. Third, they both dispatch their male units of reproduction "naked" and thus cheaply. The drones leave their colony's nest on mating flights all on their own, and the microscopic pollen grains leave their tree's flowers stuck to the hairs of bees. And fourth, because in both living systems each female unit of reproduction is many thousand times larger and costlier than each male unit, they both produce the two types in vastly different numbers. Over a summer, a vigorous bee colony will send forth only a few queen-containing swarms but many thousands of

drones, and a healthy apple tree will produce only several hundred seed-containing apples but millions of pollen grains.

This chapter aims to present a clear picture of the reproductive biology of honey bee colonies when they live in the wild. It will do so by looking at what a colony does to pass on its genes when its reproductive habits are not being manipulated by beekeepers. Most beekeepers manage their colonies in ways that hamper the colonies' production of swarms and drones—in order to boost their production of honey—so until recently we have looked at the reproductive biology of *Apis mellifera* mainly from an apicultural perspective. But this gives us a distorted picture of the subject. Fortunately, several recent studies provide detailed information about the natural reproductive habits of honey bee colonies, and we will focus on what these studies have revealed. We will also examine the inner workings of colonies whereby they adaptively regulate their production of swarms and drones. Our goal is to understand how a wild honey bee colony manages its reproductive affairs in ways that help it "to contribute a more than average number of genes to future generations."

DRONE PRODUCTION PEAKS BEFORE QUEEN PRODUCTION

Although reproductive success by a honey bee colony involves producing both fertile drones and big swarms, the two production processes do not unfold in perfect synchrony. Instead, a colony usually has a peak in its drone production approximately 30 days before the colonies in its neighborhood begin casting swarms and then sending forth virgin queens to be mated. The reason is simple. Drones have a 24-day developmental period, and they require another 12 or so days after emerging from their brood cells to reach sexual maturity. (Queens have much shorter times for development and sexual maturation: about 16 days and 6 days, respectively.) So, if a colony is to have a maximum number of sexually mature drones ready for active service at the time of year when virgin queens are most

FIG. 7.1. A honey bee swarm, with approximately 12,000 worker bees and one queen bee, resting safely inside the cluster.

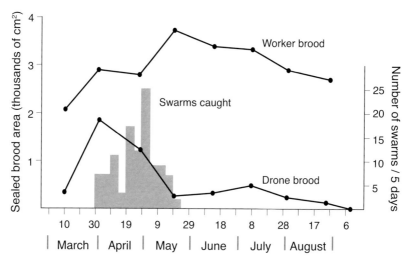

FIG. 7.2. Levels of sealed brood (pupae) in 16 test colonies in 1978, and frequency of swarming for colonies living in and around Davis, California, in 1979.

abundant—the swarming season—then it must start rearing its drones long before the seasonal peak of swarming. This is precisely what was found in a study conducted by Robert E. Page Jr. in Davis, California. As shown in Figure 7.2, the production of drone brood in Page's 13 study colonies peaked early in the spring, at the start of April, whereas the production of swarms peaked in early May, some 30 days later. The same phenomenon is shown in Figure 6.5, which shows that a colony living in a large observation hive in Ithaca began rearing its drones in late April, hence one month before it began casting its swarms in late May.

How many drones does a colony produce when it is living on its own and so is free to build as much drone comb and rear as many drones as it wants? This question can be answered using data from two papers that report measurements across a summer of the area of drone brood in colonies that were unmanaged and had a normal fraction (ca. 20%) of drone comb in their hives. The first is the study just mentioned, conducted by Robert E. Page Jr. in Davis, California, in 1978, and the second is the study conducted by Michael L. Smith and colleagues in Ithaca in 2013, mentioned in chapter 6 (see Fig. 6.5). Page measured the areas of capped drone cells (pupal brood) in 13 colonies living in standard, movable-frame hives with two frames of drone comb per 10-frame hive body. Smith et al. measured the areas of drone comb that contained brood (eggs, larvae, and pupae) in four colonies that were living in large observation hives filled with natural combs that the colonies had built the previous summer. For each study, I have converted what the authors reported—measurements of *area* of occupied drone comb on various days across the summer—to estimates of *number* of drone cells occupied on each sampling date. I then calculated the total number of drones produced per colony across the summer by calculating the total number of occupied brood cell-days per colony throughout a summer and dividing this by the relevant development time for drones: the 14-day capped brood period for the Page study and the 24-day entire brood period for the Smith et al. study. The two studies yielded similar values for the average number of drones produced over a summer by an unmanaged colony living in a hive with a normal amount of drone comb: 7,812 drones (Page) and 6,949 drones (Smith et al.). The significance of these two numbers will become clearer later in the chapter, when we use them to compare the investments that colonies make in their male (drone) and their female (queen) means of reproduction.

Because honey bee colonies benefit from starting to rear drones in early spring, and because they have filled their drone comb with honey in the previous summer and fall, they often face a problem in early spring of having many of the cells in their drone comb plugged with honey. It is not surprising, therefore, that colonies preferentially *remove* honey from their drone comb in the spring, when this comb is needed for rearing drones, and preferentially *store* honey in their drone comb in late summer and autumn, when this comb is best used for honey storage. This seasonal shift in the use of drone comb for honey storage was demonstrated recently in a study Michael L. Smith and colleagues conducted in Ithaca, in which once a month, from April to September, they installed in the hives of several colonies two frames of comb—one of drone comb and the other of worker comb—whose cells they had filled with thick sugar syrup. The two test frames installed in each colony's hive were positioned on opposite sides of two frames that contained brood; this ensured that there were nurse bees near both test frames. Fourteen days later, the

Fig. 7.3. Drone comb before (top) and after (bottom) being placed in a colony for 14 days in April, when the colony was preparing to rear drones en masse. In both images, the cells with reflections contain sugar syrup. The bottom frame shows that the bees removed "honey" from the center of the drone comb to make space for rearing drones. The study colony was living in Ithaca, New York.

investigators removed the two test frames from each hive and measured in each the area of comb that had been cleared of sugar syrup (see Fig. 7.3). They found that in April and May, the average comb area cleared was markedly *larger* for drone comb than for worker comb, and that in August and September the pattern was reversed: the average comb area cleared was noticeably *smaller* for drone comb than worker comb. Presumably, the workers in these colonies did not remove much sugar syrup from their drone comb in late summer and autumn because they knew that drone production was no longer the most important use of this special, large-celled comb.

QUEEN PRODUCTION AND SWARMING

The life cycle of a honey bee colony can be regarded as beginning in the spring when an established colony builds up its worker population and starts rearing a batch of queens in preparation for swarming. The first step in these preparations is the construction of queen cups along the lower margins of the colony's brood-nest combs. These queen cups, tiny inverted bowls made of beeswax, form the bases of the large, ellipsoidal cells in which queens are reared (Fig. 6.3). Next, the queen lays eggs in a dozen or more of the queen cups, and workers feed the hatching larvae the royal jelly that ensures their development into queens. The formation of these new queens is remarkably rapid; only 16 days pass from when an egg is laid to the moment an adult queen climbs from her cell. As the daughter queens develop, changes unfold simultaneously in the physiology of the mother queen in the colony. With each passing day, she is fed less and less by the workers. Her egg production declines, and her abdomen, no longer swollen with fully formed eggs, shrinks dramatically. Furthermore, the workers begin to shake their queen, grabbing onto her one at a time with their front legs and letting loose a volley of five or six shaking movements. These bouts of shaking, which can eventually reach a frequency of 40 to 80 per hour, appear to force the queen to keep walking about the nest. This exercise, together with reduced feeding, results in a 25 percent reduction in the queen's body weight. Shortly after the first queen cell is capped, the mother queen flies off in a swarm of some 10,000–20,000 workers, leaving behind only about a quarter of the colony's population of worker bees in the parental nest. After flying a short distance, the swarm condenses into a beard-like cluster on a tree branch (Fig. 7.1). From here the swarm's scout bees explore for nest cavities, select one that is suitable, and finally signal the swarm to break cluster and fly to the chosen home-site. The new dwelling place is rarely less than 300 meters (ca. 1,000 feet) from the bees' original residence, and can be 3,000 or more meters (more than 2 miles) away.

For about eight days following the departure of their mother queen, the workers in the parental nest are queenless, but this situation ends with the emergence of the first daughter queen. If the colony is still greatly weakened by the departure of the first swarm—what beekeepers call the prime swarm—then the remaining workers allow the daughter queen that emerges first to search through the nest to find her rival sister queens and kill them, by stinging them while they are still in their cells. Usually, however, by the time the first daughter has appeared, enough young worker bees have emerged from cells in the brood combs to restore the parent colony's strength. In this situation, the workers guard the remaining queen cells against destruction by the first daughter queen, they start shaking this queen in preparation for flight, and eventually they push her out of the nest in an afterswarm. As is shown in Figure 7.4, this process of afterswarming may be repeated with another daughter queen, and when this happens the colony is usually left weakened to the point where it cannot support further division. At this point, if there remains more than one daughter queen roaming the parental nest, the workers allow these queens to fight each other until just one remains alive. It is she who, partly by luck and partly by skill, inherits the parental nest with its rich endowments of beeswax combs and honey stores, both of which are immensely important assets that will give the colony living in the parental nest a high likelihood of survival through the coming winter.

The production of afterswarms depends strongly on a colony's residual strength—measured in workers and especially in brood—after the prime swarm has departed, so the number of afterswarms produced per episode of colony reproduction varies greatly. Fortunately, the results from several detailed studies of swarming and afterswarming by unmanaged colonies make it possible to put probabilities on the events shown in Figure 7.4. First, we know from long-term studies (described in the next section of this chapter) that, on average, the annual probability of queen turnover within an unmanaged colony living in the region of Ithaca is 0.87. There-

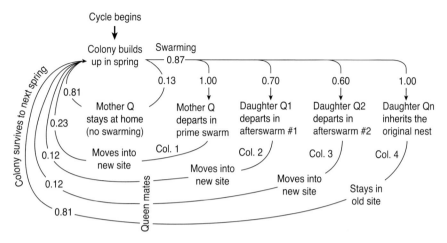

FIG. 7.4. Principal events in the life cycle of honey bee colonies, starting in the spring when a colony builds up its worker population, which sets the stage for swarming. Q = queen. Numbers along the lines denote the probabilities of the various events (e.g., the probability of a colony swarming after building up in the spring is 0.87).

fore, 0.87 is a good estimate of the probability that on any given year the mother queen in a colony will leave her nest in a prime swarm and will occupy a new nest cavity located several hundred or several thousand meters away. Second, we know from painstaking studies performed by Mark L. Winston, working in Lawrence, Kansas, and by David C. Gilley and David R. Tarpy, working in Ithaca, a great deal about the fates of the daughter queens that are produced in unmanaged colonies after the mother queen has departed in the prime swarm. Gilley and Tarpy, for example, worked with colonies living in large observation hives that enabled them—aided by a team of helpers—to monitor continuously the activities of the daughter queens in five colonies, each of which had cast a prime swarm. They maintained a round-the-clock surveillance of the daughter queens in their observation hive colonies until each one had either departed or been killed, except the one who inherited the parental nest. Taken together, the Winston and the Gilley and Tarpy studies show us that in an unmanaged colony that has produced a prime swarm, the probability that one of the daughter queens will leave in a first afterswarm is 0.70, the probability that another daughter queen will leave in a second afterswarm is 0.60, and the probability that a third daughter queen will inherit the original nest (after killing all her rivals) is 1.00.

The process of queen production and colony foundation is completed when all the surviving daughter queens have flown from their nests and mated with drones from neighboring colonies (discussed later in this chapter). At this point, the colonies that have moved into new nest sites have begun building their nests, and all the colonies—including the one occupying the old nest—are rearing brood to build up their populations and are foraging intensively to build up their honey stores to prepare for winter. The probability of surviving the coming winter is quite high, approximately 0.81, for the fortunate daughter queen that inherited the old nest and its store of honey. Sadly, for her mother queen and her sister queens, whose colonies must build new nests from scratch, the probabilities of winter survival are much lower, often less than 0.20, for reasons that will be discussed shortly.

One might wonder, why has natural selection favored mother queens who leave the old nest in a prime swarm and thereby incur a low probability of winter survival in a new nest? I think the answer is simple: by leaving in the prime swarm rather than lingering in the old nest, a mother queen dodges the high risk of being killed by one of her daughter queens when they start emerging from their cells. The danger to the mother queen of staying at home is borne out by the data on regicide committed by virgin queens, as reported by Gilley and Tarpy, and by M. Delia Allen, who worked in Aberdeen, Scotland, in the 1950s. These three investigators report the fates of 44 virgin queens who were reared in six study colonies that were living in observation hives and that swarmed. The researchers observed that within a colony, on average, one virgin queen left in an afterswarm, one virgin queen inherited the original nest, and 5.3 virgin queens died from being stung by a fellow virgin queen. Clearly, the mother queen does well to flee the killing field of her old nest before her murderous daughter queens emerge from their cells.

HOW A POPULATION OF WILD COLONIES IS PERSISTING

Under favorable conditions, a colony that is alive at the end of a summer will survive the following winter and will go on to reproduce the next summer. Colonies living on their own, however, don't always experience favorable conditions, and many perish over winter through starvation, disease, or failure to replace a senescing queen. If the rate of colony deaths exceeds the rate of colony births (through swarming), then the population of colonies in a region will shrink and may even expire. In chapter 2, we reviewed evidence that the population of wild honey bee colonies living in and around the Arnot Forest has been stable since it evolved resistance to *Varroa* mites in the 1990s. Let us now examine how this population of wild colonies can persist. We will do so by reviewing what I have learned about the patterns of colony generation and colony loss for honey bees living on their own in the wild places outside of Ithaca. What we know about these matters comes from two long-term studies that I made—in 1974–1977 and in 2010–2016—on the demography of wild colonies.

Both studies comprise two avenues of investigation: one of *reproduction* (swarming) by simulated wild colonies (SWCs) living in movable-frame hives, and one of *survival* by wild colonies living in natural nest sites. The hive-based work on colony reproduction involved setting up approximately 20 SWCs in separate, secluded places. Each SWC was established by catching a natural swarm—either by collecting it from its bivouac site or by capturing it in a bait hive—and then installing it in a 10-frame Langstroth hive (Fig. 7.5). The hive contained two frames of drone comb and eight frames of worker comb, and its entrance was reduced to a small, natural-size opening. In short, each SWC occupied a hive that simulated a natural cavity, except that its wooden walls were thinner and its entrance was lower. I labeled the queen in each colony with a paint mark so that I could detect turnovers of a colony's queen—presumably by swarming, primarily. I kept the colonies' queens labeled over the years by applying paint marks to any unlabeled queens I found during colony inspections. I

FIG. 7.5. One of the hives used for housing a simulated wild colony. The blue structure is a screen board used in getting counts of how many *Varroa* mites dropped onto a sticky board in 24 hours. Each colony's hive was permanently equipped with a screen board to make noninvasive measurements of the colony's mite load.

inspected each SWC three times each summer, in early May, late July, and late September. This meant that each colony was inspected before and after the main swarming season for the area around Ithaca (mid-May to mid-July) and before and after the secondary swarming season (mid-August to mid-September). These inspections served two purposes: to check each colony for a turnover in its queen (probably by swarming) and to check each colony for brood diseases.

The work on colony survival required finding dozens of wild colonies. I did this by hunting for bee-tree colonies in the Arnot Forest, by responding to swarm calls (which often led to discovery of the swarm's source in a tree or building near the swarm), and by putting out the word that I was looking for colonies of bees living in trees. Once I found the colonies, the rest was simple. Three times each summer—in early May, late July, and late September—I visited all the sites that I knew were (or recently had been) occupied by a wild colony of honey bees. These visits revealed whether a colony was still alive at each site. They also revealed whether a site vacated by a colony's death in the past had been recolonized by a swarm. In 1974–1977, I monitored 42 nest sites: 26 in trees and 16 in rural buildings (hunting cabins, barns, and farmhouses). In 2010–2016, I monitored 33 sites: 20 in trees and 13 in rural buildings.

The work in the 1970s and the 2010s produced surprisingly similar findings about how honey bee colonies survive and reproduce when they live on their own. They are summarized in Table 7.1. The first column in this table shows the probabilities that an established colony either did (p = 0.87) or did not (p = 0.13) swarm over a summer. We see that most, but not all, did so. The next columns in this table show the probabilities of the various events that can happen after the prime swarm has left the nest of an established colony: p = 1.00 that the mother queen occupies a new nest; p = 1.00 that a daughter queen inherits the old nest; p = 0.70 that a daughter queen leaves in a first afterswarm; and p = 0.60 that another daughter queen leaves in a second afterswarm. Table 7.1 also shows the probabilities of survival to the following year for the colony headed by the mother queen (p = 0.23) and for the colonies headed by the various daughter queens: the one that inherits the old nest (p = 0.81), the one that flies off in afterswarm one (p = 0.12), and the one that departs in afterswarm two (p = 0.12). The value of p = 0.23 for the survival of a new colony founded by a mother queen is the probability that a prime swarm will survive to the following summer. I show a more somber value, p = 0.12, for the probability of survival to the following summer for a colony founded by an afterswarm. I do so because an afterswarm, relative to a prime swarm, encounters more obstacles in its path toward surviving to the following spring: it starts building its nest later in the summer, it

TABLE 7.1. The probability of a colony swarming, and, if it does, the probabilities of various events thereafter. Also, the probabilities of colony survival to the following summer for the mother queen and the various daughter queens, and the overall probabilities of survival to the following summer for the various kinds of "offspring" colonies. "Q" = queen.

Prob. of colony swarming (SW)	Events post-swarming	Prob. of event (E)	Prob. of survival (S)	Prob. overall (SW × E × S)
0.87	mother Q occupies new nest	1.00	0.23	0.20
	a daughter Q inherits old nest	1.00	0.81	0.70
	a daughter Q leaves in afterswarm #1	0.70	0.12	0.07
	a daughter Q leaves in afterswarm #2	0.60	0.12	0.06
				1.03 colonies
Prob. of not swarming (nSW)	Event post-not swarming			Prob. overall (nSw × E × S)
0.13	mother Q stays in old nest	1.00	0.81	0.11 colonies

suffers a delay in initiating brood rearing because its queen needs about two weeks to get mated and begin laying eggs, and it risks losing its queen when she conducts her mating flight.

I draw two take-home messages from the results shown in Table 7.1. The first is that even though most (ca. 87%) of the wild colonies living around Ithaca swarm each summer, and even though the colonies that swarm often produce multiple swarms (one prime swarm and 1.3 afterswarms, on average), the net growth rate of this population of colonies is low: only 0.14 new colonies are added to the population per existing colony each year. (On average, for every colony that is alive in the spring of year one, the expected number of colonies alive in the spring of year two is 1.03 + 0.11 = 1.14.) Although low, this rate of population growth seems to be sufficient to enable this population of wild colonies to recover from a bout of heavy mortality caused by a summer with poor foraging or a winter that is especially demanding. The second take-home message is that this population appears to be close to the carrying capacity—the maximum sustainable population—of the fields and forests around Ithaca. Certainly, the low probabilities of survival for colonies moving into new nest sites suggest that it is difficult for newly founded colonies to insert themselves into the population of established colonies.

HOW LONG IS A BEE TREE ALIVE WITH BEES?

Once a swarm moves into a hollow tree or other natural nesting site, how long will this site be continuously inhabited by a honey bee colony? In other words, how long will the site be alive with bees? To answer this question, I calculated the average "life span" of an occupied nest site using the probabilities shown in Table 7.1. I did so by multiplying each possible site "age" (in years, e.g., 0, 1, 2, 3, . . . 20) by the probability of the site "death" at that age, and then summing up the 21 numbers produced by this multiplication. To this sum, I added 0.5 year. (One needs to add a half year because an individual's life span at death is, on average, one half year more than its age, expressed in whole years, at death. E.g., a person who dies at age 80 might actually have lived 80 years and 6 months.) I calculated the age-specific probabilities of site death by multiplying the probability of site survival to each age times the probability of site death at that age. In making these calculations, I assumed that the site was colonized by a swarm containing a mother queen (hence was a prime swarm), so that the site's probability of survival to age one was 0.23. For each age after that, I used 0.81 for the probability of site survival, since this is the survival probability for an established colony, whether it has swarmed that year (so a daughter queen has inherited the site) or it has not (so the mother queen remains in the site). These calculations yield an estimate of 1.7 years for the average life span of the occupation of a bee tree in the Ithaca area. It is rather short because a colony living on its own in a tree or building has a sadly low probability ($p = 0.23$) of surviving its first winter. If, however, the founding colony survives its first winter, then the probability of colony survival in this site each year thereafter is much higher ($p = 0.81$), which means that the site is likely to be alive with bees for several more years. The calculations show that, on average, a site whose founding colony has survived

FIG. 7.6. Worker bees clustered around their colony's nest entrance, high up (9.4 meters/31 feet) in a bigtoothed poplar (*Populus grandidentata*) in the Arnot Forest. Thrice-yearly inspections indicate that this bee tree was occupied continuously from when it was found on 27 August 2011 to 20 Sept 2017, hence for six years. The most recent inspection, on 8 May 2018, revealed that the colony living here died over the brutally long and cold winter of 2017–2018.

its risky first winter will be occupied continuously for another 5.2 years—and not just in theory. In my six-year (2010–2016) program of monitoring 33 nest sites occupied by wild colonies, there were eight sites that I monitored for all six years, and two were occupied throughout these years, while two others were occupied continuously for four years. Moreover, one site had achieved a seven-year record of continuous occupation as of May 2017. Figure 7.6 shows a site that had achieved a six-year record of continuous occupation as of September 2017. The entrance opening is a small knothole that sits high in a massive, big-toothed poplar (*Populus grandidentata*) and faces southwest.

INVESTMENT RATIO BETWEEN DRONES AND QUEENS

The most striking feature of the reproductive biology of honey bees is their astonishingly skewed sex ratio. As we have seen earlier in this chapter, each year a typical wild colony rears to adulthood about 7,500 drones, but it produces only 2.3 swarms: one prime swarm and (on average) 1.3 afterswarms. Evolutionary theory predicts, however, that in the simplest scenario—large breeding population, homogeneous individuals, random mating, strong outcrossing, and so forth—natural selection will favor an equal allocation of resources to the production of male and female offspring. This is because half the genes within every individual in the population (i.e., within every colony of honey bees) come via the reproductive success of males (via their sperm) and half come via the reproductive success of females (via their eggs), so male and female functions are equally effective means to achieving genetic success. Thus, it is puzzling, at least at first glance, that a honey bee colony produces thousands of drones, but only a few queens, each summer.

As a first step toward resolving what seems to be a huge disparity between theory and reality, let us consider what constitutes a honey bee colony's total investment in its reproduction via males and via females. For male offspring, this is straightforward. It is simply whatever resources the colony spends to produce its drones and support them throughout their lives. For female offspring, the situation is more complex. I believe that it is correct to calculate a colony's total investment in its reproduction via

TABLE 7.2. Calculation of the number of worker bees a honey bee colony invests, on average, in the swarms that it produces over a summer

Probability of swarming, P_{sw}	Swarm type	Probability of swarm type, P_{type}	Average no. of workers, W	Expected no. of workers invested $P_{sw} \times P_{type} \times W$
0.87	Prime	1.00	16,033	13,949
	Afterswarm #1	0.70	11,538	7,026
	Afterswarm #2	0.60	3,926	2,049
				23,024 total

females by including everything that it spends to produce its swarms. This follows from the analogy between bee colonies and apple trees in how they reproduce through females: honey bee colonies produce queens sheltered inside swarms and apple trees produce seeds buried inside apples.

This way of looking at the problem reveals a way to compare how much a honey bee colony invests in reproducing via males and via females: by determining the total dry weight of the bees that a colony produces for its reproduction via drones and comparing it to the same thing for its reproduction via swarms. Let's first calculate this dry-weight measure of investment for drones. We have seen already that measurements of the areas of drone brood in unmanaged colonies by Robert E. Page and Michael L. Smith et al. indicate that their colonies produced, on average, 7,812 and 6,949 drones over a summer, so the average is 7,380 drones per colony per summer. The dry weight of a single drone is 45 milligrams (0.0016 ounce), so the average total investment in a colony's drones (measured in terms of dry weight) over a summer is 7,380 × 45 milligrams = 332 grams (about 11.7 ounces) of dried drones.

What is the total dry weight of a colony's investment in swarm bees? The calculation is shown in Table 7.2, and it shows that, on average, a colony produces 23,024 workers (females) over a summer in support of its swarming. The dry weight of a worker bee is 17 milligrams (0.0006 ounce), so the average total investment in a colony's swarms (measured in terms of dry weight) over a summer is 23,024 × 17 milligrams = 391

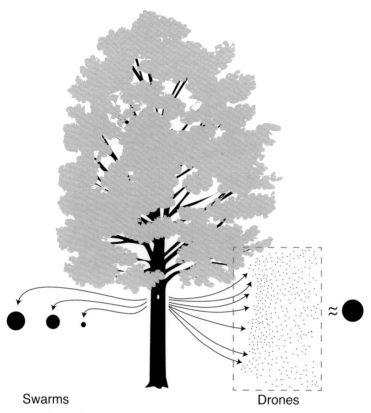

FIG. 7.7. Graphical depiction of the relative investment in female reproduction (swarms) and male reproduction (drones), on average, by a honey bee colony. The areas of the three circles representing swarms are proportional to the average investment—measured in dry weight of the bees—a colony makes in producing a prime swarm, a first afterswarm, and a second afterswarm. The area of the circle representing drones is proportional to the average investment a colony makes in producing drones. Even though the numbers and sizes of swarms and of drones differ greatly, their total dry weights are nearly the same.

grams (about 13.8 ounces of dried workers). These two values, 332 grams (of drones) and 391 grams (of workers), tell us that colonies do indeed invest approximately equal resources to building their male and female units of reproduction (Fig. 7.7). I suspect that these two estimates of a colony's investments in female and male reproduction would be even

Fig. 7.8. Worker feeding a hungry young drone even before he has climbed from his cell.

more closely matched if they included the cost of fueling the workers and drones for their contributions to colony reproduction. The fueling of the reproductive efforts of workers occurs just once, shortly before they leave in a swarm, but the fueling of the drones extends throughout their lives, starting right when they emerge from their cells (Fig. 7.8) and continuing until they die, usually a few weeks later.

OPTIMAL SWARM FRACTION

If you are a beekeeper, then you know that one of the more dismaying experiences you can have is to pop the lid off a hive during a nectar flow and discover that it is no longer brimming with bees diligently filling the storage combs with honey. Instead, the hive looks downright deserted. Curses, the colony has swarmed! Gone are most of this colony's bees, and gone with them are your prospects of a bountiful crop of honey from what was, until recently, a top-notch colony. One wonders, why does such a large fraction of the worker bee population leave in the prime swarm? In recent years, two of my colleagues, Juliana Rangel and H. Kern Reeve, and I have investigated how natural selection has tuned the "swarm fraction," that is, the percentage of a colony's workforce that leaves when the colony casts a prime swarm (the one with the mother queen). We were attracted to investigating this part of the functional design of honey bee colonies because it is a part of their biology that is not manipulated by beekeepers, so remains fully under the bees' control.

The problem that the bees must solve is this: How should the adult workers divide themselves between the new and old colonies? We know that as soon as the new colony—headed by the old queen (mother to the workers)—moves into its dwelling place, its workers begin to tackle the huge job of building a set of combs. Meanwhile the old colony—headed by a new queen (sister to the workers)—inherits an abundance of resources in the parental nest, including a full set of combs, much brood, and often a sizable store of honey. Given this marked asymmetry in the resources possessed by mother-queen and sister-queen colonies, one expects that a swarming colony will need to devote a large fraction of its workforce to the swarm, but how large a fraction is optimal?

We studied this matter in a two-part investigation. First, we built an inclusive fitness model for the optimal allocation of workers between the two colonies, based on the insight from evolutionary biology that the workers should distribute themselves between the mother-queen colony and the sister-queen colony in a way that will maximize the genetic success of the workers. The model factors together three things: 1) the genetic relatedness (r) of a worker to the offspring produced by each queen; 2) the winter survival probability of each colony, $s_{mother}(x)$ and $s_{sister}(x)$, if fraction x of the workers in the original colony departs with the mother queen; and 3) the expected reproductive success of each colony, $w_{mother}(x)$ and $w_{sister}(x)$, that has survived the winter if fraction x of the adult workers in the original colony departs with the mother queen. The variable x is called the "swarm fraction," and it refers only to the *adult* workers in the original colony, for it is only these workers that can decide how to distribute themselves between the mother-queen and sister-queen colonies.

To use this model to predict the optimal swarm fraction, we had to determine the winter-survival probabilities for mother-queen and sister-queen colonies as a function of the swarm fraction. We did so by making an artificial swarm from each of 15 colonies in June 2008. This involved removing the mother queen and some portion (0.90, 0.60, or 0.30) of each colony's workers to put in the artificial swarm. There were five colonies in each treatment (swarm-fraction) group. After we had installed each of our artificial swarms in a 10-frame hive (equipped with frames holding beeswax foundation), we left it alone to build combs, collect food, and rear brood. We did, however, check each colony once a month, from July 2008 to April 2009, to see whether it was still alive. This work yielded the following values of winter-survival probability (p) for the mother-queen colonies as a function of swarm fraction (sf): $p = 0.80$, for sf = 0.90; $p = 0.20$, for sf = 0.60; and $p = 0.00$, for sf = 0.30. For the sister-queen colonies, the corresponding winter survival probabilities were 0.20, 0.40, and 0.40.

Using these results, and assuming that the function for colony reproductive success in relation to swarm fraction ($w(x)$) is the same for both mother-queen and sister-queen colonies, we calculated the inclusive fitness of a worker bee in a swarming colony as a function of the swarm fraction. The results are shown in Figure 7.9. The model predicts that a worker bee's inclusive fitness is highest if the swarm fraction is 0.76–0.77. It also predicts that there is a considerable range of swarm fractions, from about 0.65 to 0.80, over which the inclusive fitness of a worker bee in a swarming colony is high. What are the actual swarm fractions that people have found? Three studies have reported mean values for the swarm fraction of 0.68, 0.72, and 0.75, with an overall mean of 0.72.

The fact that the model's *predicted* optimal value for the swarm fraction (0.76–0.77) is very close to the *observed* mean value for the swarm fraction (around 0.72), tells us that worker bees are indeed maximizing their genetic success (inclusive fitness) by strongly preferring to leave with the

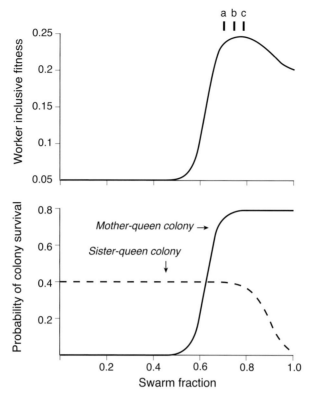

FIG. 7.9. *Top:* Worker bee inclusive fitness as a function of the fraction of workers in the mother-queen colony. It is maximized at the swarm fraction of 0.77. Bars at top indicate reported values. *Bottom:* Survival curves for mother-queen and sister-queen colonies as a function of swarm fraction. The lines are fitted to data from a field experiment.

mother queen rather than stay with a sister queen. This is partly because each worker is more related to the reproductive offspring (queens and drones) of her mother compared to those of her sister (who is probably a half sister). The strong preference of the workers to leave with their mother queen has probably also been favored by natural selection because the mother-queen colony faces the formidable challenge of establishing a new colony, and therefore needs a large workforce to have any chance of surviving to the following summer.

WILD MATING

Since the 1950s, we have known that when queens and drones soar off on their nuptial flights, they do not search randomly near their nests to find members of the opposite sex. Instead, virgin queens and young drones fly large distances to reach specific sites called drone congregation areas, where they gather—some 10–20 meters (ca. 30–60 feet) aloft—to pair in the upper air. Drones start flying to these aerial rendezvous sites at around 1300 hours, which is about one hour before the queens begin to arrive, so usually there is a horde of drones circling at each site by the time the first queen arrives. It is a striking fact that when a young queen bee sallies forth to get inseminated, she travels without a retinue of worker bees for her protection. Because a virgin queen flies alone, she is easy prey for dragonflies and other aerial insectivores, which means that her mating flight is not just the most private time in her life, it is also the riskiest. It is no surprise, therefore, that a virgin queen usually conducts just one mating flight and that she keeps it brief, mating with only 10–20 drones.

The sites where the virgin queens and sex-ready drones meet to mate appear to be stable from year after year. In the Austrian Alps, for example, one group of well-studied drone congregation areas near Lunz am See has persisted since the 1960s, hence for at least 50 years. Other drone congregation areas have also been found to persist for decades. One is on the campus of Cornell University. It was found in the 1960s by Norman E. Gary while he was conducting experiments that revealed that the main component of the queen substance pheromone, E-9-oxo-2-decenoic acid, functions as the sex attractant pheromone of honey bees. This drone congregation area fills the airspace above a small—ca. 100 × 100 meters (330 × 330 feet)—patch of lawn in an otherwise wooded, steep-sided valley just north of the College of Veterinary Medicine. On many a sunny afternoon in June, I have lain on the grass here and watched comets of drones chasing queens (or pebbles fired from my slingshot) shoot across the bright blue sky. Once, I caught a mated queen who had crashed to earth. My immediate thought was to use her to start a new colony, but then I realized that if I were to do so I would orphan a colony. So I let her fly home.

Most of what we know about drone congregation areas comes from the work of two brothers, Professors Friedrich and Hans Ruttner, who worked in Austria in the 1960s and 1970s, and their successors, Professors Gudrun and Nikolaus Koeniger, who have continued the investigations (in both Austria and Germany) to the present. These researchers have discovered that the "hookup" sites of queens and drones can have remarkably distinct boundaries. In one location, for example, the Ruttners found that when they displaced an airborne queen—confined in a cage held aloft by a hydrogen balloon—by only 30 meters (ca. 100 feet) within a drone congregation area, they often shrank by tenfold the number of drones hovering around the caged queen. They also found that in the mountainous regions where they conducted their studies, queens and drones appeared to orient to their congregation areas by flying toward low points on the horizon line, which the drones may perceive as the directions of maximal light intensity. It may be that drones continue orienting in flight in this manner until they reach a location where the intensity of light on the horizon is uniform, and there they circle. How drone congregation areas form where the countryside is flat remains a mystery. It may be that drones are distributed rather evenly over flatlands and that they congregate only when they detect the alluring scent of a queen and orient upwind to its source.

Two other inquiries about the mating habits of honey bees looked into the density of their mating sites and the distances that queens and drones will fly to reach them. One intensive search conducted near Erlangen, in southern Germany, found five drone congregation areas (DCAs) within a circular area that covered about 3 square kilometers (1.16 square miles), with a density of approximately 1.6 congregation areas per square kilometer (ca. 2.6 DCAs per square mile). The results of a similar search, conducted near Lunz am See, in Austria, are shown in Figure 7.10. The density found here is much lower than what was found near Erlangen: approximately 0.1 drone congregation area per square kilometer (ca. 0.3 DCAs per square mile).

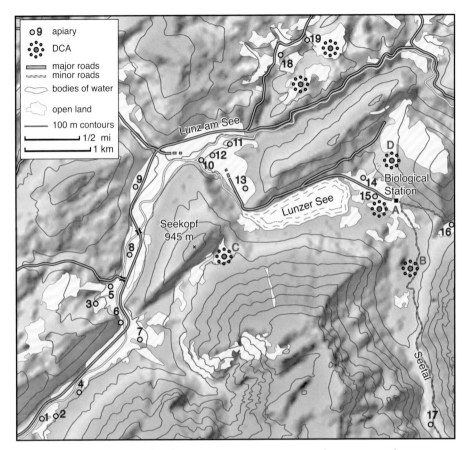

FIG. 7.10. Locations of the drone congregation areas and apiaries in the mountains around the village of Lunz am See, Austria. Mark-and-recapture studies revealed that drones from all the apiaries (except number 9) were visiting the drone congregation area C in the center of the map.

Regarding flight distances, it is clear that both queens and drones will fly great distances to reach drone congregation areas, with queens mating on average 2–3 kilometers (1.2–1.9 miles) from their homes and drones traveling 5–7 kilometers (3.0–4.2 miles) or more to find a sexually receptive queen. Perhaps the most impressive evidence of drones making long-distance mating flights comes from a massive mark-and-recapture study conducted in the Austrian Alps by Friedrich and Hans Ruttner in the mid-1960s. Their study site was near the town of Lunz am See, where in previous years they had found six drone congregation areas, and where they had access to colonies living in 19 apiaries scattered around the study site (see Fig. 7.10). They began by going to these apiaries and labeling thousands of drones, each with a colony-specific paint mark. Next, they captured drones at two of the six known drone congregation areas in the region. Their capture method worked as follows: they lofted a queen in a small plastic cage suspended from a hydrogen-filled balloon; they waited until a crowd of drones was circling around her; and then they slowly lowered her to where they could collect the queen-baited drones using a long-handled insect net. Amazingly, at drone congregation area C, which sits in a high valley in the center of their study site, they captured drones from 18 of the 19 apiaries in the region. The only apiary not represented among the drones captured at congregation area C was apiary 9, which was only 1.6 kilometers (1 mile) from this congregation area but was separated from it by the Seekopf mountain, rising more than 300 meters (ca. 1,000 feet). The Ruttners also reported how many of their captured drones came from each apiary, and from their data I have calculated the average distances flown by the drones they captured at congregation areas B and C: 3.0 kilometers (1.9 miles) and 2.3 kilometers (1.4 miles), respectively. The longest mating flight they detected was an excursion made by a drone from apiary 17 to congregation area C. He flew either a 3.9-kilometer (2.4-mile) beeline route over the mountains or (more likely) went around the mountains via an approximately 6-kilometer (3.7-mile) curved route down the long valley (the Seetal) leading to the lake.

These findings about the impressive mating flight distances of drones in the Austrian Alps are supported by what Donald F. Peer found in the 1950s working in Ontario, Canada. He studied the mating range of honey bees by introducing colonies to a region covered with vast coniferous forestlands that contained no colonies other than his experimental ones. He established an apiary stocked with 20 colonies that produced only drones carrying the Cordovan allele, a recessive color mutation. He also set out small colonies (mating nuclei), each of which had no drones but contained a virgin queen that was genetically marked by being homozygous for the

Cordovan mutation. To get data on mating flight range, he placed his mating nuclei containing virgin queens at various distances from his full-size colonies containing drones. He found that none of the 22 queens that were separated from the drone-source colonies by 19.3 or 22.6 kilometers (12 or 14 miles) mated successfully but that most of the queens separated from the drone-source colonies by 16.0 kilometers (10 miles) or less did mate successfully, and only with males carrying the Cordovan allele (hence from his apiary). Although Peer's impressive results reveal maximum mating ranges, not typical ones, the fact that honey bees have such huge mating ranges indicates that strong outbreeding is almost certainly the rule for *Apis mellifera*.

IS POLYANDRY WEAKER IN THE WILD?

Polyandry—the practice of a female mating with multiple males—is not common among insects, but it occurs at astonishingly high levels in all the species of honey bees. The level of polyandry by *Apis mellifera* queens has been measured by looking at the genotypes of the workers in colonies to determine how many sperm donors are needed to explain the genetic diversity of these workers. These investigations show that, on average, a queen mates with about 12 drones. Why are honey bee queens so promiscuous? We know that this behavior is not needed to ensure that a queen acquires a sufficient supply of sperm to last her lifetime. The average ejaculate of a drone contains about 11 million sperm, so the total number of sperm received by a queen on a mating flight can exceed 100 million. However, a queen typically stores only about 5 million sperm—and it is a random subsample of what she has acquired—in her sperm storage organ (the spermatheca). We now understand that the reason a queen mates with, and then stockpiles sperm from, a dozen or so drones is so that the fertilized eggs she lays will produce a genetically diverse workforce. Numerous studies have shown that having high genetic diversity among the workers in a colony confers many conspicuous benefits to the colony. These include improved resistance to disease, greater temperature stability in the brood nest, and enhanced acquisition of food resources through a more expansive and responsive foraging effort. The mystery of exactly how increasing the genetic diversity of a colony's workforce makes it more effective has been carefully investigated by Heather R. Mattila, starting when she was a postdoctoral student at Cornell. One line of her investigations focused on the foraging abilities of colonies, and to study this she compared the activities of individual forager bees in multiple-patriline and single-patriline colonies, that is, colonies whose queens were instrumentally inseminated with either a blend of semen collected from 10 drones or the same volume of semen collected from just 1 drone. Heather found that the multiple-patriline colonies benefited from having more social facilitators of foraging—that is, worker bees who became engaged as foragers (and waggle dancers) early in the day and thereby galvanized their colony's foraging effort. These high-participation bees belonged to only a handful of the patrilines in the multiple-patriline colonies and were often absent from the single-patriline colonies.

Until recently, all the estimates of the mating frequency of queens in the European subspecies of *Apis mellifera*—living either in Europe or North America—were based on samples of worker bees collected from managed colonies living in apiaries, that is, from colonies headed by queens that probably conducted their mating flights where drones were ultra-abundant. I wondered whether these estimates of queen mating frequency apply to queens living in the wild, where colonies are widely spaced. Do queens have lower mating frequencies when the colony density is lower and potential mates may be less plentiful? To answer this question, I worked with two colleagues—David R. Tarpy at North Carolina State University and Deborah A. Delaney at the University of Delaware—to determine the mating frequencies of the queens in colonies living in the Arnot Forest.

The first step in this study was taken in August 2011, when I and a Cornell undergraduate student, Sean R. Griffin, located 10 bee-tree colonies living in or just outside of the Arnot Forest (Fig. 7.11), by using the technique of bee lining, described in chapter 2. To collect a sample of at least 100 workers from each bee-tree colony, we captured bees from our feeding station. We only captured these bees once we had gotten within 100

184 CHAPTER 7

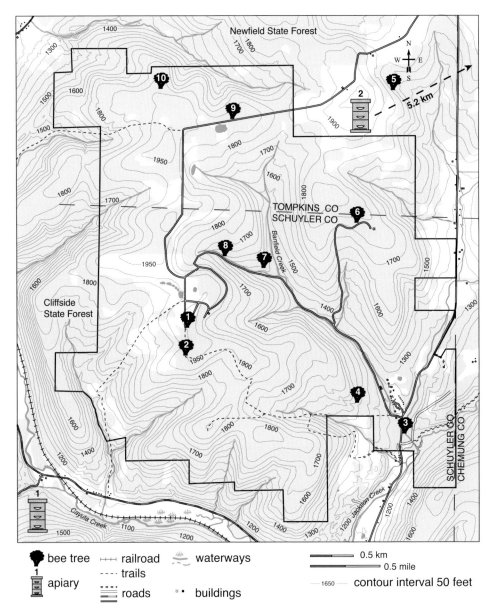

FIG. 7.11. Map of the Arnot Forest and adjacent lands showing the locations of the 10 bee-tree colonies located in August 2011 and of the two nearest apiaries outside of the Arnot Forest.

meters (330 feet) of the tree the colony occupied, at which point our feeding station was totally dominated by foragers from the nearby colony (as confirmed by subsequent genetic analyses).

  Because we wanted to compare the mating frequencies of our *wild* colonies' queens to those of the nearest *managed* colonies' queens, we also collected, in August 2011, approximately 100 workers from each of 20 managed colonies, 10 in each of the two apiaries nearest the Arnot Forest. Apiary 1 was located only 1.0 kilometer (0.6 mile) from the Arnot Forest's southwestern boundary (see Fig 7.11) and had been newly established by a commercial beekeeper in May 2011. The colonies in this apiary contained young queens purchased that spring from a queen breeder in California, so we were confident that the workers sampled from its 10 colonies provided information on the mating frequency of commercially produced queens. The other apiary (apiary 2) was located 5.2 kilometers (3.2 miles) from the northwestern corner of the Arnot Forest. The same commercial beekeeper that had established apiary 1 in May 2011 had established apiary 2 in the early 2000s, and from time to time he gave the colonies in apiary 2 new queens purchased from the same queen breeder in California that had produced the queens in apiary 1. It was handy to have these two apiaries near the Arnot Forest, for they enabled us to compare the mating frequencies of wild-colony and managed-colony queens living in the same general location. It was, however, also dismaying to find apiary 1 located only 1 kilometer from the southwestern boundary of the Arnot Forest, for its presence had the potential to alter the genetics of the bees in this forest. My dismay was short-lived, however, because a black bear destroyed apiary 1 in November 2011, and ever since then the site has been abandoned.

  When David Tarpy and Deborah Delaney performed a paternity analysis on 1,089 of the worker bees that I had collected from the 30 colonies—on average, 36.3 individuals per colony—they found no significant differences in mean number of drone fathers (sperm donors) per queen among the three groups of queens (Fig. 7.12). The mean mating frequencies of the apiary 1 queens (19.8 drones) and the apiary 2 queens (16.6 drones) were not statistically distinguishable from those of the Arnot Forest queens (15.9

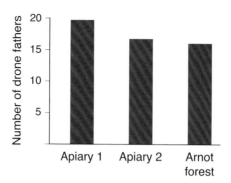

FIG. 7.12. Average mating frequencies of 10 honey bee queens sampled from the three groups: the two apiaries nearest the Arnot Forest and the population of wild colonies living in this forest. The queens in the apiary 1 colonies had been reared and had mated in California, within the operation of a large, commercial queen producer.

drones). These results show that queens living in wild colonies that are broadly dispersed do not necessarily mate with fewer drones than queens living in managed colonies that are crowded together. This is because even where colonies are spread thinly across the countryside, their queens and drones come together at a few special sites—drone congregation areas—when it is time to mate. In short, the low density of colonies in a wild population does not give rise to a low density of drones in the places where queens are inseminated. This is not so surprising when you consider that before humans started managing the lives of honey bees, their colonies lived widely dispersed, and so natural selection must have strongly favored queen bees and drones with the mysterious ability to fly off and find individuals of other colonies to tend to the affairs of sex.

8

FOOD COLLECTION

> Much have I marvelled at the faultless skill
> With which thou trackest out thy dwelling-cave,
> Winging thy way with seeming careless will
> From mount to plain, o'er lake and winding wave.
> —*Thomas Smibert, "The Wild Earth-Bee," 1851*

We generally think of a honey bee colony as a family of bees living inside a bee hive or a hollow tree. A moment's reflection will disclose, however, the important fact that during the daytime many of the bees in a colony are dispersed far and wide over the surrounding countryside, where they toil to gather their colony's food. To accomplish this, each forager bee flies as far as 14 kilometers (8.7 miles) to a patch of flowers, gathers a load of nectar or pollen (or both, Fig. 8.1), and then flies home, where she quickly off-loads her food and then heads out on her next collecting trip. On a typical day, a colony will field several thousand worker bees, or about one-third of its members, as foragers. Thus, in acquiring its food, a honey bee colony functions as a large, diffuse, amoeboid entity that can extend itself over great distances and in multiple directions simultaneously to exploit a vast array of food sources. To succeed in gathering the pollen and nectar it needs, a colony must closely monitor the food sources within its environment, and it must wisely deploy its foragers among these sources so that its food is gathered efficiently, in sufficient quantity, and with the correct

FIG. 8.1. Worker bee flying home bearing loads of yellow-green pollen on her hind legs and a load of nectar in her crop (honey stomach). That she is carrying a nectar load is indicated by the distension and translucence of her abdomen.

nutritional mix. The colony must also properly apportion the food it gathers between present consumption and storage for future needs. Moreover, it must accomplish all these things in the face of constantly changing conditions, both outside the nest as foraging opportunities come and go, and inside the nest as the colony's nutritional needs change with the seasons. In this chapter, we will see how a colony living in the wild meets these challenges.

THE ECONOMY OF A WILD COLONY

Besides fresh air, a colony of honey bees needs just four resources to sustain itself: pollen, nectar, water, and tree resin (Fig. 8.2). Pollen is the most nutritious item on the bees' short shopping list. It provides the amino acids, fats, minerals, and vitamins that immature bees need to develop

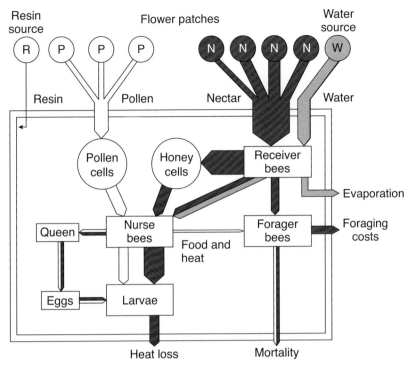

FIG. 8.2. The flow of materials in a honey bee colony on a summer day. The width of each arrow is proportional to the amount of matter flowing along its route. Matter accumulates in the growing larvae, to boost the colony's population, and in the pollen cells and honey cells, to build up its reserve supplies of pollen and honey for rainy days and winter.

properly and that adult bees need to keep their bodies working properly. The nurse bees in a colony, for example, rely on finding cells stocked with pollen near the brood nest, for if their diet lacks pollen, then their brood-food (hypopharyngeal) glands will atrophy and they will be unable to nourish their colony's young. The nurse bees' need for pollen to be stored near the brood explains why, when you peer inside a glass-walled observation hive and watch a pollen forager that has just come home with two balls of brightly colored pollen on her hind legs, you see that she looks along the margins of the brood nest to find a cell in which to off-load her pollen. You see too that once she finds a suitable storage cell, she stands for about 10 seconds with the tip of her abdomen poking into this cell, briskly rubbing her hind legs together to rake off her pollen loads. This care in unloading, performed by all the foragers engaged in pollen collection, creates a neat band of pollen-filled cells around the brood nest.

Other foragers arriving home do not carry pollen loads on their hind legs but instead land at the nest entrance with conspicuously swollen abdomens. If you snag one of these bees before she disappears inside the nest, calmly hold her by the wings and then gently compress her abdomen with forceps, you will see her disgorge a droplet of clear or pale yellow liquid. Analysis of its composition will usually reveal that it is a concentrated solution of sugars—mainly glucose, fructose, and sucrose—hence it is nectar, the bees' principal source of carbohydrates. However, not all the bees that return home with bulging abdomens are nectar foragers; a small percentage carry in cargoes of nearly pure water. These are water collectors, returning from a puddle, stream, or whatever water source is handy. Both nectar foragers and water collectors regurgitate the contents of their crops (honey stomachs) to bees working inside the nest (Fig. 8.3). To do this, the bee off-loading liquid opens her mandibles widely and disgorges a droplet of liquid that sits on the base of her folded proboscis (tongue). The bee receiving the liquid extends her proboscis to full length and quickly drinks the fluid. The recipients of the nectar and water are usually middle-aged bees. Those that accept nectar (nectar receivers) distribute some of this fresh food to others for immediate consumption, but they process most of it into honey for future consumption. Those that accept water (water receivers) will either spread it over the combs to cool the nest or distribute it among the nurse bees. The collection of water for the nurse bees occurs most often in early spring, when the colony is subsisting on honey and stored pollen, and the colony's nurse bees need water to maintain their water balance as they prepare suitably dilute food for the colony's larval brood.

Pollen, nectar, and water are the substances most commonly gathered by a colony's foragers. But during late summer and early fall, if you keep a close watch at a hive's entrance, you will also spy a few bees returning

Fig. 8.3. A nectar forager with bulging abdomen (right), having returned to her nest, regurgitates her nectar load to a food-storer bee (left), which has inserted her tongue between the mouthparts of the forager.

home with shiny brown loads of tree resin stuck in their pollen baskets (Fig. 5.15). As discussed in chapter 5, the bees jam this gluey material into cracks and small holes in the walls of their nest cavity, making their home more weathertight and easier to defend. We also saw that they use this resin to coat the walls of their nest cavity because it has antimicrobial properties that promote colony health.

What follows now is a quantitative look at a colony's collection of nectar and pollen, with emphasis on the total amounts of these two materials that a colony consumes over a year. This overview shows that a colony's foraging operation is a massive enterprise, even for a relatively small colony living in the wild. We will focus here on the nectar and pollen sectors of a colony's economy, because the resin sector has been examined already in chapter 5, and the water sector will be explored in chapter 9, when we look at nest thermoregulation.

Most studies of nectar and pollen consumption by honey bee colonies are based on colonies managed for honey production, but because this book focuses on the lives of wild colonies, we will focus on what I have learned about these matters by studying unmanaged colonies, or simulated wild colonies. These were colonies that I housed in hives the size of natural nest cavities and then monitored—mainly by weighing them once a week—for three years. As I explained in chapter 6, I did this work in the early 1980s, when I was living in the city of New Haven, Connecticut, which is about 400 kilometers (250 miles) east of Ithaca. The populations of my simulated wild colonies ranged from a minimum of about 8,000 adult bees in late winter (March) to a maximum of some 30,000 adult bees in spring (May–June), just before they produced swarms. These population numbers represent biomasses of roughly 1 and 4 kilograms (2.2 to 8.8 pounds), respectively. Because honey bee colonies have considerable mass and because they consume large quantities of food, I could track the food consumption of my study colonies during most of the year by monitoring changes in the total weight of a colony, its food reserves, and its nest. (Hereafter, I will refer to the sum of these three weights as "hive weight.") Between late September and late April, these colonies collected little or no food, so their hive weights dropped steadily as they consumed their honey and pollen stores. As was discussed in chapter 6, on average, the total mass of food eaten over a winter by each of these colonies was approximately 25 kilograms (55 pounds), of which approximately 1 kilogram (ca. 2 pounds) was pollen and the rest was honey.

Determining a colony's total consumption of honey and pollen for the summer—late April to late September, in New York and New England—is more complicated than for the winter, because resources now flow into the colony, counterbalancing losses in colony weight due to food consumption. Fortunately, extended periods of cool, rainy weather occurred in the summer, during which the bees did not forage. Losses in hive weight at these times indicated the summertime rate of resource use. (Note: these weight losses must underestimate the rate of food consumption. This is because honey and pollen resources that are consumed but are then converted into brood are not represented in the losses of hive weight.) The drops in hive weight during rainy weather ranged from 1.0 to 4.0 kilograms (2.2 to 8.8 pounds) per week and averaged about 2.5 kilograms (5.5 pounds) per week. Given a summer season of 22 weeks (late April to late September), I estimate the total mass of resources consumed over the summer by a wild colony in New York or New England is about 2.5 kilograms/week × 22 weeks = 55 kilograms (120 pounds). The pollen portion of this total can be estimated by noting that it requires about 130 milligrams (0.004 ounces) of pollen to produce a bee, and that the average colony population across the summer is about 30,000 bees. Given that a worker bee lives about one month in summer, we can estimate that a wild colony in New York or New England rears about 150,000 bees each year over the five-month summer season. Consuming about 130 milligrams of pollen per bee reared, a colony needs about 150,000 × 130 milligrams = 20 kilograms (44 pounds) of pollen each summer to support its brood rearing.

To summarize, I estimate that the yearly food consumption of a wild colony where I live is approximately 20 kilograms (44 pounds) of pollen and 60 kilograms (132 pounds) of honey (25 kilograms/55 pounds in winter plus 35 kilograms/77 pounds in summer). These are, of course, only rough estimates; the precise values will vary depending on colony size, local climate, and forage abundance. The comparable figures for colonies managed for honey production in Europe and North America are all considerably higher. These colonies have been estimated to rear 150,000–250,000 bees annually and to consume 20–35 kilograms (44–77 pounds) of pollen, and 60–80 kilograms (132–176 pounds) of honey each year.

The number of foraging trips required to procure the materials consumed by a wild colony and the efficiency of this foraging work are both rather easily calculated. With respect to pollen, a typical load weighs about 15 milligrams (0.0005 ounce), so the collection of 20 kilograms (44 pounds) of pollen requires approximately 1.3 million foraging trips. Given an average total flight distance—out and back—of 4.5 kilometers (2.8 miles), a flight cost of 6.5 joules per kilometer, and an energy value for pollen of 14,250 joules per gram, the total cost of flying to collect this pollen is about 3.8×10^7 joules (1.3×10^6 trips × 4.5 kilometers per

Fig. 8.4. A forager collecting pollen from New England aster (*Symphyotrichum novae-angliae*) flowers.

trip × 6.5 joules per kilometer), and the pollen energy value is nearly 2.9×10^8 joules. These numbers show that worker honey bees achieve an approximately 8:1 ratio of energy return in collecting pollen (Fig. 8.4). Parallel calculations can be made about the number of collecting trips required to produce the 60 kilograms (132 pounds) of honey that a colony consumes in a year. Knowing that nectar is on average a 40 percent sugar solution, while honey is an 80-plus percent sugar solution, and that a typical nectar load weighs about 40 milligrams, we can calculate that the production of 60 kilograms of honey requires approximately 3 million foraging trips. More number crunching indicates that worker honey bees achieve an approximately 10:1 ratio of energy return when they collect nectar.

These numbers make it clear that the collection of food each year by a wild honey bee colony living in a cold climate region is an enormous undertaking. Each such colony can be thought of as an organism that weighs

1–5 kilograms (ca. 2–10 pounds), rears 150,000 bees, and consumes some 20 kilograms (44 pounds) of pollen and 60 kilograms (132 pounds) of honey each year. To collect this food, which comes as tiny, widely scattered packets inside flowers, a colony must dispatch its workers on some 4 million foraging trips, with these foragers flying some 20 million kilometers (12 million miles) overall. Given these facts, we can expect that honey bee colonies have been under strong natural selection for great skill in the acquisition and use of their food.

A VAST SCOPE OF OPERATION

Among the more mind-boggling attributes of a honey bee colony is its ability to conduct its food-collection operation over an immense area around its nest extending more than 100 square kilometers (40 square miles). A colony can exploit food sources over such a broad area because each of its foragers can fly to and from flower patches located 6 or more kilometers (3.6 miles) from home. A flying bee cruises along at about 30 kilometers per hour (18 miles per hour), so a 6-kilometer trip takes only about 12 minutes, which does not sound so special, but if you consider the small size of a bee, you will realize that a foraging range of this scale is truly impressive. A 6-*kilo*meter flight performed by a 15-*milli*meter- (0.6-inch-) long bee is a voyage of 400,000 body lengths. A comparable performance by a 1.5-meter- (5-foot-) tall human would be a flight of some 600 kilometers (360 miles), such as from Boston to Washington, D.C., or Zurich to Berlin, or Los Angeles to San Francisco.

Among my fondest memories from when I was a novice beekeeper—some 50 years ago—is lying in the grass among my hives in late summer and gazing up at the foragers crisscrossing the blue sky like shooting stars. Naturally, I wondered how far my colonies' foragers were flying to reach their work sites. Several years later, when I was a college student, I read scientific papers that reported studies of the scope of a colony's foraging operation. In some of these studies, the investigators had used the familiar mark-and-recapture method: foragers were labeled in their hives—with paint, fluorescent powders, sugar syrup containing radioactive isotopes, or a genetic color marker—and then the fields around their hives were searched for the labeled bees. In another study, one that I found much more exciting than those just mentioned, the scientists moved colonies to the semidesert badlands of northwestern Wyoming, where there were no sources of nectar except in two irrigated areas, separated by 28 kilometers (17.4 miles), in which alfalfa (*Medicago sativa*) and yellow sweet clover (*Melilotus officinalis*) were being grown. The researcher, John E. Eckert, placed colonies along "the old winding stagecoach road" that connected the two irrigated areas, and then he determined which of his colonies discovered and exploited the two agricultural oases. In still other studies, Norman E. Gary and colleagues devised a clever advancement on the usual mark-and-recapture method: install a row of magnets over the entrance of a study colony's hive; capture bees from flowers in crop fields all around; weakly glue a steel identification disk on each captured bee's abdomen; and carefully record where each ID disk was deployed. When the foragers from the study colony return to their hive, the magnets over the entrance automatically recapture the ID disks. By referencing the records of which ID disks were deployed in which fields, researchers can determine quite precisely where the bees retuning home to this colony with tags were foraging.

The studies using one or another of these three approaches reported that most of the foragers from the colonies studied were traveling to flowers within 2 kilometers (1.2 miles) of their hives, but that the bees would fly as far as 14 kilometers (8.7 miles) to reach flowers if none were closer (the situation of some colonies in Wyoming). None of these studies, however, provided a picture of the spatial distribution of the foraging efforts of a colony living in nature. This is partly because these studies were conducted in artificial settings—the study colonies were placed either alongside crop fields or in a barren, semidesert habitat—and partly because their findings reflected not only where the colonies' foragers went to find flowers but also where the researchers went to find bees. Inevitably, the researchers did not go everywhere the bees went. Furthermore, most of the studies made with mark-and-recapture methods were conducted in settings where the forage was unusually plentiful, such as alfalfa fields and almond orchards in full bloom.

In the spring of 1979, I started a study with a friend, Kirk Visscher, the aim of which was to get an accurate, bird's-eye view of the foraging activities of a colony living under natural conditions. Our approach was to map out, day by day, the places being visited by the foragers of a full-size colony that we would establish near the center of the Arnot Forest. We did so using a technique that had been pioneered by one of Karl von Frisch's students—Dr. Herta Knaffl—in 1948–1950: spy on the waggle dances performed by the foragers in a colony living in an observation hive. The attraction of this approach is that it reveals where a colony's foragers are going even if they are exploiting multiple flower patches and each patch is many kilometers (or miles) away. This technique, however, does not reveal *all* the places where a colony's foragers are working each day, because on any given day it is only the bees returning home from the most profitable sites that advertise their work sites with recruitment dances. Bees that are working flower patches that merit continued exploitation, but not additional foragers, do not advertise these sites by performing waggle dances, so these sites are invisible to anybody monitoring the dances performed within a colony. Even so, this dance-surveillance method produces an accurate picture of the spatial scale of a colony's foraging operation, because every major flower patch that a colony exploits is advertised by waggle dances produced during the initial, buildup stage of its use by the colony.

The first step in our investigation was to build an observation hive that was roomy enough to house a full-size colony of bees (Fig. 8.5). Ours had a volume of 40 liters (10.6 gallons), the median volume of the nest cavities of the wild colonies in the area (Fig. 5.3). This hive held four large combs, whose total area (1.35 square meters/14.5 square feet) matched what is found in the nests of the colonies living in hollow trees around Ithaca. Because we wanted to be able to see all the waggle dances performed in our hive, we guided all the foragers in our study colony to enter the hive on the front side of its huge wall of comb. We also discouraged these just-returned foragers from crawling to the back side of this comb by closing

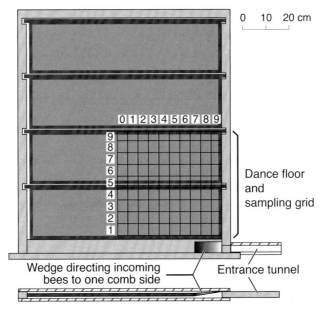

FIG. 8.5. The large observation hive used for sampling and reading the dances of the foragers in a full-size colony to determine where the colony's foragers were gathering their food.

off with beeswax all the passageways between the two sides of the comb that were near the hive's entrance. Most foragers perform their waggle dances soon after entering their nest (or hive), so by directing the traffic of incoming foragers to one side of the comb, we created a well-defined "dance floor" near the entrance on the front side of our hive's wall of comb. We drew a sampling grid on the glass over this dance-floor area, so that we could sample the dancing bees at random when we began our data collection. At this point, we installed a colony of approximately 20,000 worker bees and a queen (and some drones) in the hive, and a few days later we moved it to the Arnot Forest, where we installed it in a special hut that we had positioned in the forest's center. An important feature of this hut was its roof, built of translucent fiberglass panels, which provided diffuse sunlight for our observations of the bees' dances.

A few days later, we began collecting data from the bees performing waggle dances in our large observation hive. This work involved sitting beside the hive from 8:00 A.M. to 5:00 P.M. and manually recording data from one (randomly chosen) dancing bee at a time. For each bee, we recorded four items: 1) the angle (relative to vertical) of her waggle runs, 2) the duration of her waggle runs, 3) the color of her pollen loads (if any), and 4) the time of day of her dance. Using these four pieces of information, we could estimate where she was working, and we could determine whether pollen was available at her work site or only nectar. Finally, we plotted each dancer's recruitment target on a map. This gave us a synoptic picture of the colony's richest foraging opportunities—those being advertised by its waggle-dancing foragers—on each day.

Figure 8.6 shows the distribution of distances to the food sources that our study colony exploited in the Arnot Forest. It is based on observing 1,871 dancing bees during four nine-day periods spread over the summer of 1980. We see that this colony's foragers conducted some of their work within several hundred meters of the hive, but that they also flew to many flower patches several kilometers from their home. The modal (most common) distance from homesite to flower patch was 0.7 kilometers (about 0.4 miles), the median distance was 1.7 kilometers (1.1 miles), the average distance was 2.3 kilometers (1.4 miles), and the maximum distance was 10.9 kilometers (6.8 miles). It was fascinating, too, to see that the average distance flown by the bees to reach their food sources varied across the summer. For example, during the second period of data collection, 9–17 July 1980, the average distance flown was ca. 2 kilometers (1.2 miles), but during the third period, 28 July to 5 August, the average distance flown was ca. 5 kilometers (3 miles). Evidently, the foragers in our study colony had difficulty finding good food within the Arnot Forest during late July and early August; their dances showed us that throughout this time they collected most of their food by flying 5–6 kilometers (3.0–3.6 miles) to the Pony Hollow valley to the north. We knew there were several farms in this valley, and a check of their fields revealed honey bees busily collecting nectar and pollen on alfalfa plants in bloom.

The most noteworthy feature of the distribution shown in Figure 8.6 is the location of the 95th percentile, which falls at 6 kilometers (3.7 miles).

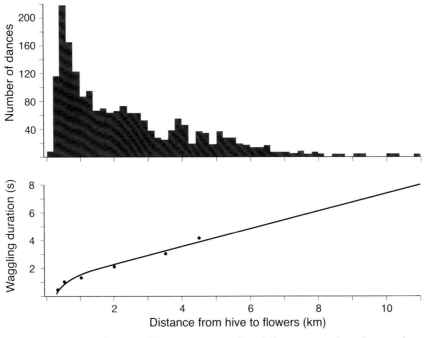

FIG. 8.6. *Top:* Distribution of distances to a colony's forage sites, based on analysis of 1,871 waggle dances. *Bottom:* Coding of distance information in waggle dance: the duration of each waggle run is proportional to the length of the outbound flight.

This shows that if we were to draw a circle around the hive that was large enough to enclose 95 percent of this colony's food sources, then it would have an area greater than 113 square kilometers (43 square miles)! One might wonder, though, was our study colony exceptional in conducting such widespread foraging? I believe that the answer is no, for several reasons. The first is the biological reality that the bees' system of recruitment communication—the waggle dance—has been tuned by natural selection so that successful foragers can guide their hive mates to rich food sources that are 10-plus kilometers (6 or more miles) from their home (see Fig. 8.6). This shows us that their communication system has evolved to coordinate foraging over great distances. Second, there are several other studies that have reported honey bees flying 10 or more kilometers from their hives to capitalize on rich foraging sources. The first is the 1953 study by Herta Knaffl, the postdoctoral student of Karl von Frisch who pioneered the method of spying on the waggle dances performed in a colony's nest to investigate the distances that its foragers are traveling. She reported that 95 percent of the 2,456 dances that she decoded in colonies living in small observation hives—all located within the city of Graz, Austria— advertised horse-chestnut trees (*Aesculus hippocastanum*) and other food sources less than 2 kilometers (1.2 miles) away. However, she also reported observing dances that announced discoveries of bountiful food sources 5–6 kilometers (3.0–3.6 miles) and even 9–10 kilometers (5.4– 6.0 miles) away.

The most impressive demonstration of a honey bee colony's capacity for long-range foraging comes from a study conducted by Madeleine Beekman and Francis L. W. Ratnieks, with a colony living in an observation hive located within the city of Sheffield, in England. They video-recorded and then decoded 444 waggle dances performed by their colony's foragers on three days in the middle of August 1996, a time when vast patches of a major nectar source, ling heather (*Calluna vulgaris*), were in full bloom on the moors in the Peak District west of Sheffield. Their data analysis revealed that their study colony foraged in two areas, one relatively close to the hive, less than 2 kilometers (1.2 miles) away, and one much farther away, 5–10 kilometers (3.1 to 6.2 miles). Overall, 50 percent of the dances performed in their observation hive represented sites more than 6 kilometers (3.6 miles) from the hive, and 10 percent represented sites more than 9.5 kilometers (5.9 miles) away! Clearly, there are times when a honey bee colony benefits greatly from having workers with a long maximum flight range, for this endows its food-collection operation with a vast scope.

TREASURE HUNTING BY THE BEES

To benefit fully from its immense foraging range of 6-plus kilometers (3.7-plus miles), a honey bee colony must be able to discover the richest flower patches within the ca. 100-square-kilometer (ca. 40-square-mile) region around its nest. Furthermore, it must be able discover these first-class food sources soon after they come into bloom, before they are dominated by competing colonies. These considerations raise an important question: How good is a wild colony's surveillance of its environment for rich flower patches popping up in the swamplands, woodlands, and fields that form the landscape around its home? I addressed this question in the mid-1980s by performing an experiment in which I conducted two treasure hunts with four colonies of honey bees. In both trials of this experiment, I created lush patches of flowering buckwheat plants, dispersed within a large forest, and I measured each colony's success in finding these prize food sources.

The forest was the 3,213-hectare (7,840-acre) Yale Myers Forest in northeastern Connecticut. It is a beautiful study site that has much in common with the Arnot Forest in New York. Both are large wooded areas surrounded by state and private forestlands. Also, both have a history of human use, which includes abandonment from agriculture by the 1870s and then regrowth to a lightly managed, mixed-species hardwood forest that has some stands of eastern hemlock and eastern white pine trees. The Yale Myers Forest, however, has one habitat feature not found in the Arnot Forest: large, open wetlands created by beaver activity. These sunny expanses—brimming with cattail and purple loosestrife plants—provide much food for the bees. Before conducting this experiment, I searched the southeastern corner of the Yale Myers Forest and found four wild honey bee colonies living in hollow trees, which told me that the Yale Myers Forest, like the Arnot, is a prime habitat for wild colonies of *Apis mellifera*. Indeed, this is as true now as it was in the 1980s. I had no difficulty finding honey bees on flowers deep in this forest when, in August 2017, I spent a day there bee hunting.

The layout of my experiment consisted of four clustered hives of bees and six dispersed patches of buckwheat (Fig. 8.7). The buckwheat patches were identical in size—100 square meters (1,078 square feet)—but they differed in distance from the hives: 1.0–3.6 kilometers (0.6–2.2 miles). I timed the planting of these flower patches so they would come into bloom when little other forage would be available—either in late June, after the

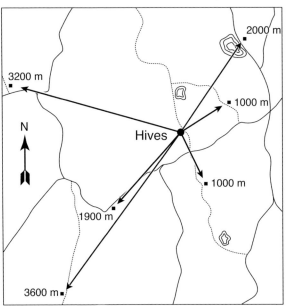

FIG. 8.7. *Top:* One of six patches of buckwheat planted in the Yale Myers Forest to assess the extent of a honey bee colony's reconnaissance for rich food sources. *Bottom:* Map of the array of experimental buckwheat patches in the forest.

nectar flows from raspberry (*Rubus* spp.) and sumac (*Rhus* spp.), or in mid-August, before the nectar flow from goldenrod—hence at times when colonies would be eager to exploit my buckwheat flowers. Once the patches were fully in bloom, I went to each one, daubed paint of a patch-specific color on 150 of the approximately 200 bees busily foraging in each patch, and then scooted back to the four hives to watch each one's entrance for foragers bearing my paint marks. If I saw bees entering or leaving a hive with paint marks of a certain color, then I knew that the colony inhabiting this hive had discovered the corresponding patch. I learned from this experiment that the four colonies had high probabilities of discovering the patches at 1,000 meters/0.6 miles ($p = 0.70$) and at 2,000 meters/1.2 miles ($p = 0.50$), but that—to my surprise—none found the patches at 3,200 and 3,600 meters/2.0 and 3.7 miles ($p = 0.00$).

I suspect that these probabilities understate the actual surveillance ability of a honey bee colony, if only because my method for determining which colonies discovered each patch probably did not detect all the discoveries made by my four colonies. If a colony had only a handful of foragers visiting a flower patch, then probably I would have failed to see that it had found this patch. Nevertheless, my treasure-hunt experiment revealed an impressive ability by colonies to monitor their environment for rich food sources. Even though a 100-square-meter (1,100-square-foot) patch of flowers—about half the size of a tennis court—represents less than 1/125,000 of the area enclosed by a circle with a 2-kilometer (1.2-mile) radius, my four colonies had a probability of 0.5 or higher of discovering any flower patch of this size located within 2 kilometers of its hive.

CHOOSING AMONG FOOD SOURCES

For a colony to be skilled at extracting food from the myriad flowers in the woods, swamps, fields, and gardens around its nest, it must couple its impressive ability to discover productive flower patches with an ability to selectively dispatch its foragers to the richest ones. In other words, it must be skilled not only at exploring its options, but also at choosing among them. Our first sign that a wild colony does indeed possess an impressive collective intelligence in choosing among its broadly dispersed food sources emerged from the study described earlier in this chapter in which Kirk Visscher and I monitored the waggle dances of a full-size colony living in an observation hive in the middle of the Arnot Forest. As explained above, we took data each day from a random sample of the bees performing waggle dances in our hive, and we plotted the locations of the food sources that these bees had advertised with their dances. We now know that it is only the bees that have visited top-quality forage sites that perform long-lasting waggle dances—the ones that Kirk and I were most likely to watch—so we are confident that each day's map of the colony's recruitment targets revealed the whereabouts of the colony's most attractive food sources for the day.

These maps, four of which are shown in Figure 8.8, revealed that the spatial distribution of a colony's richest, most exciting food sources can change dramatically from one day to the next. Indeed, on each of our 36 days of spying on the colony's waggle-dance communications, we found that the colony's dances revealed a new spatial distribution of its primary recruitment targets. The following day-by-day review summarizes these dynamics for the four-day period represented in Figure 8.8.

> **13 June 1980.** Good weather. The main foci of recruitment were clearly indicated: sites 0.5 kilometers (0.3 miles) SSE and SSW of the hive, producing yellow and yellow-gray pollen, and a large site 2–4 kilometers (1.2–2.4 miles) SSW of the hive, producing mainly nectar. Two other sites profitable enough to elicit long-lasting dances were one producing orange pollen 1 kilometer (0.6 miles) to the NE and one producing yellow-gray pollen 4 kilometers (2.4 miles) to the NE.
>
> **14 June 1980.** Good weather. The source of yellow-gray pollen 4 kilometers (2.4 miles) to the NE now aroused few, if any, dances; none were "hit" in our sampling of the dances. A nectar source 0.5 kilometers (0.3 miles) to the NW became extremely profitable, stimulating many bees to advertise it with persistent dances. The pol-

len sources 0.5 kilometers (0.3 miles) SSE and SSW of the hive and the nectar source 2–4 kilometers (1.2–2.4 miles) SSW remained highly attractive.

15 June 1980. Good weather. The large nectar source 2–4 kilometers (1.2–2.4 miles) to the SSW has become less exciting, as has the nectar source 0.5 kilometers (0.3 miles) to the NW. The two sites producing mainly yellow and yellow-gray pollen 0.5 kilometers (0.3 miles) to the SSE and SSW remained highly desirable. A source of brown pollen ca. 0.5 kilometers (0.3 miles) to the SW was advertised for the first time.

16 June 1980. Cool and intermittent rain. The bees foraged relatively little and only rather close to the hive. The source of yellow-gray pollen 0.5 kilometers (0.3 miles) to the SSW remained attractive, but the nearby source of yellow pollen in the SSE stimulated no dancing. A source of orange pollen 0.5 kilometers (0.3 miles) to the NW was exciting. The richest source of nectar on this damp day was a new site 0.5 kilometers (0.3 miles) S of the hive.

A more detailed picture of a colony's ability to choose among a dynamic array of foraging opportunities comes from an experiment performed several years later, in which instead of simply monitoring changes in the waggle dances advertising different foraging sites, I undertook the technically greater challenge of measuring changes in the number of foragers engaged in exploiting different foraging sites. To accomplish this, I worked with a colony in which all 4,000 of its workers were painstakingly labeled for individual identification. After labeling these bees at my bee laboratory in Ithaca over a two-day period, I moved this special colony 240 kilometers (150 miles) north to one of my favorite study sites, the Cranberry Lake Biological Station (CLBS). This remote field station is in the northwestern corner of the 24,400-square-kilometer (9,375-square-mile) Adirondack Park in northern New York State. For at least 10 kilometers (6 miles) in all directions, the landscape around the CLBS is one of forests, ponds,

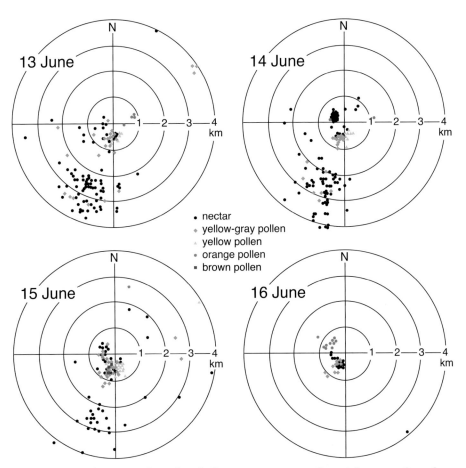

FIG. 8.8. Daily maps of a colony's foraging sites, as inferred from reading the waggle dances performed by the colony's foragers. Each dot represents the location indicated by one bee's dance. Black dots denote sites yielding nectar; all other dots denote sites yielding pollen of the color shown. On the four days represented here, only a small fraction (2%) of the dances indicated foraging sites beyond 4 kilometers (2.4 miles), and most of these sites are not shown.

bogs, and lakes, so it contains no rich food sources for the bees. Indeed, the bee forage here is so scanty that no wild colonies of honey bees live in the area, though there are some colonies of bumble bees. This meant that there was no risk of foragers from wild honey bee colonies intruding upon my experiment and no risk of natural food sources luring my meticulously labeled bees away from the two sugar-water feeders that lay at the heart of my experiment.

This experiment began when my assistants and I trained 10 bees to collect sugar water from each of two feeders located north and south of the study colony's hive. Both were 400 meters (0.25 miles) from the hive. During this training period, both feeders were filled with a dilute (30 percent) sugar solution that was attractive enough to motivate the bees visiting each feeder to continue foraging from it but was not rich enough to stimulate them to recruit hive mates to it. The critical observations began at 7:30 on the morning of June 19, which followed a 10-day period of cool, rainy weather, during which the bees could not leave their hive. We loaded the north and south feeders with 30 percent and 65 percent sucrose solutions, respectively, and began recording the number of different individuals visiting each feeder. By noon, the colony had generated a strong pattern of differential exploitation of the two foraging sites: 91 bees were bringing home food from the richer feeder to the south, but only 12 bees were doing so from the poorer one to the north (Fig. 8.9). We then swapped the positions of the richer and poorer feeders for the afternoon, and by four o'clock the colony had switched the primary focus of its foraging from south to north. The colony's ability to choose between forage sites was again demonstrated the following day in a second trial of the experiment. Thus, we saw that when we gave our study colony choices between two food sources that differed in profitability, the colony consistently focused its collection efforts on the more profitable site. The net result was that the colony steadily tracked the richest foraging site within a changing array. Perhaps the most impressive feature of these experimental results is the high speed of the colony's tracking response. Within four hours of the noon reversal of the positions of richer and poorer sites, the colony had completely reversed the distribution of its foragers. These results show us that wild colonies are highly skilled at coping with the dynamics in foraging opportunities that they experience in the wild, such as those that Kirk Visscher and I detected when we spied on the waggle dances performed inside a colony living in the Arnot Forest.

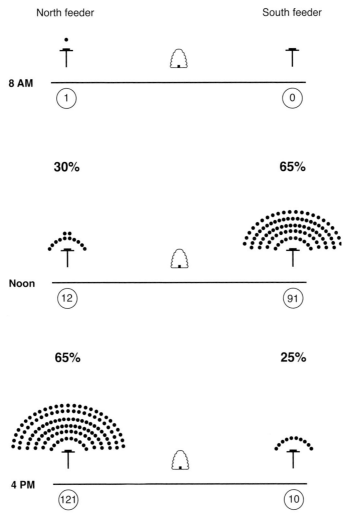

FIG. 8.9. Diagram showing the ability of a colony to choose between two foraging sites (sugar-water feeders) with 30 percent and 65 percent sucrose solutions. The number of dots above each feeder denotes the number of different bees that visited the feeder in the 30-minute period preceding the time shown on the left (8:00 A.M., Noon, 4:00 P.M.) on 19 June. For several days prior to the start of the experiment, a small group of bees was trained to go to each feeder (12 bees to the north feeder and 15 to the south); thus on the morning of 19 June the two feeders had essentially equivalent histories of low-level exploitation by the study colony. The two feeders were located 400 meters (0.25 mile) from the hive and were identical except for the concentration of the sugar solution.

HONEY ROBBING IN THE WILD

Beekeepers have long known that honey robbing can be severe among colonies in their apiaries, but until recently nothing was known about the importance of robbing among colonies in the wild. We can be sure that in both settings the bees have a powerful incentive to steal. A pound of honey is about 80 percent sugar, whereas a pound of nectar is (on average) only about 40 percent sugar. These numbers tell us that when a worker bee flies home with her crop stuffed with stolen honey, she brings home twice as much energy for the colony than if she were returning with her crop filled with nectar. Furthermore, a crop filled with honey can be far less costly to obtain than one filled with nectar; a honey robber visits just one spot to load up (see below) whereas a nectar forager typically visits dozens, if not hundreds, of flowers to gather a full load. A robber, however, risks being killed by defenders of the honey she tries to steal. But if this risk is low, as it is when the colony being robbed is weak or even dead, then the energetic advantages of robbing honey relative to collecting nectar strongly favor theft.

Robbing is a fascinating part of the biology of honey bees given its significance as a form of *intra*specific kleptoparasitism, or parasitism by theft (usually of food). I suggest, though, that the primary significance of robbing lies in its potential to promote *inter*specific parasitism by fostering the transmission of parasites and pathogens between colonies. For example, if a colony is robbed while it is dying from an infestation of *Varroa* mites and associated viruses, then the robbers are likely to become infested with these mites and carry them home. Robbers can also carry home certain pathogens from dead colonies, including the highly virulent spores of *Paenibacillus larvae* (American foulbrood) and the less dangerous spores of *Ascosphaera apis* (chalkbrood). To understand the importance of robbing in spreading parasites and pathogens between colonies, we need to know how long these agents of disease can survive in dying and dead colonies, and how quickly such colonies are found by robbers. Previous studies of robbing have focused on robbing in the artificial situation of colonies living crowded in apiaries, so until recently, the importance of robbing among colonies living dispersed in the wild remained unknown.

To begin to fill this gap in our knowledge of the natural history of *Apis mellifera*, one of my PhD students, David T. Peck, and I investigated how rapidly unguarded honey is discovered by robbers in two settings: wild colonies in the Arnot Forest and managed colonies in my five apiaries at Cornell. As was explained in chapter 2, the colonies in the Arnot Forest have a density of 1 colony per square kilometer, so in this setting the average distance to a colony's nearest neighbor is about 1 kilometer (0.6 miles). The colonies in my apiaries are arranged in pairs on hive stands, so here a colony's nearest neighbor is less than 1 meter (ca. 3 feet) away. We provided unguarded honey in these two scenarios (forest and apiary) by putting out "robbing test boxes" (RTBs), which were made by cutting old Langstroth hive bodies in half and adding a wall, a floor, and a lid. The entrance opening of each RTB was a hole 2.5 centimeters (1 inch) in diameter drilled into one end. We equipped each RTB with one frame of capped honey and three frames of dark, aromatic comb. The most revealing trial of this experiment was conducted in October 2015, when frosts had eliminated most of the flowers so the bees were keen to find colonies to rob. We set out simultaneously 10 RTBs in the Arnot Forest and 10 RTBs in my apiaries near Ithaca, about 25 kilometers (15 miles) from the Arnot Forest. We deployed the 10 RTBs assigned to the forest by hanging them from tree limbs to make them safe from black bears (see Fig. 2.15). We also spaced these boxes approximately 1 kilometer (0.6 mile) apart, to mimic the spacing of the nests of wild colonies. We deployed the 10 RTBs assigned to the five apiaries in two ways: either suspended from the limb of a tree (as in the Arnot Forest) near the apiary or set atop a cement block within the apiary.

The results for our experiment are shown in Figure 8.10. Every robbing test box that was hung near or placed within an apiary was discovered by the end of the first day of good weather. Every robbing test box hung in the Arnot Forest was robbed too, but these boxes were discovered more slowly. It took 10 days of good weather for the robbers to find them all.

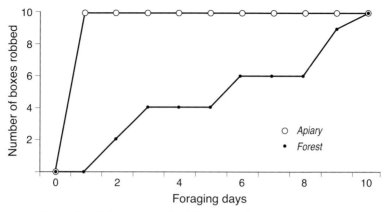

Fig. 8.10. Records of discovery of robbing test boxes in Arnot Forest and in apiaries, 8 October to 12 November 2015. Time scale is foraging days, i.e., days in which weather conditions were good enough for foragers to make flights.

Thus, we found that in an apiary setting, during a nectar dearth, every box containing unguarded honey was discovered rapidly by robbers. This result was expected. What was not expected, however, was the test result that all the boxes dispersed in the Arnot Forest were also discovered, though relatively slowly. It may be that our test boxes are more conspicuous, and thus are more easily found by robbers, than natural nest cavities. Nevertheless, the complete success of robber bees in discovering our test boxes in both scenarios suggests that even in the wild, where dying (or dead) colonies must be relatively hard for robbers to find, there are opportunities for parasites and pathogens to be transmitted between colonies by robbers bringing home the honey stores of colonies weakened (or killed) by disease.

By chance, David Peck and I also managed to watch closely the behavior of robber bees while they were loading up on honey inside an unoccupied hive. This happened in early July 2017, after I noticed bees robbing from a bait hive that I keep on the sloping roof of a shed attached to my barn at home. A swarm had moved into this hive in June 2016, but it had died during the winter of 2016–2017. In early May 2017, I opened the hive to clean out the dead bees, and in doing so I found that the hive contained two frames still nearly full of honey. I left them in place thinking that they might help attract another swarm. When I came home from work on 30 June 2017, I was excited to see bees zipping in and out of this hive. Were they scouts from a swarm checking it out? With a ladder, I climbed to the hive and gently removed its cover to have a look inside. What I saw was robbers loading up on honey. Cool! This opportunity to watch robbers plunder another colony's nest was too good to pass up, so I had to do something that would allow me to watch them closely. Soon I was back with a two-frame observation hive, which I set on sawhorses arranged inside the shed. I transferred from the bait hive to the observation hive the two frames of honey-filled comb, and then I hid the bait hive. Almost immediately, the robbers were investigating the observation hive.

For the next two days, David Peck and I watched the robbers at work inside the observation hive. We noticed that, as a rule, a robber bee does not uncap a honey cell to have a personal honey source. Instead, she walks across dozens or hundreds of capped honey cells and eventually joins a knot of fellow robbers extracting honey from one of a handful of already opened cells. We also saw that if a robber *does* open a capped cell of honey, she cuts with her mandibles just a pinprick-size hole in the cell's capping. This opening is just large enough to admit her tongue. In time, it will be enlarged by other bees that join her in drinking from "her" cell. This economy of the robber bees in uncapping honey cells means that they spend most of their time inside the robbed nest standing head-to-head, sometimes pushing each other back and forth slightly, but usually standing surprisingly still. We timed several robber bees from entry to exit and found that on average they spent 12 minutes in the hive. Seeing these bees standing nearly motionless, and crowded around a few partially uncapped cells, revealed to us that when robbers invade a dying colony they provide *Varroa* mites with excellent opportunities to climb onto them. One prior study has shown that *Varroa* mites are extraordinarily nimble creatures; they can climb onto a feeding worker bee, via a leg or her tongue, in the blink of an eye. Another study has shown experimentally that when *Varroa* mites find themselves in highly infested colonies they no longer avoid foragers when they scramble onto worker bees.

It is now clear from studies of the ability of robbers to find dead or dying colonies, together with our little study of the behavior of robbers inside a dead or dying colony's nest, that robbing by honey bees gives *Varroa* mites excellent opportunities for moving from a dying colony to a healthy colony, even where colonies are widely spaced across a landscape. Fifteen years ago, I was deeply surprised to find that *Varroa* mites had managed to infest the entire population of wild colonies in the Arnot Forest. But now, knowing how skilled robber bees are at finding the nests of dead and dying colonies, and knowing how robbing bees stand nearly motionless for several minutes when loading up on honey inside the nests of these colonies, I will be surprised to ever find a wild colony without *Varroa* mites.

9

TEMPERATURE CONTROL

No warmth, no cheerfulness, no healthful ease,
No comfortable feel in any member—
No shade, no shine, no butterflies, no bees,
No fruits, no flowers, no leaves, no birds,
November!
—*Thomas Hood, "November," 1844*

Thomas Hood's poem, written in London in 1844, expresses well the isolation that we can feel when winter sets in and nature goes quiet. No butterflies, no bees. On these days, though, we can take heart from knowing that the bees are merely out of sight, not gone. Inside a bee hive or bee tree, there are still thousands of active bees, huddling and shivering, creating a warm refuge that is well stocked with heating fuel—their honey stores. In this regard the honey bee is unique, for it is the only insect living in seasonally cold climates that creates year-round a cozy microclimate in which it stays active. From late autumn to midwinter, the time when honey bee colonies are without brood, the temperature of the bees clustered inside a hollow tree or man-made hive stays well above freezing. The core temperature of a broodless winter cluster rarely falls below 18°C (64°F), and the temperature of its mantle—its outermost layer—usually stays above 7°C (44°F), even when the ambient temperature is −20°C (−4°F) or colder. Then from late winter to early autumn, the stretch of time when colonies are rearing brood, the temperature in the nursery region of a honey bee colony's nest is maintained between 34.5°C and 35.5°C (94°–96°F), varying by less than 0.5°C (1°F) across a day, even if the temperature outside drops below 0°C (32°F) in late winter or it soars above 40°C (104°F) on the hottest days of summer.

Another measure of the remarkable thermal stability of the nursery region of a honey bee colony's nest is the high sensitivity of honey bee brood to small deviations from the normal, 34.5°–35.5°C temperature range of the broodnest. When Jürgen Tautz and colleagues reared capped brood—pupae and late-stage larvae—in incubators set to 32°C, 34.5°C, and 36°C (90°F, 94°F, and 97°F), labeled the young bees upon emergence with paint marks, and then introduced them to a foster colony living in an observation hive, they found clear differences in behavior among the temperature-treated bees when they foraged from a sugar-water feeder 300 meters (980 feet) from the hive. On average, a bee raised at 32°C performed only 10 circuits of the waggle dance upon return to the hive, whereas those raised at 34.5° or 36°C performed 50 circuits. Also, the waggle dances performed by the 32°C-reared bees indicated the feeder's 300-meter distance with less precision than those of their sisters reared at the two higher temperatures. Subsequent work looked for temperature-induced effects on the worker bees' brains, and found that the connections between neurons in the mushroom bodies—the centers of information integration in worker bees' brains—were highest in bees that matured at the normal brood-nest temperature (34.5°C) and were significantly lower in bees raised at temperatures just 1°C (less than 2°F) above or below normal.

The temperature of any living system reflects the relative rates at which it gains heat and it loses heat, so to understand how a honey bee colony maintains a stable, and elevated, temperature in its brood nest, we must examine how it adjusts both its *production* of heat through metabolism and its *loss* of heat through various means, including nest ventilation and evaporative cooling. These processes are the same for managed colonies living in hives and wild colonies living in trees, but how hard the bees need to work to heat and cool their nests often differs greatly for the two types of colo-

TEMPERATURE CONTROL 217

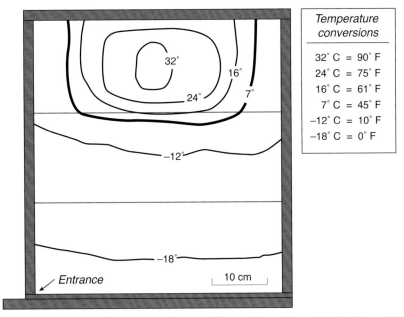

FIG. 9.1. Isotherms of a winter cluster of honey bees living in Madison, Wisconsin. Data were collected at 1700 hours on 25 February 1951, from a colony housed in a Langstroth-type, movable-frame hive that consisted of three medium-depth hive bodies. The 7°C (45°F) isotherm marks the outer surface of the bees' cluster. Note the pocket of relatively warm air in the upper half of the hive. It is this microenvironment around the cluster that is its direct thermal environment.

nies. As we shall see, colony thermoregulation—in both summer and winter—is generally easier for wild colonies because the thick wooden walls of their tree-cavity homes provide better insulation than do the thin lumber walls of most hives, and because cracks in the walls of the wild colonies' homes are filled with propolis, which makes them less drafty. These differences in insulation and draftiness are important because they strongly influence the microenvironment inside a colony's nest cavity, and it is the temperature inside the nest cavity, not the temperature outside it, that is the direct thermal environment of a bee colony (Fig. 9.1). Increasing the insulation and decreasing the draftiness of a nest enclosure slows the heat flow between the microenvironment of the nest and the macroenvironment of the world outside. This means that on a day when the air is cold, a wild colony living inside a well-insulated and well-sealed tree cavity needs to produce relatively little heat to keep its brood nest warmed to the ca. 35°C (95°F) set point, because the microenvironment around it is so well isolated from the cold macroenvironment outside. It also means that on an extremely hot day, when heat will tend to flow into a colony's nest cavity, a wild colony living in a thick-walled tree hollow may need to do relatively little cooling to prevent its brood nest from overheating, because the microenvironment inside the nest cavity is so well isolated from the high temperatures outside.

EVOLUTIONARY ORIGINS OF COLONY THERMOREGULATION

The ability of a honey colony to maintain a warm microclimate inside its nest is ultimately derived from the adaptions of honey bees for flight. Being insects, honey bees fly by flapping their wings—the most energetically demanding mode of animal locomotion—and the flight muscles of insects are among the most metabolically active of tissues. A worker bee in flight expends energy at a rate of about 500 watts/kilogram (230 watts/pound). In comparison, the maximum power output of an Olympic rower is only about 20 watts/kilogram (9 watts/pound). Therefore, whenever a bee is airborne, she not only consumes the energy in her fuel at a prodigious rate, she also generates a great deal of heat. The efficiency of a bee's flight apparatus in converting metabolic fuel to mechanical power is about 10–20 percent, so more than 80 percent of the energy expended in flight appears as heat in the muscles. The rate of heat loss from a worker bee's hairy thorax is sufficiently low that during sustained flight her thorax temperature is typically 10°–15°C (18°–27°F) above the ambient temperature.

We see, therefore, that in honey bees an elevated thorax temperature is an inevitable consequence of flight, but what is critical for understanding the origins of the colony thermoregulation abilities of honey bees is the fact that an elevated thorax temperature has become *essential* for their flight. Workers must maintain a thorax temperature above about 27°C

(81°F) to be able to fly. Flight muscles cooler than this simply cannot generate the high wingbeat frequency and power output per stroke needed for takeoff and flight. This high minimum thorax temperature for flight reflects two "design" constraints on the bees' flight-muscle enzymes: 1) they must withstand the high thoracic temperature produced by flight, but 2) when built with sufficiently strong intramolecular bonds to resist degradation at high temperatures, they are too rigid to operate efficiently at low temperatures. So, when honey bees evolved flight muscles adapted to high temperatures, they also evolved the ability to conduct preflight warm-ups of these muscles, without which they would remain grounded at temperatures below 27°C (81°F). Bees warm up their flight muscles by simultaneously activating the wing-levator and the wing-depressor muscles in the thorax. This causes these muscles to contract isometrically, which produces much heat but few or no wing vibrations.

This preflight warm-up behavior evidently set the stage for the evolution of nest thermoregulation by honey bees, since they use the same mechanism of isometric muscle contractions for warming their flight muscles and for heating their brood combs. Recordings of the thorax temperatures of foragers preparing to leave on foraging flights and of nurse bees heating cells of capped brood show identical patterns of a 2°–3°C (4°–5°F) per minute rise in thorax temperature. And in both settings the wings of a bee warming herself remain motionless, folded over her abdomen. Sometimes the bees that are heating brood stand perfectly still while pressing their thoraces onto the caps of cells containing pupae, but other times they enter empty cells amidst cells of sealed brood and then remain in them for up to 30 minutes with their thoraces heated to 41°C (106°F), to warm the pupae in the adjacent cells (Fig. 9.2).

The colony-level thermoregulation abilities of *Apis mellifera* evolved in tandem with the evolution of this bee's social life. In part, honey bee colonies gained their sophisticated control of nest temperature when they evolved into large groups, simply because a group has a greater capacity for heat production than an individual. After all, a colony of 15,000 bees can generate heat some 15,000 times more powerfully than can a single

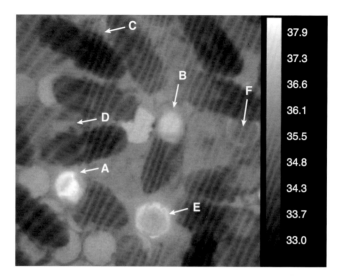

FIG. 9.2. Thermogram taken with an infrared camera of worker bees on a comb containing capped brood cells and empty cells. Capped cells appear gray with no outline; empty cells are recognized by the hexagonal shape of their rims. A: worker with hot (ca. 38°C/100°F) thorax that is about to enter an empty cell adjacent to three sealed brood cells. B: worker that has just left the warm (ca. 37°C/98°F) open cell in the center of the image. C and D: workers not producing heat; each has a cool thorax. E and F: cells containing workers producing heat; in each, the cell interior glows around the dark silhouette of the cool abdomen of the heater bee within the cell.

bee. A second advantage in thermoregulation enjoyed by groups relative to individuals is reduced heat loss per individual, especially when the group's members crowd together into a tight cluster. The surface area of an isolated worker honey bee—a cylinder 14 millimeters (0.55 inch) long and 4 millimeters (0.16 inch) in diameter—is about 3.8 square centimeters (0.6 square inches), but the surface area of 15,000 bees, when contracted into a dense cluster 18 centimeters (7 inches) in diameter, is only about 1,000 square centimeters (155 square inches). So, when a bee is huddling in a cluster, her effective surface area is reduced to only 0.067 square centimeters (0.01 square inches), some 60 times smaller than when she is standing alone.

BENEFITS OF TEMPERATURE CONTROL

Honey bee colonies benefit greatly from being able to both cool and heat themselves. With thousands of bees crowded together inside a nest cavity that has only one rather small entrance opening, a strong colony faces a risk of disastrous overheating when the temperature outside the nest cavity rises above about 30°C (86°F) and stays there all day. Sustained temperatures over 37°C (99°F) inside a nest will disrupt larval metamorphosis. Also, if the temperature inside a nest rises above 40°C (104°F), then the beeswax combs can soften dangerously, and those laden with honey can collapse. Moreover, the adult bees can survive only a few hours at temperatures of 45°–50°C (113°–122°F), which is just 10°–15°C (18°–27°F) above their optimum temperature for full activity (35°C/95°F). In contrast, honey bees can survive indefinitely at 15°C (59°F). This shows that worker bees, like most organisms, possess a narrower range of heat tolerance above their optimum than below it. The low tolerance of high temperatures by honey bees reflects the fact that they have not evolved enzymes more stable than are normally necessary. This makes sense, because an enzyme that would be stable at temperatures far above this bee's normal range would be too rigid to function efficiently at its usual temperatures.

The adaptive significance of avoiding nest overheating is obvious, but what selective forces favored the evolution of nest warming? The main benefit during the warm months is probably acceleration of brood development. Speedy brood development enables rapid colony growth, which is valuable whenever a colony's population has dropped sharply, such as at the end of winter, after swarming, and following heavy mortality from predation. Significant deceleration of brood development occurs when brood is cooled just slightly. Vern G. Milum found, for example, that brood located on the perimeter of a colony's brood nest, where the temperature averaged about 31.5°C (89°F), required 22–24 days between egg laying and adult emergence, whereas brood in the nest center, where it was about 3°C (4.5°F) warmer, required only 20–22 days to complete development.

In addition to fostering rapid colony growth, the elevated temperature of a colony's brood nest helps it cope with disease. Work done by Anna Maurizio, in the 1930s, showed that chalkbrood, a disease of honey bee larvae caused by the fungus *Ascosphaera apis*, is blocked if a colony keeps its larvae warm (at 35°C/95°F), but that letting the larvae cool to 30°C (86°F) for just a few hours is all that is needed for a successful infection of these larvae by this fungus. Recently, Phil Starks and colleagues have shown that honey bee colonies have a brood-comb fever response when exposed to chalkbrood spores. Specifically, colonies that were fed a 50 percent sugar solution containing ground sporulating chalkbrood mummies—dead larvae covered with the fruiting bodies of the fungus—raised their brood-comb temperatures by nearly 0.6°C (1.0°F) (Fig. 9.3). Given that the normal range of the brood-comb temperature is just 2°C, from 34°C to 36°C (93°F to 97°F), a 0.6°C increase is a sizable elevation, and it appears to have been effective in preventing infection. None of the colonies that were fed the spores acquired the disease. Honey bee colonies also suffer from at least 15 viral and two bacterial diseases, but the effects of high brood-comb temperature on a colony's vulnerability to viral and bacterial infections remain unknown. Studies with other insect viruses have found that they do not cause infections when their hosts are reared at temperatures like those found in the brood nest of a honey bee colony, so the elevated temperatures found in honey bee colonies may provide resistance to viral diseases, too.

Of course, another important benefit of the honey bee colony's ability to create a warm microclimate within its nest is greater resistance to cold temperatures out in the general environment. Through its advanced techniques of social thermoregulation, *Apis mellifera* has greatly expanded its thermal niche, living today in geographic locations where colonies would otherwise perish over winter. As was discussed in chapter 6, the honey bee is basically a tropical insect that has expanded its range into cold-temperate regions through various adaptations, especially its ability to maintain a warm cluster throughout long, freezing winters.

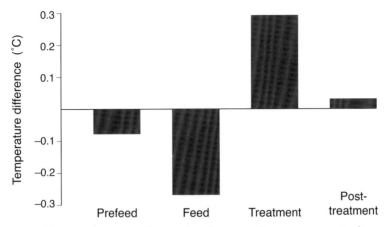

FIG. 9.3. Differences between observed and expected temperatures in the center of the brood comb in three small colonies living in two-frame observation hives, on days when they were fed sugar syrup that was pure (Feed interval) or contained chalkbrood (*Ascosphaera apis*) spores (Treatment interval), and on days before (Prefeed) and after (Post-treatment) the feeding. Brood-comb temperatures decreased during the Feed period, because some bees left the brood comb to collect the sugar syrup. Despite the cooling effect of feeding, the colonies had relatively high brood-comb temperatures when the sugar syrup was inoculated with chalkbrood spores.

WARMING THE COLONY

The primary problem in thermoregulation that is faced by a honey bee colony living in a place with long, cold winters is that of staying warmer than the surrounding environment. As mentioned already, the internal temperature that a colony strives to maintain varies depending on whether it is or is not rearing brood. If it is, then the brood-nest region is kept at 34°–36°C (93°–97°F). If it is not rearing brood, then it turns down its thermostat and maintains its core temperature above about 18°C (64°F) and its mantle temperatures above about 8°C (46°F). These two temperatures are critical lower limits. Bees chilled below about 18°C (64°F) cannot generate the neuronal activity that is needed to activate their flight muscles to produce more heat, and bees cooled below about 8°C (46°F) become immobilized and enter a sort of chill coma. Whether a bee survives such hypothermia depends on its duration; chilling to 10°C (50°F) or colder kills most bees within 48 hours.

A colony maintains a suitably warm microclimate inside the part of its nest that it occupies—the combs that it covers—by controlling the rates of heat production within and heat loss from this region. Figure 9.4 depicts the ways that a colony in a winter cluster loses heat to the environment inside its nest cavity: by conduction, convection, evaporation, and thermal radiation. It loses heat by conduction through the ceiling of the cavity and through the combs, and it loses heat by convection through air currents moving through the porous cluster and within the nest cavity. The colony also loses heat from respiratory evaporation of the adult bees, surface evaporation from the moist bodies of larval bees, and surface evaporation from any damp combs inside the cluster. Finally, the colony emits thermal radiation toward all objects around it: the unoccupied combs and the nest cavity's walls. The isotherm lines in this figure show us that the loss of heat from the cluster created a pocket of relatively warm air outside the cluster in the top of the nest cavity. Note that the temperature in the bottom of the nest cavity was the same as outside, $-21°C$ ($-6°F$), but that in the top of the cavity, and all around the cluster, the temperature was much warmer, $-1°C$ ($30°F$).

The heat that is lost from the cluster warms not only the air around it but also the nest cavity's ceiling and walls, and they in turn lose heat to the environment outside by conduction and thermal radiation. If wind outside causes air currents to pass in and out of the nest cavity through the entrance, then there also can be loss of heat to the outside through convection. Wild colonies reduce heat loss from their nest cavities through convection by filling cracks and small holes with seams of propolis, as discussed in chapter 5. In late summer, some colonies will also reduce their nest's entrance opening by building a curtain of propolis across most of it, as shown in Figure 5.7.

The take-home message from Figure 9.4 is that a honey bee colony has two ways to minimize its loss of heat to the environment: 1) by reducing

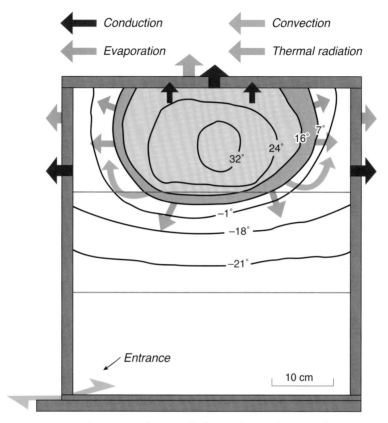

FIG. 9.4. Anatomy of a winter cluster of a honey bee colony inside a Langstroth hive when the ambient temperature was −21°C (−6°F), at 0700 hours on 25 February 1951. It shows the ways in which a cluster loses heat to the nest cavity's air and walls: through conduction, convection, evaporation, and thermal radiation. It also shows how the nest cavity's air and walls lose heat to the outside environment: though conduction, convection, and thermal radiation. Dark shading indicates the dense, insulating mantle of the cluster. The colony's small brood nest is inside the 32°C (90°F) isotherm.

heat loss from the colony to the nest cavity, and 2) by reducing heat loss from the nest cavity to the environment outside of whatever is housing the colony (hereafter, called the "nest enclosure"). In both stages of a colony's heat loss—from cluster and from cavity—the rate of heat transfer from inside to outside increases approximately in proportion to the difference in temperature between the inside and the outside (T_I minus T_O). This temperature difference is the driving force for the heat loss. When there is no heat transfer due to evaporation, Newton's law of cooling applies:

$$\text{Rate of heat transfer} = C \times (T_I - T_O)$$

The coefficient C, the thermal conductance of a structure, is a measure of how readily heat can move from the inside to the outside of the structure (whether a bee cluster or a nest enclosure). A bee cluster or a nest enclosure with a low C has a high resistance to loss of heat, hence it has a high insulation value. Let's now review what is known about how a honey bee colony adaptively adjusts the $C_{cluster}$ for its mass of bees, and how a honey bee colony acquires an impressively low $C_{enclosure}$ for the walls of its dwelling place by occupying a thick-walled cavity in a tree.

A colony of honey bees can strongly reduce its heat loss by lowering its thermal conductance ($C_{cluster}$). It does so when the bees press together tightly to form a compact, roughly spherical cluster. This process begins when the temperature inside the nest cavity dips below about 14°C (57°F). If the temperature outside the cluster falls further, the bees press together more tightly, thereby shrinking the cluster's size, though at about −10°C (14°F), cluster contraction reaches its limit. Between 14°C and −10°C, the volume of the cluster shrinks roughly fivefold. The structure of a winter cluster was closely investigated in 1951, by Charles D. Owens, in Madison, Wisconsin, while he was working for the U.S. Department of Agriculture. He did this by taking careful measurements of the temperatures throughout a colony's cluster on a day when the ambient temperature was −14°C (7°F), then killing the colony with hydrogen cyanide gas, and finally carefully dissecting the dead colony. Figure 9.4 shows the two-part internal organization that Owens found. There was an *outer zone*, between what had been the 7°C and 16°C (44° and 61°F) isotherms, consisting of several layers of densely packed bees oriented with their heads pointed inward. They filled all the empty cells in the combs and had pushed themselves as close together as possible in the spaces between the combs to form an insulating mantle. Bees in the *inner zone*, however, had room to crawl about, feed on the honey stores, fan with their wings, and tend the brood.

Why do bees form a tight cluster at low ambient temperatures? Clustering reduces a colony's heat loss in part by shrinking its surface area for heat loss by thermal radiation. Clustering also reduces heat loss by convection (air currents), since by pressing together the bees reduce the porosity of their cluster and therefore its loss of heat by air currents. Most important, the dense outer layer of a cluster forms an effective blanket of insulation that reduces its heat loss by conduction. Based on measurements of a colony's metabolic rate as a function of ambient temperature, Edward E. Southwick has estimated the heat conductance from a 17,000-bee (hence, a 2.2-kilogram/4.9-pound) winter cluster to be: 0.10 watts per kilogram per °C. Rather amazingly, this low heat conductance matches, or is even below, that of birds and mammals of the same body weight. Evidently, the insulation effectiveness of the dense outer layer of bees in a winter cluster is as good as, or better than, that of the feathers of birds or the fur of mammals.

A honey bee colony living in the wild also reduces its heat loss by occupying a thick-walled tree cavity. The thick wooden walls have low heat conductance (low $C_{enclosure}$) so they impede heat loss by conduction from the nest cavity to the general environment. At present, nobody has investigated whether nest-site scouts assess the thickness of a prospective nest cavity's walls and factor this into their assessments of the overall quality of a prospective homesite. I bet they do. What has been studied, however, is how strongly the total heat conductance of a honey bee colony's nest enclosure differs between a thick-walled tree cavity and a conventional, thin-walled wooden hive. This is the work of Derek Mitchell in England. The thick walls of tree cavities differ from those of wooden hives in both heat *capacity* and heat *conductance*, but Mitchell decided to study the effects of differences in heat conductance per se, so he worked with models of tree cavities that he constructed using sheets of polyisocyanurate foam, a material that has very little heat capacity. He built a foam model of a tree cavity whose walls had the same thermal conductance per unit height (watts per °C per meter) as the thick-walled tree cavities that Roger Morse and I had reported on in the 1970s, in our study of the nests of wild colonies. He also built his tree-cavity model so that its size (40 liters/10.6 gallons), shape (a tall cylinder), and entrance properties—a 15-centimeter- (6-inch-) long passageway, 5 centimeters (2 inches) in diameter—matched the average values of these variables for the tree-cavity nest sites that we had studied. Furthermore, Mitchell worked with various kinds of wooden hives, including British National and Warré. Inside his tree-cavity model and each of his hives, he installed an array of temperature sensors, and he suspended in its upper region a heating element for simulating the heat production of a colony of bees. Finally, he powered everything up and let the heating elements run and the temperature sensors take data for several hours, until the conditions inside each structure came to equilibrium. What he discovered was that his model of a thick-walled tree cavity had a value of total (lumped) enclosure heat conductance ($C_{enclosure}$) that was only about 0.5 watts per °C, whereas his British National hives (for example) had values of $C_{enclosure}$ that were approximately five times higher, around 2.5 watts per °C.

What are the effects of such a strong difference in heat conductance between the walls of natural tree cavities and conventional wooden hives? One is that the bees in a strong colony inhabiting a well-insulated tree cavity can stay mobile inside their nest well into winter, possibly even throughout it. Mitchell's analysis shows that a 1-kilogram (2.2-pound) colony of bees, producing heat at what is a normal rate of 20 watts at a nest-cavity temperature of 20°C (68°F), will not need to form a tight, well-insulated cluster until the temperature outside the cavity drops below −30° or −40°C (−22° or −40°F), if it is living inside a thick-walled, well-insulated tree cavity. In contrast, a colony of the same size that is living in a standard, thin-walled wooden hive will need to go into a tight cluster soon after the temperature outside the hive drops below about 10°C (50°F), because its nest cavity is so poorly insulated (Fig. 9.5). This striking difference may translate into better winter survival for colonies residing

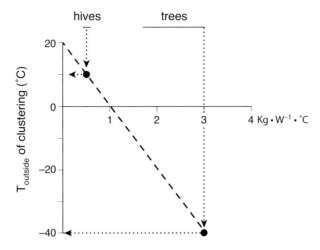

FIG. 9.5. Graphical analysis of the threshold temperature at which a 1.0-kilogram (2.2-pound) colony will need to form a cluster to stay warm depending on the total heat conductance of its nest enclosure: a thin-walled wooden hive (British National) vs. a thick-walled tree cavity. Tree cavities have relatively low total heat conductance, so they have a high ratio of colony mass to nest enclosure heat conductance (kg·W^{-1}·°C). Therefore, a kilogram colony of bees living in a tree cavity will experience a warm, 20+°C (68+°F) microclimate inside its nest cavity until T$_{outside}$ drops below −40°C (−40°F), but a kilogram colony of bees living in a British National hive will experience such a warm microclimate only when the outside temperature is above about 10°C (50°F).

in tree cavities relative to those in hives, because colonies overwintering inside trees can stay mobile on their combs for longer and so are probably better at staying in contact with their honey stores. This subject requires much more investigation, however, because the thick walls of tree cavities have high heat capacities—not just high insulation values—relative to the thin walls of conventional hives. This means that once the walls of a tree cavity get cold over winter, these cold walls will delay the warming of the cavity air to the temperature, ca. 14°C (59°F), at which a colony living inside the tree cavity can break its cluster. Therefore, colonies living in thick-walled tree cavities may be delayed in becoming active in spring relative to colonies living in thin-walled wooden hives.

To explore further the microclimate differences that exist between a thick-walled tree cavity and a thin-walled hive, I recently started a study with two colleagues, Robin Radcliffe and Hailey Scofield, to describe the microclimates inside both types of structure, without any bees, throughout a year. We began by building two nest cavities, one made of standard pine lumber and the other cut with a chain saw into a large sugar maple tree (Fig. 9.6). The two cavities are identical in shape (tall and narrow), volume (50 liters/13.2 gallons), and entrance size (5 centimeters/2 inches in diameter), but they differ greatly in wall thickness: ca. 2 centimeters (0.75 inches) vs. 36 centimeters (14 inches). Therefore, both cavities have the shape and size of a natural nest cavity, but only one has the wall thickness of a natural nest cavity. They are located side by side, and each contains two temperature sensors/recorders, positioned in the center of the cavity, 20 centimeters (8 inches) from cavity top and bottom. There is also a temperature sensor/recorder mounted in a shady spot between the two structures for measuring ambient temperature. Our aim is to compare the temperature dynamics inside both cavities across two years, for one year with each cavity unheated and for one year with each containing a 40-watt heating element, installed inside to simulate the heating it would have if it were occupied by a 2-kilogram (4.4-pound) colony of honey bees.

The graph in Figure 9.6 shows the temperature readings from the two cavities and the ambient temperature sensor for two weeks in April 2018. It shows that during this early-spring time period, the readings for the tree cavity are far more stable than those for either the ambient temperature or the hive box. We also see that the latter two sets of readings are nearly identical, though on sunny days the hive box interior became warmer than the air outside it (ambient). Furthermore, we see that during every night, the tree cavity was warmer than the box. It is too early to know what this study will tell us overall, but already it has revealed a stunning difference in the stability of the nest-cavity temperatures experienced by wild colonies that live deep inside massive trees vs. managed colonies that reside in thin-walled hives.

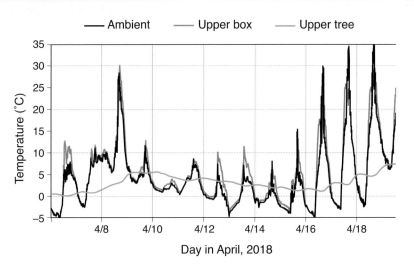

FIG. 9.6. *Top:* The two artificial nest cavities being used to compare, over a year, the microclimate inside a thick-walled tree cavity and a thin-walled wooden box of the same dimensions and with the same size entrance opening. *Bottom:* Sample of temperature recordings from inside the thick-walled tree cavity and inside the thin-walled wooden box and from outside both structures.

The best insulation in the world will not, however, keep a living system warm unless the system generates heat, so it is not surprising that the adaptations of honey bee colonies for heat retention are complemented by an impressive capacity for heat production. A respectable amount of heat is generated by the resting metabolism of brood and adults: about 8 and 20 watts per kilogram (3.6 and 9 watts per pound), respectively, at the brood-nest temperature of 35°C (95°F). However, by isometrically contracting their flight muscles—so they burn energy without producing wing movements—adult bees can boost greatly their metabolic heat production, up to about 500 watts per kilogram (230 watts per pound)! It is, therefore, the adult worker bees' metabolic rate that is varied to adjust a colony's heat production. Workers engaged in heat production for the colony are almost indistinguishable from resting bees; both stand motionless on the combs. However, one can sometimes observe a worker press her thorax firmly onto the cap of a cell containing a pupa and then stand motionless in this posture for several minutes. She is generating heat (with her flight muscles) to raise her thorax temperature to approximately 40°C (104°F). This will boost the temperature of the brood cell by a few degrees. Also, measurements of single bees or small groups of bees confined in a controlled-temperature respirometer show clearly that worker bees will resist chilling by dramatically raising their metabolic rate. For example, whereas a group of 10 workers at 35°C (95°F) showed little elevation of thorax temperature (36°C/97°F) and had a low metabolic rate (29 watts per kilogram/13 watts per pound), bees in a group held at 5°C (41°F) boosted their metabolic rate to 300 watts per kilogram (136 watts per pound) and so maintained a thorax temperature of 29°C (84°F), far above the ambient temperature.

In nature, where each bee works together with thousands of fellow colony members in resisting cold, the process of the workers increasing their heat production operates together with the process of them reducing their heat loss by clustering. Figure 9.7 shows how the bees coordinate these two thermoregulatory mechanisms. Heat production increases as the ambient temperature drops from 30°C (86°F) to about 15°C (59°F) and

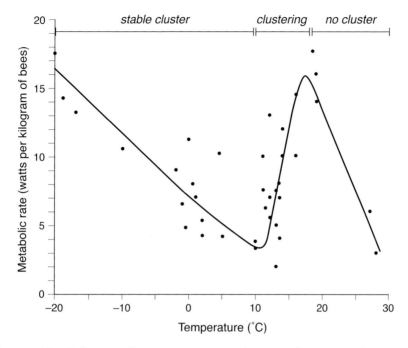

Fig. 9.7. Metabolic rate of an overwintering colony in relation to ambient temperature. The colony was inside a controlled-temperature cabinet that served as a metabolic chamber. Each data point shows the minimum metabolic rate of the colony during a 24-hour run at a fixed temperature.

does so again when it falls below about 10°C (50°F), but it decreases sharply when the ambient temperature declines from about 15°C to 10°C. As discussed above, this is the temperature range over which a colony coalesces into a well-insulated cluster and can resist deepening cold entirely by reducing its heat loss. Because clusters continue to shrink down until temperatures reach about −10°C (14°F), reducing heat loss continues to play a role in colony thermoregulation down to this temperature, though as we can see in Figure 9.7, it is accompanied by rising heat production. Presumably the reason colonies do not initiate their cluster formation at higher temperatures, and thereby reduce their nest-heating costs, is that coalescing into a tight cluster disrupts other colony operations, such as foraging and food storage.

COOLING THE NEST

A honey bee colony sometimes needs to cool its home to keep the temperature in the brood-nest region below about 36°C (97°F). Prolonged temperatures just 2°–3°C (about 3°–5°F) higher than this will disrupt the development of immature honey bees undergoing metamorphosis—that is, the transformation of a larva to an adult. Some of the danger of nest overheating arises from the colony's inevitable production of metabolic heat. The resting-level heat production by brood and non-incubating adults is much lower than that of actively incubating bees; nevertheless, even non-incubating bees produce enough heat that, when the air temperature outside a colony's nest rises above about 27°C (80°F), the colony can face a threat of its brood nest overheating. This danger is more common for managed colonies residing in thinly insulated hives fully exposed to direct sunlight than for wild colonies inhabiting well-insulated tree cavities shaded by neighboring trees. But whenever overheating of the brood nest arises, the bees deploy mechanisms for keeping their nests cool that are as effective as the ones they use to keep them warm. For example, when Martin Lindauer placed a colony of bees living in a thin-walled, wooden hive in full sunlight on a lava field near Salerno, in southern Italy, the colony's maximum temperature inside its hive never exceeded 36°C (97°F), even though the temperature of the air at hive level outside rose to 60°C (140°F) and the temperature inside an unoccupied hive nearby rose to 41°C (106°F). To prevent nest overheating, colonies deploy a battery of cooling mechanisms in a graded response. They start with the adults spreading out inside the nest and partially evacuating it (to reduce internal heat production and foster heat loss by convection), followed by fanning (to create forced convection), and finally spreading water on the combs (to turn on evaporative cooling).

The dispersal of adult worker bees inside their nest cavity as its temperature rises is an extension of the cluster expansion that starts when the temperature inside the nest cavity rises above about −10°C (14°F). The air temperature outside the nest at which fanning behavior begins to supplement this dispersal by the bees is variable and depends on such factors as the sun exposure of the nest structure (tree or hive), the insulation of the nest cavity's walls, and the strength of the colony (which influences its overall rate of heat production). What matters to the colony, ultimately, is keeping the internal nest temperature below 36°C (97°F), and many observers have reported bees initiating strong ventilation of their nest when its internal temperature approaches this critical limit for the brood-nest temperature. The fanning bees deploy themselves throughout the nest, aligning themselves in chains to drive the air along existing (unidirectional) currents. Additional fanning bees stand outside the entrance opening with their abdomens pointing away from it, pulling air out of the nest.

Recently, Jacob Peters and colleagues at Harvard University have measured the *velocity* of airflow at the entrance and have reported that it can be as high as 3 meters per second (ca. 10 feet per second), hence 10.8 kilometers (6.7 miles) per hour. This occurs when the temperature of the air that is being expelled is over 36°C (97°F), indicating that the colony's brood-nest temperature is dangerously high. An earlier investigator, Engel H. Hazelhoff, working in the Netherlands, measured the *volume* of airflow through a nest created by the fanning bees. He constructed a hive with two openings, one at the top, connected to an airspeed indicator (anemometer), and one at the bottom, serving as the hive's entrance. With this hive, he could accurately measure the airflow through the hive produced by the fanning bees. Once, when there were 12 bees fanning steadily, all spaced evenly across the 25-centimeter- (10-inch-) wide entrance of his hive, Hazelhoff measured the rate at which cool, fresh air flowed in through the top opening: up to 1.0–1.4 liters (0.26–0.37 gallons) per second!

Hazelhoff also discovered that a high level of carbon dioxide in the air inside a nest, not just a high temperature, will stimulate strong fanning by bees for ventilation. This means that nest ventilation by honey bees functions not only in social thermoregulation but also in social respiration. Remarkably, the carbon dioxide content of the air inside a colony's nest cavity when it is not being actively ventilated by the bees is 0.7%–1.0%, which is 20–30 times above the normal percentage of carbon dioxide of atmospheric air (0.03%–0.04%). The ability of honey bee colonies to thrive under such "stuffy" conditions shows us how remarkably these social bees are adapted to living in large numbers inside tree cavities with small openings.

In chapter 5, we saw how the tree-cavity nests of wild colonies often have just one entrance opening and that often its size is quite small, only 10–20 square centimeters (1.5–3.0 square inches). This raises the question of how the bees achieve sufficient airflow through just one small entrance opening to prevent buildup of excess heat (and carbon dioxide) in the nest. The work of Jacob Peters and his colleagues shows us how the bees solve this problem. The fanning bees distribute themselves asymmetrically around the entrance opening so that air continuously enters and exits at different locations around its perimeter (Fig. 9.8). The clustering of the fanning bees also has the benefit of reducing the fluid friction of the air, which increases the ventilation efficiency. Evidently, the fanning bees achieve this efficient partitioning of the inflow and outflow airstreams by sensing the air temperature—which is highest where the air is shooting out—and aligning themselves with the direction of the airflow wherever the air is hottest. In other words, the fanning bees use the airflow itself, not direct interactions with the other fanners, to efficiently partition the outflow of hot air and the inflow of cooler air through their nest's entrance.

When the bees cannot cool their nest adequately by means of worker dispersal and nest ventilation, they bring the power of evaporative cooling to the problem. Water absorbs a great deal of heat (energy) when passing from a liquid into a gas, so water evaporation is a wonderfully powerful means of cooling objects. We all know this from our experience in sweating when our bodies are overheating. A vivid demonstration of the power of evaporative cooling in the context of beekeeping comes from a report of a catastrophe by P. C. Chadwick, a beekeeper in southern California. One day in June 1916, when the midday air temperature rose to 48°C (118°F), his bees brought large amounts of water into their nests and pre-

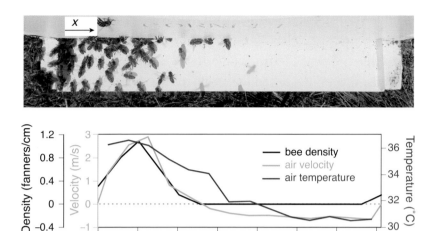

FIG. 9.8. *Top:* Worker bees ventilating at the entrance of a hive, in a cluster on the left side of the entrance opening. *Bottom:* Air velocity (green), bee density (black), and air temperature (red) across the hive entrance. The outflow and inflow velocities are indicated by positive and negative values, respectively.

vented the melting of their combs. By 9:00 P.M., the temperature had dropped to 29.5°C (85°F), but at midnight a hot breeze blowing in off the desert raised the air temperature back up to 38°C (100°F). The supplies of water in the colonies were soon exhausted, no more water could be collected until daylight, and the combs of many of Chadwick's colonies softened and collapsed during the night.

    The details of the acquisition, handling, and storage of water in nests, and the regulation of its collection, have received close attention in recent years. For example, it is now clear that some of a colony's workers that are old enough to work outside the hive will specialize in water collection, and that they can travel to water sources up to 2 kilometers (1.2 miles) away, even though they must make their return flights with their crops (fuel tanks) filled with water. That some of the workers in a honey bee colony are water-collector specialists makes sense, because colonies need water daily, sometimes for evaporative cooling, other times for diluting stored honey and producing glandular secretions to feed brood, and still other times to humidify their nests to prevent desiccation of the brood. There are times, too, when the adult bees need water simply to relieve personal thirst—that is, to maintain osmotic homeostasis in their bodies.

I witnessed bees in a broodless colony responding to powerful sensations of personal thirst one January day in Ithaca, when deep snow covered the land but a strong sun had warmed the air enough that some of the bees living in the observation hive in my office were making flights outside. At first, I figured that these bees were simply conducting cleansing flights, but then I noticed several bees inside the hive performing extremely vigorous waggle dances for a site just outside the building. These were water collectors! They had found puddles of snowmelt in the parking lot, and upon entering the observation hive they were being mobbed by thirsty bees. One of these water collectors produced the most exuberant and persistent waggle dance that ever I have seen; it lasted for 339-plus dance circuits without a pause. I must write "339-plus," because I did not see the start of her eye-catching performance.

At the time, I thought that the extraordinary wintertime thirst of these bees was not natural. I figured it was an artifact of their living in an observation hive located in a heated room, an arrangement that prevented their metabolic water from condensing on the walls of their nest cavity (the observation hive). Since then, I have learned that even bees living outdoors in cold wooden hives can get extremely thirsty in winter. Ann Chilcott, a beekeeper in northern Scotland, has recorded bees collecting water in January and February, even under cloudy skies, as long as the air temperature is above 4°C (39°F) (Fig. 9.9). In a related study, Helmut Kovac and colleagues at the University of Graz, in Austria, used an infrared camera to measure the thorax temperatures of wintertime water collectors while they were drinking from a water source near their hives. They discovered that when a wintertime water collector is at a water source, busily loading up on water, she activates her flight muscles (i.e., she shivers) to keep her thorax temperature *always above 35°C (95°F)*, even when the air temperature is as low as 3°C (37°F)! Evidently, this ensures that the water collector can complete a short flight home before her thorax temperature falls—

Fig. 9.9. Worker bee loading up at a moss-filled water source on a chilly January morning, near Inverness, in northern Scotland.

from the wind chill during flight—below the critical 25°C (77°F) needed to produce a wingbeat frequency high enough for flight.

These reports of bees desperately collecting water at low temperatures in winter make me wonder whether some condensation on the walls inside a hive or tree cavity might be beneficial to a colony during winter. Perhaps this helps explain why colonies in the wild avoid nest cavities with top entrances, which allow warm, moist air to escape. Derek Mitchell has explained that if a nesting cavity is well insulated and lacks a top vent hole, then overwintering bees will *not* have cold condensation dripping down on them, because the temperatures of the ceiling and walls above the bees will be above the dew point. There will be condensation on the cooler walls below the bees, but this may be beneficial, because it provides them with a ready source of fresh water. This may help explain, too, why bees living in natural nest cavities coat their walls with propolis: the condensed water is not lost to the bees by soaking into the wooden walls of their home.

During the warm months, most honey bee colonies have ready access to water sources, so they can meet their water needs without maintaining large water reserves in their nests. Reliance on external water sources, however, requires activation and deactivation of a colony's water collectors as conditions change. Recently, Madeleine M. Ostwald, a Cornell undergraduate student, worked with me and one of my PhD students, Michael L. Smith, to investigate how a colony's water collectors are activated and deactivated. To do so, we moved a colony living in a glass-walled observation hive into a greenhouse, where we could control the bees' access to water. We provided the bees with just one water source, which was set on scales; by measuring its drop in weight, we could measure precisely the colony's water collection during our experiments. We then stimulated the colony's water collection by heating the observation hive with an incandescent lamp, and when bees began visiting the water source, we labeled these water-collector bees for individual identification. By closely watching the behavior of the water collectors inside the hive on days when we again gave the colony a heat stress, we saw how the colony's water collectors sprang into action (Fig. 9.10). These bees did so about one hour after the heat stress began, which showed that the water collectors were not responding to the high temperatures in their nest. Instead, these bees were stimulated to resume their work by experiencing either greater personal thirst or more frequent beggings for water, or both. A previous study, by Susanne Kühnholz and myself, found that once a water collector has sprung into action, she keeps informed about her colony's water need by paying attention to what she experiences when she returns to the hive and seeks to unload her water to other bees (water spreaders). The higher the colony's water need, the fewer unloading rejections she experiences and the faster she delivers her water load.

FIG. 9.10. Changes in behavior of water collectors in a colony when it experienced brood-nest overheating and was temporarily deprived of water collection. Experimental treatment: brood nest was not heated for first hour, then was heated and water was withheld for two hours, and finally water was provided *ad libitum* outside the hive. Six focal water collectors were observed repeatedly over the five hours, to determine the proportion of time these bees spent standing,

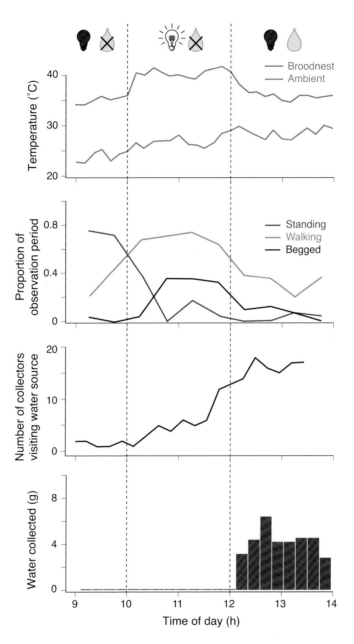

walking, or being begged for water. A roll call was made every 15 minutes of the water collectors (labeled with paint marks for individual ID) that visited the water source. Amount of water collected was measured by weighing the water source every 15 minutes.

While it is true that honey bee colonies do not maintain large water stores in their nests, beekeepers have reported finding some water (amounts not specified) stored in the combs of colonies during droughts in Australia and South Africa. A second means of water storage is for worker bees to hold water in their crops (honey stomachs). O. Wallace Park reported finding clusters of reservoir bees—workers with swollen crops containing dilute nectar—in a colony in Iowa in early spring, when the weather was chilly and the bees were rarely able to obtain water for rearing brood. He also observed water collectors flying out en masse from a colony living in an observation hive. These bees gathered water from grass blades and puddles, and upon return to the hive they transferred their loads to other bees that served as the water reservoirs. Madeleine M. Ostwald, Michael L. Smith, and I also found water-filled bees standing quietly on the combs in our observation hive in the evening after the study colony had experienced a day of heat stress combined with water deprivation. Moreover, we found water stored in the comb cells located just inside the hive entrance. It now seems clear that temporary water storage in reservoir bees, and in comb cells, is an important part of the social physiology of honey bee colonies.

10

COLONY DEFENSE

> Life consists with wildness.
> The most alive is the wildest.
> —Henry David Thoreau, "Walking," 1862

Every living system faces a legion of predators, parasites, and pathogens, each of which is equipped with a sophisticated tool kit for penetrating the defenses of its prey or host. In the case of a honey bee colony, there are several hundred species, ranging from viruses to black bears, whose members are forever trying to breach the bees' defenses. What makes a bee colony *so* attractive to *so* many is, of course, the store of delicious honey and the horde of nutritious brood that lies inside its nest. In summer, the combs inside a bee hive or a bee tree typically hold 10 or more kilograms (20-plus pounds) of honey, plus thousands of immature bees (eggs, larvae, and pupae). Moreover, these brood items are neatly packed together in the warm center of the bees' nest, making them an absolute bonanza for any viruses, bacteria, protozoa, fungi, and mites that succeed in infecting or infesting this host of developing bees. Clearly, a colony of honey bees is an immensely desirable target. It is also a perfectly stationary target. Because a colony's beeswax combs are a huge energetic investment, and because these combs are often filled with brood and food, a honey bee colony cannot afford to find safety by fleeing its home when threatened. Instead, it must cope with its foes by standing its ground, and usually it succeeds, by drawing on a sophisticated arsenal of biochemical, morphological, and behavioral weapons.

Given that honey bees have a 30-million-year history, it is likely that most of the relationships between *Apis mellifera* and its predators and agents of disease are long-established. We can expect, therefore, that colonies living undisturbed in the wild possess defense mechanisms that usually prevent pathogens and parasites from multiplying sufficiently to cause severe disease. Indeed, it is likely that wild colonies of honey bees have perpetual, endemic infections of parasites and pathogens, and it is also likely that symptoms of disease arise in these colonies only when they are weakened by adverse environmental circumstances, such as food shortages or damage to their nests. We will see, however, that the balance of power between the bees and their pathogens and parasites can be upset by intrusive beekeeping practices, and that these practices can lead to severe losses of colonies. For example, a common feature of several diseases of honey bees—including American foulbrood and chalkbrood—is that the causative pathogens persist over winter as resting stages on the combs. In nature, these combs would be cleaned by the bees in live colonies or destroyed by scavengers of the nests of dead colonies (e.g., the greater wax moth, *Galleria mellonella*). Either action would control the infections. But ever since beekeepers switched from using hives with fixed combs (such as skeps) to working with hives with movable combs (such as Langstroth hives), they have recycled the honey storage combs after extracting the honey from them. Beekeepers generally store the emptied combs away from the bees for the winter, which means that these combs are not naturally cleansed by the bees. Then in the spring, when they are returned to the bees, the combs can reinoculate the colonies with pathogens.

A sad example of how beekeeping practices have disrupted some of the balances that once existed between honey bee colonies and their agents of disease is the recent rise in the virulence of the deformed wing virus. The history of this problem begins in the late 1800s, when beekeepers in the Russian Empire began exporting colonies of the western hive honey bee

(*Apis mellifera*) from Europe to eastern Asia, specifically to the Primorsky region of the Russian Far East. This intercontinental transplanting of colonies occurred first by sailing ships but later via the Trans-Siberian Railway, which links Moscow and Vladivostok. Eventually, somewhere in the Primorsky region, the ectoparasitic mite *Varroa jacobsoni* underwent a host shift from colonies of the eastern hive honey bee (*Apis cerana*), which is native to eastern Asia, to colonies of the western hive honey bee (*Apis mellifera*), which had been introduced to eastern Asia. Then, after making this host shift, the population of *Varroa* mites living on the *Apis mellifera* colonies in the Primorsky region speciated into *Varroa destructor*. Because *Varroa* mites feed on the hemolymph (blood) of adult and juvenile honey bees, they are highly efficient vectors of the bees' viruses, and this has created major health problems for colonies of *Apis mellifera*, especially those that live crowded in apiaries, where the mites and the viruses they carry spread easily among colonies. As will be explained shortly, the dispersal of *Varroa* mites and the deformed wing virus among unrelated colonies—so-called horizontal transmission—has favored the evolution of at least one highly virulent strain of the deformed wing virus.

What follows is a look at how the lives of wild colonies and managed colonies of European honey bees differ in ways that affect the endless problem of colony defense. We will see that even though colonies living in the wild receive no human protection from predators, parasites, and pathogens, they tend to experience fewer problems—and incur lower costs—of colony defense relative to colonies living in apiaries.

LIVING WITHOUT VS. WITH TREATMENTS FOR *VARROA DESTRUCTOR*

Since the 1970s in Europe and the 1990s in North America, most beekeepers have found that they must routinely treat their colonies with miticides; otherwise they will die after a year or two from severe infections of viruses spread by *Varroa* mites, most importantly the deformed wing virus. In the last 10 or so years, however, several beekeepers and research biologists have reported populations of European honey bee colonies that are thriving without receiving treatments with miticides.

Primorsky Kray, Russia

One of these populations is a commercial line of honey bees available in the United States that is based on stock imported by the U.S. Department of Agriculture from the Primorsky Kray (region) of eastern Russia. As mentioned above, colonies of *Apis mellifera* were brought here from western regions of the Russian Empire starting in the late 1800s, and some time thereafter the mite *Varroa jacobsoni* made a host shift from *Apis cerana* to *Apis mellifera*. This host shift by the mites must have happened only rarely (possibly just once), because the gene exchange between the mites living on *A. cerana* and those living on *A. mellifera* was rare enough for these two populations of *Varroa* mites to diverge in morphology and behavior and eventually to live as two distinct species. The result is *Varroa destructor*, a mite that is, alas, superbly adapted for life as an ectoparasite of *Apis mellifera*.

Once *Varroa* mites began to parasitize colonies of the European honey bees that had been introduced to eastern Russia, these bees began to undergo natural selection for resistance to these mites. Eventually, the colonies of *Apis mellifera* living in eastern Russia evolved a stable host-parasite relationship with *Varroa destructor*. The resistance mechanisms of these honey bees have been studied closely by a team of researchers led by Thomas E. Rinderer, at the Honey Bee Breeding, Genetics, and Physiology Research Laboratory in Baton Rouge, Louisiana. A series of studies started in the late 1990s has revealed that workers in colonies of Russian bees—compared to workers in colonies of stocks used commercially in the United States—are superior at grooming *Varroa* mites from their bodies, at removing mite-infested pupae from their cells, and perhaps also at biting the legs off the mites. The net result is that populations of *Varroa* mites grow much more slowly in colonies of Russian bees than in colonies of U.S. domestic stocks (Fig. 10.1).

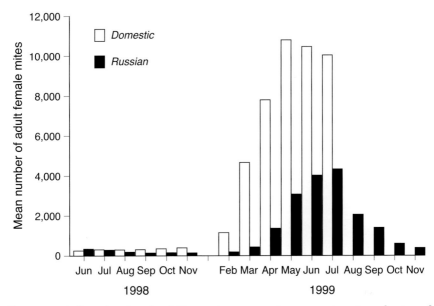

FIG. 10.1. Infestation levels of *Varroa destructor* mites over time in colonies of Primorsky, or Russian, bees (black bars) and domestic colonies of North American stock (white bars). The two types of colonies lived together in two shared apiaries.

Gotland, Sweden

The Russian bees demonstrate unequivocally that a population of *Apis mellifera* colonies of European origin can evolve mechanisms for survival despite being infested with *Varroa destructor*. Unfortunately, nobody tracked this evolutionary change, so we do not know how rapidly the bees' resistance to *Varroa* mites evolved in eastern Russia. In the early 2000s, however, an answer to the question of how quickly resistance to *Varroa* mites can evolve emerged from a remarkable experiment that was conducted in Sweden under the leadership of the late Ingemar Fries, then a professor at the Swedish University of Agricultural Sciences. The goal of the experiment was to see whether *Varroa* mites "will eradicate European honey bees in an isolated area under Nordic conditions, where no mite control or swarm control of honey bee colonies are implemented."

FIG. 10.2. One of seven apiaries that were set up in isolation on the south end of the island of Gotland, Sweden.

This experiment began in 1999, when a population of 150 genetically diverse colonies—headed by *A. mellifera mellifera*, *A. m. ligustica*, and *A. m. carnica* queens from various sources in Sweden—was established in isolation on the southern end of Gotland, a 3,200-square-kilometer (1,200-square-mile) island in the Baltic Sea, about 90 kilometers (56 miles) east of the Swedish mainland. These 150 colonies were distributed among seven apiaries (Fig. 10.2). Each colony was housed in a two-story Swedish hive, which is like a Langstroth hive, with two full-depth, 10-frame hive bodies. Once in place, the colonies were basically left alone and allowed to swarm. The only management was the feeding of a few colonies whose honey stores were insufficient for winter survival. Each colony started out free of *Varroa* mites but was inoculated with approximately 60 mites when the investigators gave each colony 1,000 bees they had collected from mainland colonies heavily infested with the mites. Also, each colony's queen was given a paint mark, so the investigators could detect turnovers in the colonies' queens, usually due to swarming. The colonies were disturbed only four times each year, when inspections were made to check for colony survival in late winter, colony size in early spring, colony swarming over summer, and mite infestation level in late October. Swarms from the colonies were collected during regular checks of the apiaries by a local beekeeper and by putting up bait hives. These swarms were installed in vacant hives in the apiaries.

What happened to the colonies in this live-and-let-die experiment? Figure 10.3 shows that going into the first winter (October 1999) the average mite infestation level (mites/100 bees) was low for these colonies, and that over the first winter (1999–2000) the level of colony mortality was low (around 5%). We see too that the colonies were strong enough to swarm heavily in summer 2000, which kept the number of colonies almost at the starting level, but that by October 2000 their mite infestations had grown strongly and this was followed by relatively high colony mortality (nearly 30%) over the winter of 2000–2001. The state of the population deteriorated further in the summer of 2001, with less swarming across the summer, higher mite infestations in October, and high colony mortality (nearly 80%!) over the winter of 2001–2002. By the third summer, in 2002, there were few colonies still alive, and those that remained were so weak that none swarmed. In October 2002 there remained only 21 colonies, and on average they had high mite infestations and high colony mortality (57%) over the winter of 2002–2003. In the fourth summer, in 2003, there were signs of improvement. Although the number of colonies was lower than ever—only eight of the 120 hives were inhabited in October 2003—one of these survivor colonies was strong enough to swarm, the mean level of mite infestation had begun to decline, and the colonies had impressively low mortality (12%) over the winter of 2003–2004. These improvements gained momentum in the fifth summer, in 2004, when more than half the colonies produced swarms, their mite infestations were far lower than in the previous four years, and their mortality over the winter of 2004–2005 was again fairly low (only 18%). Since the spring of 2005, this population of colonies has been left alone, so investigators could see what would happen over time, and over the following 10 years (through 2015), it consisted of 20–30 self-sustaining colonies.

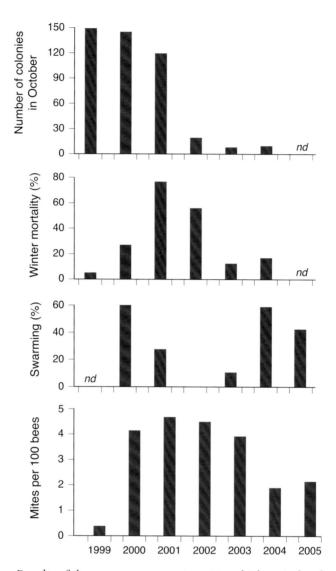

FIG. 10.3. Results of the seven-year experiment in which an isolated population of 150 honey bee colonies was established on the southern tip of Gotland, an island in the Baltic Sea. Colonies were infested with *Varroa destructor* mites and were left unmanaged and untreated for mites. Colonies were monitored for winter mortality, swarming over the summer, and *Varroa* infestation level in October.

The genetics of this experimental population of Swedish bees on Gotland was not studied closely, so we do not know what genetic changes occurred within it, but there can be little doubt that these bees experienced harsh natural selection for traits that confer resistance to *Varroa* mites and the viruses they transmit. What traits might these be? A cross-infection study of mite population growth, using mites from the Gotland survivor population and from elsewhere in Sweden, together with bees from the Gotland survivor population and from elsewhere, found that the Gotland survivor colonies had an 82 percent lower rate of mite population growth, regardless of the mites' source. This shows that the improvement in the survival of the Gotland bees was based on the evolution of increased resistance by the bees, not reduced virulence of the mites.

There are two stages at which honey bees can take actions that will suppress the reproduction of *Varroa* mites: 1) when the female mites are phoretic—moving about and feeding on adult bees, and 2) when the female mites are sealed in brood cells containing pupae. There is no evidence that the Gotland bees are skilled at attacking and killing the phoretic mites by biting off their legs and antennae. There is, however, compelling evidence that the Gotland bees are skilled at disrupting the female mites when they are sealed in cells containing pupae. Two of the Swedish investigators, Barbara Locke and Ingemar Fries, found in one study that only about 50 percent of the mites in the Gotland survivor bee colonies produced viable and mated daughter mites, whereas nearly 80 percent of the mites in control (mite-susceptible) colonies did so. This may be, at least in part, because the Gotland bees have stronger *Varroa*-sensitive hygienic (VSH) behavior—that is, the behavior in which bees uncap brood cells with reproducing mites and remove the mite-infested pupae from these cells. It may also be, at least in part, a result of the Gotland bees having a heightened tendency to chew open, and then recap, worker brood cells, especially those that are mite-infested. A recent study has shown—by comparing the proportions of nonreproductive female mites in cells that have and have not had their caps removed by experimenters—that the mere uncapping and recapping of brood cells is effective at reducing the reproductive success of mites. These manipulations of the brood cells' cappings are not lethal to the pupae developing inside the uncapped (and eventually recapped) cells, so this mechanism of mite control appears to be less costly to a colony than the VSH behavior, which kills bee pupae while disrupting the mites.

The Gotland colonies also tend to be smaller, more likely to swarm, and more restrained in their rearing of drones relative to mite-susceptible colonies living elsewhere in Sweden. It is likely that these colony-level traits of the Gotland bees also help reduce the growth of the mite populations in their colonies. Swarming creates a break in the brood rearing of a colony and thus disrupts the reproduction of the mites. Rearing only a few drones probably also hinders the mites' reproduction by limiting the supply of their preferred hosts.

Arnot Forest, United States

The honey bees living in the Arnot Forest and other woodlands immediately south of Ithaca are a third example of a population of honey bee colonies that have not been treated with miticides to control *Varroa destructor* and that possess good defenses against this mite. When did these wild colonies become infested with the mites? As explained in chapter 2, I first noticed *Varroa* mites in the colonies at my laboratory back in 1994. This fact, together with what we know now about the astonishing skill of *Varroa* mites in climbing onto worker bees while they are foraging or robbing honey, makes it likely that these mites spread widely soon after they reached the Ithaca area in the early 1990s. I believe, therefore, that it is likely that the colonies in the Arnot Forest were infested with *Varroa* mites by the mid-1990s.

It was also explained in chapter 2 that by 2004, just 10 years after discovering *Varroa* mites in my managed colonies, I had compelling evidence that the wild colonies in the Arnot Forest already possessed mechanisms for controlling the populations of *Varroa* mites living in them. Therefore, the bees living in the Arnot Forest provide us with a third example of a population of untreated colonies that has (evidently) rapidly evolved, through strong natural selection, potent defenses against *Varroa* mites. A feature of this third example is that it provides insights into the genetic changes that arose in this population of colonies following the arrival of the *Varroa* mites.

Our investigations of the impact of *Varroa* on the honey bee colonies living in the forests around Ithaca began in the summer of 2011, when a Cornell student, Sean Griffin, and I conducted a third survey of the wild colonies in the Arnot Forest using the same methods of bee hunting as I had used there in 1978 and 2002. Our first goal was to locate as many of the wild colonies living in this forest as possible and to collect from each colony that we found a sample of at least 100 worker bees. Our second goal was to locate the nearest apiaries outside the Arnot Forest and to collect from the colonies in them samples of 100 worker bees per colony. We would then give these precious samples to two colleagues, Deborah A. Delaney and David R. Tarpy, who would perform genetic analyses that would tell us whether the population of wild colonies living in the Arnot Forest is self-sustaining or instead is persisting based on immigration of swarms from the nearest managed (and treated) colonies.

From late July to early September, Sean and I bee hunted over half the Arnot Forest. We located nine colonies living inside hardwood trees within this forest, plus one more colony living inside a handsome white pine some 500 meters (0.3 miles) from the forest's northeastern corner. (Note: these numbers indicate a density of wild colonies in the Arnot Forest of at least 1 per square kilometer/2.5 per square mile, which matches the estimates of colony density from the 1978 and 2002 censuses.) We collected a sample of 100 worker bees from each these 10 bee-tree colonies. We then hunted for managed colonies living within 6 kilometers (3.7 miles) of the Arnot Forest's boundaries and found just two apiaries, each containing about 20 colonies. One was a new apiary sitting 1 kilometer (0.6 miles) from the southwestern boundary line of the Arnot Forest. The other was a long-standing apiary sitting 5.2 kilometers (3.2 miles) from the forest's northeastern corner (see Fig. 7.11). Both belonged to the same commercial beekeeper. I reached him by phone and explained my project and requested permission to collect 100 bees from each of 10 colonies in his two apiaries. The phone was silent on his end for about 30 seconds. Then he asked me to say again how many bees I would remove from each colony, and I repeated the figure. Another long silence occurred, and then he said, slowly, "OK, take what you need." Great! I soon collected the critical samples of worker bees from both apiaries. My geneticist collaborators then analyzed DNA extracts of individuals from each colony at 12 variable microsatellite loci and found large genetic differences between the wild colonies in the Arnot Forest and the managed colonies in the two apiaries. These findings showed that the nearest managed colonies living outside the Arnot Forest had had little influence on the genetics of the wild colonies living in this forest.

The next step in the genetic investigation of the Arnot Forest bees started late in the summer of 2011, when a former Cornell undergraduate student and friend, Alexander (Sasha) Mikheyev, visited me. Sasha is a professor at the Okinawa Institute of Science and Technology (the "MIT of Japan"), where he heads the Ecology and Evolution Unit. When I told Sasha about the recent sampling of worker bees from the wild colonies living in this forest, he asked if I had samples of worker bees collected from the wild colonies in this forest *before* the arrival of *Varroa destructor*. If so, then he could use sophisticated genetic tools—whole-genome sequencing—on both the old and the new specimens to look for genetic changes caused by the introduction of the *Varroa* mites. Happily, I had just the samples he had in mind. Back in 1977, when I was monitoring several dozen wild colonies living around Ithaca to learn about their longevity (discussed in chapter 7), I had collected samples of worker bees from 32 of these colonies, mounted the bees on insect pins, and archived them in the Cornell University Insect Collection. The label on each specimen indicated its collection site and collection date. Some of these voucher specimens came from colonies that I had caught that summer in bait hives in the Arnot Forest, but others had come from wild colonies living in trees, barns, and farmhouses south of Ithaca. So, to make a proper comparison between these 1977 bees and the 2011 bees, I needed to collect samples

FIG. 10.4. Map of the locations of the wild honey bee colonies from which samples of 100 worker bees were collected in 2011. Their distribution in the forested hills south of Ithaca, including the Arnot Forest, essentially matches that of the wild honey bee colonies from which worker bees were previously sampled, in 1977.

of bees from wild colonies living in trees, barns, and farmhouses south of Ithaca again in 2011. This was easy to do, for at the time I was repeating the work that I had done in the 1970s on wild colony survival and longevity, so I had already a list of sites occupied by wild colonies. In a few days, I had collected worker bees from 22 of these sites to accompany the 10 samples of worker bees collected from bee trees in the Arnot Forest (Fig. 10.4). In early October, I shipped to Sasha 64 samples of worker bees that I had collected from 64 wild colonies living around Ithaca, mostly in forested places to the south of this small city. I was delighted that 32 of these samples were collected *some 20 years before* and the other 32 were collected *some 20 years after* the arrival of *Varroa destructor* in this region of New York State. A pleasing symmetry!

What did Sasha's analysis reveal? First, when he looked at the mitochondrial DNA of the bees, he found a striking loss in its diversity between 1977 and 2011: nearly all the old mitochondrial lineages had gone extinct (Fig. 10.5). He also found that the mitochondrial lineages that were per-

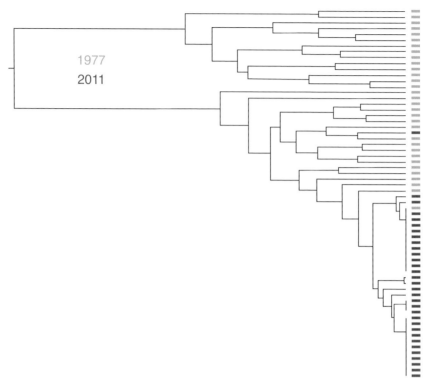

FIG. 10.5. Phylogeny of mitochondrial genomes of the wild colonies of honey bees living in the forests south of Ithaca, New York, in the 1977 population (blue) and the 2011 population (red). Most of the mitochondrial genetic diversity in the old population has been lost in the modern population. Evidently, the arrival of *Varroa destructor* was associated with massive colony mortality and intense selection acting on the bees. The modern population has descended from a relatively small number of queens.

sisting in the wild colonies were not present in the commercial stocks of honey bees. (I had also sent Sasha samples of the worker bees collected from colonies in the two apiaries closest to the Arnot Forest.) These findings tell us that the population of wild colonies living near Ithaca went through a bottleneck, but did not suffer extinction, sometime between 1977 and 2011. Evidently, this population of wild colonies experienced a collapse following the arrival of *Varroa destructor* like the one seen in

Sweden (Fig. 10.3) and in several other populations of honey bees—in southern France, Norway, Louisiana, Texas, and Arizona—that were monitored before, during, and after the arrival of *V. destructor*. These findings also tell us that even though the density of wild colonies is the same in the 2010s as it was in the 1970s—in the Arnot Forest, at least—most of the wild colonies living near Ithaca today are descendants of a small number of queens.

A second set of insights emerged when Sasha and his team looked at the nuclear DNA of the old and new bees living in the woods near Ithaca. First, it revealed that since 1977 there has been some introgression of genes from Africa (see Fig. 1.4), presumably via queens and colonies moved by beekeepers from Florida to the area around Ithaca. This work also revealed some clues about the mechanisms by which the colonies of the modern bees can tolerate infestations of *Varroa destructor*. These are signatures of selection—significant changes in the bees' genes—scattered among 634 sites across the honey bee genome, about half of which seem to relate to the bees' development. Evidently, the colonies that are surviving now in the wild possess resistance mechanisms that are based, at least in part, on changes in the development programs of their members. Consistent with the findings of changes in genes for development, Sasha found that the worker bees collected in 2011 are markedly smaller in body size—that is, in head width and distance between the bases of the wings—than those collected in 1977, but just how this contributes to resistance to *Varroa* remains a mystery.

Starting in 2015, one of my PhD students, David T. Peck, began to investigate the mechanisms of mite-resistance possessed by the worker bees in the colonies living in the Arnot Forest. There are two general possibilities: attack the female mites during the *phoretic phase*, when they are clinging to adult bees, or do so during the *reproductive phase*, when they are sealed in brood cells. When mites are in the phoretic phase, worker bees can groom mites from themselves or nest mates, and then damage them by chewing off the mites' legs, antennae, and the structure called the dorsal plate. When mites are in the reproductive phase, worker bees can disrupt the mites' reproduction either by uncapping mite-infested cells and removing the pupae inside—*Varroa*-sensitive hygienic (VSH) behavior— or by simply uncapping and recapping the cells of pupating bees.

David has tested the Arnot Forest bees with respect to both general ways for worker bees to suppress the population of *Varroa* mites in a colony. He began by capturing swarms of honey bees in the forest using bait hives that were hung out of reach of black bears (see Fig. 2.14) and then moving the swarm-founded colonies into hives in an isolated apiary outside of the Arnot Forest (for safety from bears). He then tested the Arnot Forest bees for the grooming/chewing responses that kill the mites in the phoretic phase and for the disruption responses (VSH behavior and cell uncapping and recapping) that kill them in the reproductive phase. He has found that the workers in his colonies of Arnot Forest bees, relative to workers in control colonies headed by queens of lines not selected for *Varroa* resistance, have both a stronger grooming/chewing response and a stronger VSH response. Also, he has found that some of his Arnot Forest bee colonies have a high percentage (over 40%) of brood cells that had been uncapped and recapped, which we are learning is another way that the bees can disrupt the mites' reproduction. What I think is the most important insight gained from David's work is that the worker bees in wild colonies living in the Arnot Forest possess *multiple* behavioral mechanisms for suppressing the mite populations in their colonies. In short, they are deploying a diverse set of behavioral resistance weapons against *Varroa destructor*, not just a single silver bullet. Bee breeders, take note.

LIVING WITH COLONIES FAR APART VS. CLOSE TOGETHER

When we humans switched from hunting wild colonies to keeping managed colonies, we imposed on the colonies under our supervision a fundamental change in their ecology: an enormous reduction in colony spacing. We have seen already that colonies in the wild usually live spaced far apart, often with a kilometer (0.6 mile) or more of woodland between their bee-tree homes (see Figs. 2.6, 2.12, and 7.11). Colonies managed by beekeepers, however, almost always live crowded together. The clustering of colonies in apiaries is certainly beneficial for beekeepers because it makes beekeeping practical, but it is not altogether beneficial for the bees. Relative to wild colonies, managed colonies experience greater competition for forage, a higher risk of having their honey stolen, and more problems with reproduction, such as when a young queen coming home from a mating flight enters the wrong hive and is killed by worker bees standing guard against intruders.

I suspect, though, that the greatest harm that we inflict on honey bee colonies by forcing them to live jam-packed in apiaries has come through the evolution of highly virulent strains of their parasites and pathogens. We cause this by facilitating the spread of disease agents between *unrelated* colonies (so-called horizontal transmission). This mode of disease transmission favors virulent strains, because horizontally transmitted parasites and pathogens do not need their hosts to stay healthy. Instead, natural selection favors the strains that reproduce rapidly in a host, for although this damages the host, it boosts the parasite's, or pathogen's, odds of being transmitted to another host. This way of life works well for parasites and pathogens that can easily spread to other potential hosts. Among the many parasites and pathogens of honey bees, there are three that are often transmitted horizontally (between unrelated colonies), and all three have highly virulent strains: *Varroa destructor*, American foulbrood, and the deformed wing virus. It is not surprising that these are the agents of bee disease most hated by beekeepers.

Horizontal transmission of honey bee diseases can occur when beekeepers move combs bearing infected bees and brood between colonies and when bees rob honey from a colony that has been weakened by disease. I suspect, though, that the most common mechanism of disease transmission between unrelated colonies within an apiary is drifting—adult bees returning to the wrong hive by accident. The frequency of this mistake depends on how the hives are arranged in the apiary, and it can be greatly reduced by increasing their spacing, painting them different colors, and having them face different directions. A typical arrangement of hives, however, is lined up in a straight row, spaced about 1 meter (ca. 3 feet) apart, painted the same color, and facing the same direction. In this situation, it is common for 40 percent or more of the bees to drift from their natal colony to a neighboring colony. This much drifting means that pathogens and parasites spread easily between unrelated colonies.

When I started thinking about the implications for disease ecology of the enormous difference in colony spacing between wild and managed colonies, I searched widely for articles by biologists and beekeepers about the health effects of clustering colonies in apiaries, but I found essentially nothing. So, I decided to perform an experiment that I hoped would shed light on how colony spacing affects the challenge that honey bee colonies face in avoiding infections with pathogens and parasites from their neighbors.

I began, in June 2011, by setting up two groups of 12 small colonies in one of Cornell University's designated natural areas. The one I used is an expanse of flat, brushy, abandoned farmland near a large beaver pond north of Ithaca (Fig. 10.6). In one group, the colonies were clustered in an apiary in which the hives were arranged in a row, with less than 1 meter (ca. 3 feet) between neighboring colonies (Fig. 10.7). In the other group, which was a few hundred meters from the first group, the colonies were dispersed among clearings in and around a long field, so on average each colony was 34 meters (110 feet) from its nearest neighbor.

In both groups, each colony occupied a hive that consisted of two full-depth Langstroth hive bodies. To measure the levels of bee drift among hives, I installed in 10 of the 12 colonies in each group a Golden Italian queen. This is a queen that is homozygous for a mutant form (called Cordovan) of the gene that provides instructions for making an enzyme (tyrosine) involved in the synthesis of melanin, the substance that gives honey bees their normal black or leathery-brown color. By installing a Golden Italian queen in 10 of the 12 colonies, I insured that all the drones produced in these 10 colonies would have a bright yellow coloration. Meanwhile, I installed in the other two colonies in each group a Carniolan queen, which ensured that all the drones produced in these two colonies had a dark brown, or even black, coloration. In each group, the two colonies with a Carniolan queen were placed in the center of their group.

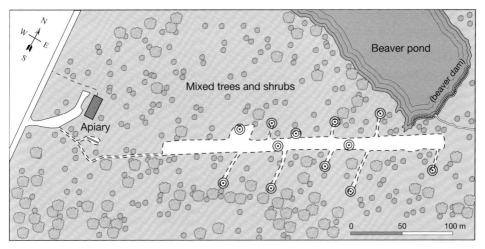

FIG. 10.6. Map of the study site used in investigating the effects of crowding colonies in apiaries on the mixing of bees and the spread of diseases among colonies. The 12 hives of the colonies crowded in the apiary are indicated by the black bar beneath the word "Apiary" on the left. The 12 hives of colonies dispersed around the nearby field are indicated by black squares surrounded by circles.

FIG. 10.7. The 12 crowded colonies arranged in the apiary. The two hives with bricks standing upright on top are the ones that had Carniolan queens and produced only dark drones.

Every colony's queen was labeled with a paint dot, so I could detect turnovers in queens (usually from swarming) when I made monthly inspections of the colonies from May to October. The study colonies received no treatments to control the *Varroa* mites throughout the two-year period of the experiment, June 2011 to May 2013. The colonies started out as two-frame nucleus colonies, so they all spent the summer of 2011 building up strength, but they did not swarm. They all went into the winter of 2011–2012 in good health.

Then, over the summer of 2012, I observed several striking differences between the lives of the crowded and the dispersed colonies. First, when 7 out of 12 colonies in each group swarmed, the crowded colonies had poorer success than the dispersed colonies in getting requeened after swarming: only two successes out of seven vs. five out of seven, respectively. I attribute most of this difference to the young queens in the crowded group entering the wrong hive when they came home from their mating flights; twice I found a dead queen lying outside the hive entrance of a non-swarming colony in the crowded group. Second, there was vastly more drifting of drones among the crowded colonies than among the dispersed colonies. In counts made in September 2011 and April 2012—before any colonies had swarmed, so when each had its original queen, either Golden Italian or Carniolan—I found that 46 percent (in September) and 56 percent (in April) of the drones flying into the two black-bee (Carniolan) colonies in the crowded group were bright yellow (Cordovan) drones! Counts made at the same time at the two black-bee colonies in the dispersed group were much lower, just 1 percent and 3 percent. It is likely that the difference between the two groups in their levels of intercolony mixing of drones applies also to intercolony mixing of workers. Third, late in the summer of 2012, when the colonies were finishing their second summer, there were conspicuous surges in the *Varroa* levels in the two crowded colonies that had swarmed and requeened but *not* in the five dispersed colonies that had swarmed and requeened. All of these seven colonies had low mite counts through June and July, presumably an effect of their swarming in June, but this healthy, low-mite condition persisted into August in only the five dispersed colonies. I cannot say for sure why the mite counts surged in August in the crowded colonies, but I suspect that foragers from these two colonies unintentionally brought home lots of mites while robbing honey from the nearby colonies that were collapsing. This experiment ended in May 2013, and my final inspection revealed that none of the 12 crowded colonies were alive but that 5 of the 12 dispersed colonies were still alive. Indeed, they were thriving.

This was a small-scale experiment, and it was performed just one time and in just one place, so we cannot draw broad and rock-solid conclusions from it, but I feel that it is a valuable step toward a better understanding of how the challenge of colony defense differs between colonies living far apart in the wild vs. close together under management.

LIVING IN SMALL VS. LARGE NEST CAVITIES

Packing colonies together in apiaries is not the only way that beekeepers enlarge the problems of colony defense for managed colonies relative to wild colonies. They also do so by keeping their colonies in hives that provide vastly more living space than the bees usually have in nature. We saw in chapter 5 that wild colonies living in the woods around Ithaca occupy tree cavities with volumes that are usually in the range of 30–60 liters (ca. 8–16 gallons), whereas most managed colonies in the United States are housed (over summer) in hives with volumes of 120–160 liters (ca. 32–42 gallons) (see Fig. 5.3). Beekeepers give their colonies such roomy homes so the bees have plenty of space for their honey stores. Roughly speaking, giving a colony an additional 100 or so liters (ca. 26 gallons) of nesting space enables it to store up an additional 50 kilograms (ca. 100 pounds) of honey. Beekeepers also like to give their colonies roomy homes so that their bees are less likely to become crowded and produce swarms. We have seen in chapter 7 that when a colony casts a prime (first) swarm, it sheds nearly 75 percent of its workforce (see Fig. 7.9), which is good for the bees as they strive for colony reproduction but is disastrous for the beekeeper who wants his or her bees to focus on honey production.

Biologists have long recognized that housing honey bee colonies in large hives makes perfect sense for the beekeeper but is utter nonsense for the bees, as it hampers their reproduction. The arrival of *Varroa destructor* has complicated this already awkward situation, because a big, strong colony is likely to produce a big, strong crop of deadly *Varroa* mites, not just a mammoth crop of honey. This has been demonstrated in a two-year experiment that I conducted with two students, J. Carter Loftus and Michael L. Smith, in which we compared two groups of 12 colonies that lived in two apiaries that were spaced only 60 meters (200 feet) apart. In one group, each colony occupied a small (42-liter/11.1-gallon) hive that consisted of one full-depth, 10-frame Langstroth hive body. These were our small-hive colonies; they simulated wild colonies of honey bees. In the other group, our large-hive colonies, each colony occupied a large (168-liter/44.4-gallon) hive and was managed in the usual ways that help maximize a colony's honey production. Each was given two full-depth hive bodies for a brood chamber plus another two full-depth hive bodies (honey supers) for honey storage. We also inspected these large-hive colonies bimonthly for queen cells—a sign of a colony preparing to swarm—and removed all that we found, as does any beekeeper who desires a good honey crop from his or her bees.

We set up these 24 colonies in May 2012 and then tracked them until May 2014. Once a month, from May to October in 2012 and 2013, we took measurements of each colony's brood and adult bee populations, mite infestation level, presence of disease, and signs of swarming (queen turnover). We also recorded the annual honey production and survival of these 24 colonies over this two-year period. Every colony started out (in May 2012) the same size: a five-frame nucleus colony consisting of two frames of comb covered with bees (one comb filled with brood and the other partially filled with pollen and honey), plus three more frames of comb with food but no brood. Also, every colony started out with a young queen purchased from one queen breeder in California. There was, however, one thing that we did not do for these colonies: we withheld miticide treatments from all of them throughout the two years of the experiment.

We predicted that the small-hive colonies would survive better than the large-hive colonies over the two-year test period, because the small-hive colonies would be smaller and would swarm more often than the large-hive colonies, and therefore would be less apt to suffer the dangerously high infestations of *Varroa* mites that lead to high levels of virus-based diseases of the colonies' brood. We based our prediction on several findings from previous studies. One key piece of background information was that when a colony casts a swarm it sheds about 35 percent of its adult *Varroa* mites. This happens because about 70 percent of a colony's adult bees leave when a colony casts a swarm (see chapter 7) and because approximately 50 percent of the *Varroa* mites in a colony are on the adult bees (the rest are in the cells of sealed brood). A second key piece of background information was that when a colony casts a swarm it experiences a break in its brood rearing. This occurs because it takes time for the new queen in the colony to finish her development, kill off her rivals, get mated, and finally begin laying eggs. *Varroa* mites cannot reproduce in colonies that lack brood. We expected, therefore, that having a broodless period would help shrink the population of *Varroa* mites in a colony by imposing an interruption of their reproduction and by depriving the mites of their main hideouts—sealed cells of brood. What we did not know in advance, though, was whether the more intense swarming by the small-hive colonies would remove a sufficiently large fraction of the adult mites in a colony and would reduce sufficiently the reproduction and survival of the mites that remained behind, so that the small-hive colonies would survive better than the large-hive colonies.

Figure 10.8 shows what we learned about the dynamics of the adult bee populations and of the *Varroa* mite infestations in the colonies of the two treatment groups. We see that the average number of adult bees per colony was the same for the two groups when the experiment got started in 2012, but that the population censuses for the two groups diverged markedly over the summer of 2013. On average, the populations in the small-hive colonies did not grow much above 10,000 bees, whereas those in the large-hive colonies climbed above 30,000. We also see that the average

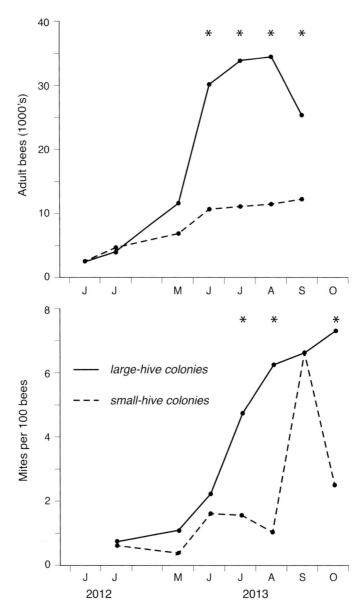

Fig. 10.8. Dynamics in the adult honey bee populations (*top*) and in the *Varroa* mite infestation rates on adult bees (*bottom*) in colonies housed in large hives and small hives, from June or July 2012 to September or October 2013. Asterisks denote significant differences.

infestation level of *Varroa* mites started out the same for the two groups, but then it diverged markedly over the summer of 2013. On average, the infestation stayed at a safe, low level (two mites per 100 bees) in the small-hive colonies until September, at which time it shot up to the dangerously high infestation rate found in the large-hive colonies (six-plus mites per 100 bees) and then dropped down. It should be noted that this temporary bump in the average infestation rate of the small-hive colonies happened in a most curious way. Three of the 12 small-hive colonies experienced eye-catching spikes in their mite count for September: a surge to 15–17 mites per 100 bees. This proved very telling, as we shall see.

What are the causes and the effects of the differences in mite infestation levels between the two treatment groups across the summer of 2013? First, although none of the colonies in either treatment group swarmed in 2012, in 2013 nearly all (10 out of 12) of the small-hive colonies swarmed, but almost none (2 out of 12) of the large-hive colonies swarmed. I believe this explains why in 2013 the mite infestation rates of the small-hive colonies were so much lower than those of the large-hive colonies (except in September). Specifically, 10 of the 12 small-hive colonies exported mites in swarms and experienced a break in brood rearing, but only 2 of the 12 large-hive colonies did so. It was not at all surprising, therefore, to see that over the winter of 2013–2014, the large-hive colonies suffered far heavier mortality (10 out of 12 colonies) than the small-hive colonies (4 out of 12).

What is perhaps the most interesting outcome of this experiment is what happened unexpectedly: the mite infestation levels in three of the small-hive colonies surged temporarily to 15–17 mites per 100 bees in mid-September 2013. (Note: the surges in the mite counts for these three colonies are what caused the spike in mite counts for the small-hive colonies shown in Fig. 10.8). These explosions in the mite counts for these three colonies coincided with the collapse of one of the large-hive colonies in the other apiary, only 60 meters (200 feet) away. When I inspected this collapsing colony, I found a pile of dead bees in front of its hive (and *only* this colony's hive), and I found virtually no bees, almost no brood, no stored honey, and very few mites (all dead) inside its hive. The floor of this hive was littered with dead bees (most with shriveled wings) and rough-edged wax flakes of the sort that robber bees create as they carelessly uncap cells holding honey. It was obvious that the colony had collapsed from a high infestation of mites and then its honey stores had been robbed, but it was not clear what had happened to the mites. I suspect that most of the horde of *Varroa* mites that killed this colony had climbed onto robber bees and then had been airlifted to the homes of these bees. I suspect, too, that many of the robbing bees came from the three small-hive colonies whose mite counts temporarily skyrocketed in mid-September. Incidentally, of the four small-hive colonies that died over the winter in 2013–2014, three were the three colonies whose *Varroa* mite populations had spiked in September 2013. The fourth was a healthy colony whose queen became a drone layer in July 2013. Lacking female brood, the colony could not replace her, so eventually it contained only drone brood and gradually died out.

I am sorely tempted to repeat this experiment, to be doubly sure that the striking finding of this experiment—simply housing colonies in smaller hives greatly helps the bees cope with *Varroa destructor*—is correct. If I do, I will separate the small-hive colonies more widely from the large-hive colonies to minimize the flow of mites between the two groups that can happen when robber bees plunder the honey stores of a colony that has collapsed.

LIVING WITH SMALLER VS. LARGER COMB CELLS

One method of controlling the parasitic mite *Varroa destructor* that has been much discussed and debated among beekeepers is to reduce the size of the worker cells in a colony's nest. The idea is that smaller cells might cause higher mortality of immature mites, because the mites develop directly beside immature bees in cells, so having a smaller space between the developing bee and the cell wall might hamper the movements of the immature mites. Specifically, it might reduce the ability of the immature mites to reach their feeding site on the abdomen of the pupal bee within the cell. We saw in chapter 5 that the mean wall-to-wall dimensions of the worker-comb cells of three wild colonies living in the woods near Ithaca were 5.12, 5.19, and 5.25 millimeters (0.201, 0.204, and 0.206 inch). Thus, on average, their worker cell size was 5.19 millimeters. For comparison, in my managed colonies, which have built their combs on standard beeswax comb foundation purchased from various manufacturers, the average wall-to-wall dimension of the worker cells is larger: 5.38 millimeters (0.212 inch). This raises the question, are the smaller comb cells in the nests of the wild colonies living in the woods around Ithaca helping these bees defend themselves from *Varroa destructor*?

It is highly doubtful that this is the case. Three recent studies conducted in the southeastern United States (Florida and Georgia) and in Ireland have experimentally tested the idea that giving colonies of European honey bees small-cell combs reduces their susceptibility to *Varroa* mites. These studies compared the growth of mite populations in colonies with either small-cell (4.91 millimeters/0.193 inch) or standard-cell (5.38 millimeters/0.212 inch) combs and found no sign that providing colonies with small-cell combs impedes reproduction by the mites. Sean R. Griffin, one of my students, and I have also tested this idea in Ithaca. We established seven pairs of equally strong colonies that started out equally infested with mites. In each pair, one hive contained only standard-cell (mean width 5.38 millimeters/0.212 inch) comb and the other hive contained only small-cell (mean width 4.82 millimeters/0.190 inch) comb. Because we used manufactured plastic combs for our small-cell treatment, and because we removed from all our hives any drone comb that the bees built, there was no doubt that the colonies in our small-cell treatment group had only small cells and that the colonies in our standard-cell treatment group had only standard cells in their combs. Despite this large and unambiguous difference in cell size between the two treatment groups, we found no difference (across an entire summer) between the two treatment groups in level of mite infestation, measured as number of mites per 100 worker bees.

If it were the case that *Varroa* mite reproduction on European honey bees is impeded when these mites parasitize bees living on combs with smaller cells, then over the summer we should have seen a decline in the mite levels in the small-cell colonies relative to those in the standard-cell colonies, but we did not. I think the reason that nobody has found a clear effect of cell size on mite reproduction is that even in small cells the mites have plenty of space to move around on the surface of a pupa. When Sean and I, as well as a team of bee researchers in Ireland (John McMullan and Mark F. T. Brown), measured the fill factor—the ratio of bee thorax width to cell width, expressed as a percentage—for both the standard-cell combs and the small-cell combs used in our studies, we both found that bees reared in our standard-cell combs had a fill factor of 73 percent, whereas the bees reared in our small-cell combs had a fill factor of 79 percent. Because the fill factors were low for both groups of bees, and only slightly higher for the bees living on small-cell combs, I'm not surprised that we found no indication that small-cell combs hinder the reproduction of *Varroa* mites on European honey bees.

LIVING WITH A HIGH VS. A LOW NEST ENTRANCE

Perhaps the most obvious, but least understood, difference between wild colonies inhabiting natural cavities and managed colonies occupying manmade hives is the difference in the heights of their homes. We like to place hives at ground level for our convenience, of course, but when the bees can choose where to live they select living quarters with lofty entrances (see chapter 5). Why? We don't know exactly, but there are several possibilities. One is to make it safer for the bees to take cleansing flights in winter. Bees exiting and entering a nest opening far above ground level are probably less apt to crash-land on snow and become stranded with chilled flight muscles. Another possibility is to reduce the likelihood of the nest entrance becoming buried in snow. Still another possibility is to gain exposure to the sunnier and warmer microclimate of the forest canopy (Fig. 7.6) rather than the shadier and cooler microclimate near the forest floor. In chapter 5, we saw that colonies with sunny, south-facing nest entrances have greater success in winter survival than those with shady, north-facing ones; and beekeepers in the Northern Hemisphere generally agree that south-facing colonies, relative to north-facing ones, produce more honey.

Perhaps, though, the primary benefit to wild colonies of nesting high in trees is making them less conspicuous to terrestrial predators, especially bears. This remains to be studied experimentally, but two things that I have learned from working in the Arnot Forest have convinced me that wild colonies acquire substantial protection from black bears by choosing nesting cavities whose entrance openings are high off the ground. The first is what I learned from finding bear tracks in the Arnot Forest (Fig. 10.9, top) in 2002: black bears were roaming this forest. The second is something I learned from monitoring a set of bee trees in the Arnot Forest. In 2002, I found 8 bee trees in the Arnot Forest, and in 2011, I found 10 more. Ever since I found these 18 bee trees, I have been checking up on them three times a year, as described in chapter 7. These bee trees have been intermittently occupied by wild colonies as old colonies die out and new ones move in. Over the past 16 years, I have accumulated a total of 51 colony-years of observations of Arnot Forest trees occupied by bees. This means that I have had many, many chances to detect attacks by bears on the nests of the wild colonies living in this forest. The amazing thing is, though, that I have detected just one attack of a bee-tree colony by a black bear, and it occurred under special circumstances: the tree, a red oak atop Irish Hill, was blown down during a winter storm. This dropped the entrance height of the nest in this tree from 10.9 meters (36 feet) to 1.2 meters (4 feet).

It was when I made my round of bee-tree checkups in May 2009 that I discovered this bee tree lying on the ground. I also discovered then that the colony inhabiting it was still alive, indeed it was thriving, for bees were pouring in carrying loads of pollen. I further discovered that the bark all around the knothole that was the bees' nest entrance was stripped off and that the bare wood around the entranceway was inscribed with claw marks. There could be no doubt: at least one black bear had found this colony and had endeavored mightily, but unsuccessfully, to get into its nest. The most important part of this story, though, is that *over the years, the black bears have not found the colonies in "my" 17 other bee trees in the Arnot Forest*. If they had, I

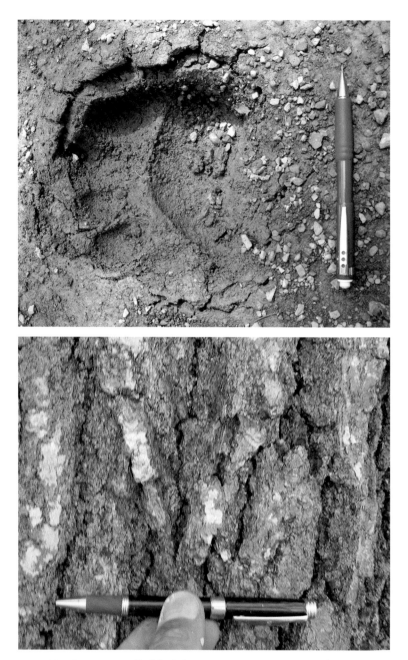

FIG. 10.9. *Top:* Paw print of a black bear (*Ursus americanus*) in the Arnot Forest. *Bottom:* Claw marks of a black bear in the bark of an oak tree in the Arnot Forest.

am sure that I would have detected it, because every time a black bear has found (and attacked) a colony in one of my bait hives in the Arnot Forest (Fig. 2.14), it has left behind not only a busted-up bait hive lying on the ground but also a bunch of conspicuous calling cards: claw marks in the bark of the tree in which I had mounted the bait hive (Fig. 10.9, bottom). Because I have never found claw marks in the bark of the 17 other bee trees that I keep monitoring, I am convinced that the bears have never discovered the colonies in these trees. Why haven't they? My reference book on the mammals of the eastern United States tells me that black bears have acute senses of smell and hearing but "only adequate vision." Adequate for some things, but evidently not for spotting small bees coming and going from dark knotholes and slender cracks high up in trees.

LIVING WITH VS. WITHOUT A PROPOLIS ENVELOPE

We have seen already in chapter 5 that wild honey bee colonies living in tree cavities collect antimicrobial plant resins and use them to coat the ceilings, walls, and floors of their nest cavities, to create propolis envelopes around their nests. It is, therefore, quite telling that honey bee colonies do not form thick coatings of propolis on the inner surfaces of manufactured hives, though they will fill with propolis the crevices between the frames as well as any cracks around a hive's lid. Evidently, the stimuli that elicit propolis deposition are small cracks and crevices (see Fig. 5.4). Surfaces that are rough and porous are ideal for bacterial growth, for they give bacteria abundant substrates to cling to as well as a supply of nutrients and moisture. And if left alone, bacteria living in these places will build biofilms that can protect them from being dislodged. It is not at all surprising, therefore, that honey bees work hard to coat the rough and porous surfaces of their nest cavities with plant resins.

Many laboratory-based studies have tested the effectiveness of propolis against the growth of the causative agents of two brood diseases of honey bees: American foulbrood (a bacterial disease caused by *Paenibacillus larvae*) and chalkbrood (a fungal disease caused by *Ascosphaera apis*). All these studies report strong inhibitory effects of propolis on these pathogens. It is not entirely clear, however, how the findings from these in vitro studies conducted in laboratories relate to disease control in colonies, especially because it is not clear that honey bees consume propolis and incorporate its compounds in the food they feed to their brood.

Marla Spivak and her colleagues at the University of Minnesota have assessed the health benefits to honey bees of living in a nest that is surrounded by a propolis envelope by listening to the bees themselves. One of their experiments involved comparing the transcription levels of immune-related genes in worker bees living in two types of hives. The hives in the experimental group had their inside surfaces coated with ethanol extracts of propolis, while those in the control group had their inside surfaces coated with straight ethanol. After seven days, individual bees living in the propolis-enriched hives, relative to bees in the control hives, showed lower bacterial loads and lower levels of activity of genes involved in insect immune responses. Evidently, coating the insides of hives with propolis extracts lowered the levels of the immune elicitors (i.e., the levels of bacteria and fungi) in the treatment hives, relative to the control hives.

Another experiment performed by the Minnesota group shows even more strongly the benefits of a propolis envelope to the health of a honey bee colony. This time, the researchers compared colonies with and without *true* propolis envelopes, and they ran their study for *two years*. Each year they measured multiple indicators of health in 24 colonies, 12 with and 12 without propolis envelopes. They stimulated half the colonies to construct propolis envelopes by fixing sheets of plastic "propolis trap" material to the smooth inner walls of their hives, and the bees responded by coating these plastic sheets with propolis (Fig. 10.10). They left the inner walls of the hives of the other 12 colonies bare, and these colonies did not construct propolis envelopes.

The experiment yielded two main findings. First, during the summer and autumn months, the expression (activity) levels of several genes involved in insect immunity were consistently lower and steadier in the worker bees living in hives with a propolis envelope, compared to workers

Fig. 10.10. View of propolis deposited on the inner walls of a hive after the removal (at end of experiment) of the sheets of plastic propolis-trap material that investigators had stapled to the walls inside the hive. Left behind are the blobs of brown propolis that filled the slots in the propolis-trap sheets.

in hives without a propolis envelope. This difference in level of immune gene expression is functionally significant, because when a bee's immune system is working hard to fight infections, it can be the costliest part of her physiology. Lowering immune system activity, therefore, enables bees to allocate more energy to other tasks, such as brood rearing, wax production/comb building, and foraging. The second main finding is even more telling: colonies with propolis envelopes had better survival in one of the two years (Fig. 10.11), more brood in May in both years, and *better nutritional status of its young workers in both years*. The nutritional status of the young worker bees was assessed by measuring their levels of activity of the gene Vg. This is the gene that young worker bees activate to produce vitellogenin, their primary storage protein. Indeed, in healthy young nurse bees, vitellogenin constitutes approximately 40 percent of the protein in their body fluids, for it is what these bees draw upon to produce protein-

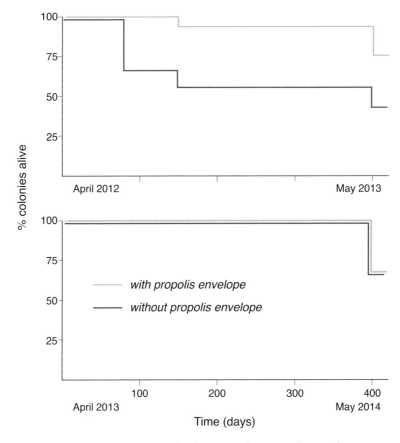

Fig. 10.11. Survivorship records of colonies with a propolis envelope (*blue line*) and without a propolis envelope (*magenta line*), in two trials of the experiment. In each trial, there were 12 colonies in each treatment group. There was a significant difference in colony survivorship at the end of the first trial but not the second trial.

rich royal jelly for feeding young larvae. Healthy, well-nourished nurse bees are the foundation of a colony's brood production and growth, so what this ingenious experiment has revealed is that when a colony builds itself a propolis envelope, it is making a far-reaching investment in its future growth and ultimate success.

11

DARWINIAN BEEKEEPING

> Beekeeping today is still as it has always been: the exploitation
> of colonies of a wild insect; the best beekeeping is the
> ability to exploit them and at the same time to interfere
> as little as possible with their natural propensities.
> —Leslie Bailey, *Honey Bee Pathology, 1981*

In the first 10 chapters of this book we have reviewed the interwoven topics of annual cycle, colony reproduction, nest building, food collection, thermoregulation, and colony defense that form the natural history of the honey bee, *Apis mellifera*. We have also examined the cultural history of beekeeping and the ways in which this unique form of animal husbandry is built upon, but also disrupts, the natural lives of these marvelous insects. This final chapter aims to integrate these two general themes, by applying what we have learned in recent decades about how honey bees live in the wild, to revise some of our beekeeping practices in ways that are mutually beneficial for bees and beekeepers. Our goal is to respect the natural lives of honey bees but also to enjoy the benefits of their hard work as honey makers and crop pollinators. We will work toward this goal in two stages. First, we will review the most important differences in living conditions between wild colonies and managed colonies, an exercise that will highlight the many ways in which standard beekeeping practices alter the lives of honey bees and often put these insects under stress. Second, we will look at ways in which beekeepers can revise their practices to make the lives of their six-legged partners less stressful and therefore more healthful. We will see that the essence of doing this is to manage colonies of honey bees in ways that enable them to live, as much as possible, under conditions like those in which they evolved and thus to which they are adapted. We will also see that this often requires putting the needs of the bees before those of the beekeeper.

WILD COLONIES VS. MANAGED COLONIES

Again and again, throughout this book, we have seen that there are stark differences between the original environment that shaped the biology of wild honey bee colonies—their environment of evolutionary adaptation (EEA)—and the current living conditions of managed colonies. Wild and managed colonies live under different conditions because beekeepers, like all farmers, modify the environments of their livestock in order to boost their productivity. Unfortunately, these changes in the living conditions of agricultural animals often make them more prone to parasites and pathogens. Table 11.1 lists 21 differences between the living conditions in which wild colonies have lived (and often still live) and the living conditions in which managed colonies of honey bees now live.

Difference 1: Colonies are genetically adapted vs. are not genetically adapted to location.

The process of adaptation by natural selection produced the differences in worker-bee color, morphology, and behavior that distinguish the 30 subspecies of *Apis mellifera* that live within the species' original range of Europe, western Asia, and Africa. The colonies in each subspecies are well adapted to the climate, seasons, flora, predators, and diseases in their native region of the world. Moreover, within the geographic range of each subspecies, natural selection has produced ecotypes—populations that are fine-tuned to their local conditions. Perhaps the best-documented example of this geographical adaptation is the ecotype of *A. m. mellifera* (the dark European honey bee) that is adapted for living in the Landes region of southwestern France. The rhythm of its annual cycle of colony growth is

TABLE 11.1. Comparison of the environments in which honey bee colonies have lived as wild colonies and those in which they live currently as managed colonies

Environment of evolutionary adaptation	Current circumstances
1. Colonies are genetically adapted to their location	Colonies are not genetically adapted to their location
2. Colonies live widely spaced in the landscape	Colonies live crowded in apiaries
3. Colonies occupy small nest cavities	Colonies occupy large hives
4. Nest cavity walls have a propolis coating	Hive walls have no propolis coating
5. Nest cavity walls are thick	Hive walls are thin
6. Nest entrance is high and small	Nest entrance is low and large
7. Nest has 10%–25% drone comb	Nest has little (< 5%) drone comb
8. Nest organization is stable	Nest organization is often altered
9. Nest-site relocations are rare	Hive relocations can be frequent
10. Colonies are rarely disturbed	Colonies are frequently disturbed
11. Colonies deal with familiar diseases	Colonies deal with novel diseases
12. Colonies have diverse pollen sources	Colonies have homogeneous pollen sources
13. Colonies have natural diets	Colonies sometimes have artificial diets
14. Colonies are not exposed to novel toxins	Colonies exposed to insecticides and fungicides
15. Colonies are not treated for diseases	Colonies are treated for diseases
16. Honey not taken, pollen not harvested	Honey taken, pollen sometimes harvested
17. Combs not moved between colonies	Combs often moved between colonies
18. Honey cappings are recycled by bees	Honey cappings are harvested by beekeepers
19. Bees choose larvae for queen rearing	Beekeepers choose larvae for queen rearing
20. Drones compete fiercely for mating	Queen breeder may select drones for mating
21. Drone brood not removed for mite control	Drone brood sometimes removed and frozen

attuned to the massive bloom of ling heather (*Calluna vulgaris*) in August and September. Colonies of honey bees that are native to this region have an unusual, second strong peak of brood rearing in August. This gives them a second burst of colony population growth that helps them exploit this late summer heather bloom. Colony transplant experiments have been performed in which colonies from the region around Paris (which lacks a heather bloom) and from the Landes region were moved to each other's location and then their brood-rearing patterns were recorded. These experiments show that the difference in annual brood cycle between these two ecotypes has a genetic basis. This example shows us that shipping mated queens, and trucking whole colonies, to places hundreds or thousands of miles away—for instance, from Hawaii to Maine or Italy to Sweden—is likely to force colonies to live in environments to which they are ill suited.

Difference 2: Colonies live widely spaced in the landscape vs. crowded in apiaries.

This shift makes beekeeping practical, but it also creates a fundamental change in the ecology of honey bees. Crowded colonies experience greater competition for forage, greater risks of being robbed, and greater problems in reproduction: swarms combining when leaving their hives, and queens entering the wrong hive when returning from their mating flights. Probably the most harmful consequence of crowding colonies, though, is boosting pathogen and parasite transmission between colonies. This facilitation of disease transmission boosts the incidence of disease among colonies and favors the virulent strains of the bees' pathogens and parasites.

Difference 3: Colonies occupy small nest cavities vs. large hives.

This modification also profoundly changes the ecology of honey bees. Colonies in large hives have the space to store huge honey crops, but they also swarm less often because they are not as space-limited. This weakens natural selection for strong, healthy colonies, since fewer colonies reproduce. A more immediate problem of keeping colonies in large hives is that colonies suffer greater problems with brood parasites, such as *Varroa*, because large, non-swarming colonies provide these parasites with a vast and steady population of their hosts: the larvae and pupae of honey bees.

Difference 4: Colonies live with vs. without a nest envelope of antimicrobial plant resins.

Living without a propolis envelope increases the physiological costs of colony defense against pathogens. It is now known, for example, that workers in colonies without a propolis envelope invest more in costly immune-system activity—such as synthesis of antimicrobial peptides—relative to workers in colonies with a propolis envelope.

Difference 5: Colonies have thick vs. thin nest-cavity walls.

This amounts to a difference in nest insulation, which strongly affects the energetic cost of colony thermoregulation. This effect is especially strong when the colony is rearing brood and so is heating or cooling the brood nest to keep its temperature within the narrow, optimal range of 34.5°–35.5°C (94°–96°F). The rate of heat transfer from (or to) a wild colony living in a typical, thick-walled tree cavity is four to seven times lower than for a managed colony living in a standard, thin-walled wooden hive.

Difference 6: Colonies have small and high vs. large and low nest entrances.

This difference renders managed colonies more vulnerable to robbing and predation, because larger entrances are harder to guard. It may also lower their likelihood of winter survival, because low entrances are often blocked by snow, which prevents bees from making cleansing flights, and because low entrances make it more likely for bees to crash or land on the snow while trying to make cleansing flights. Bees that settle on the snow are often trapped there, unable to warm their flight muscles sufficiently to fly back to their nest.

Difference 7: Colonies live with vs. without plentiful drone comb.

Depriving a colony of drone comb will inhibit it from rearing drones, and this will boost its honey production and slow the reproduction of the *Varroa* mites in the colony. But it will also hamper natural selection for colony health, because it will prevent the healthiest colonies from being the best at passing on their genes via drones.

Difference 8: Colonies live with vs. without a stable nest organization.

Disruptions of nest organization for the purposes of beekeeping may hinder colony functioning. In nature, honey bee colonies organize their nests with a consistent, three-dimensional spatial structure: a region of cells densely filled with brood that is surrounded by cells containing pollen, and a region in the periphery of cells filled with honey, mostly in the upper part of the nest. This spatial organization helps ensure that the brood-nest region is kept at its proper temperature. It also helps increase the work efficiency of the nurse bees by ensuring that they (the primary consumers of the pollen) have a ready supply of pollen near the brood. Beekeeping manipulations that alter this nest organization—such as inserting empty combs to reduce congestion in the brood nest—hamper brood-nest thermoregulation and may disrupt other aspects of colony functioning such as egg laying by the queen, brood-food production by the nurse bees, and nectar storage by the food-storer (middle-aged) bees.

Difference 9: Colonies experience infrequent vs. sometimes frequent relocations.

Whenever a colony is moved to a new location, as occurs in migratory beekeeping, the foragers must learn the landmarks around their hive and must discover new sources of nectar, pollen, and water. One study found that colonies moved overnight to a new location had smaller weight gains in the week following the move relative to control colonies already living in the location.

Difference 10: Colonies are rarely vs. frequently disturbed.

We do not know how often wild colonies experience severe disturbances (e.g., attacks by bears, skunks, or yellow jacket wasps), but it is probably rarer than for managed colonies, whose nests are often cracked open, smoked, and manipulated. In one experiment that was conducted during a honey flow, comparisons were made of the weight gains of colonies that were and were not inspected. It found that colonies that were inspected gained 20–30 percent less weight (depending on the degree of disturbance) than control colonies on the day of the inspections.

Difference 11: Colonies deal with familiar vs. novel diseases.

Historically, honey bee colonies dealt only with the parasites and pathogens with whom they had long been in an arms race. Therefore, they had evolved means of surviving with their agents of disease. Humans changed all this when we spread the ectoparasitic mite *Varroa destructor* from eastern Asia, small hive beetle (*Aethina tumida*) from sub-Saharan Africa, and chalkbrood fungus (*Ascosphaera apis*) and tracheal mite (*Acarapis woodi*) from

Europe. The near-global spread of *Varroa destructor* alone has resulted in the deaths of millions of honey bee colonies, both wild and managed.

Difference 12: Colonies have diverse vs. homogeneous pollen sources.

Many managed colonies are placed in agricultural ecosystems—for example, vast almond (*Prunus amygdalus*) orchards or huge fields of oilseed rape (*Brassica napus*)—where they experience low-diversity pollen diets and relatively poor nutrition. The effects of pollen diversity have been studied by comparing nurse bees given diets with monofloral pollens and those given polyfloral pollens. In tests that used nurse bees infected with the microsporidian parasite *Nosema ceranae*, it was found that the bees fed with the polyfloral blend of pollens lived longer than those fed with monofloral pollens.

Difference 13: Colonies have natural diets vs. are fed artificial diets.

Some beekeepers feed their colonies protein supplements (pollen substitutes) to stimulate colony growth before pollen is available. This is done to help meet the colony-size requirements of pollination contracts and to produce larger honey crops. The best pollen supplements/substitutes do stimulate brood rearing, though often not as well as real pollen. Colonies that are nutritionally stressed by a lack of pollen (or by a poor artificial diet) produce workers that have reduced longevity, an early onset of foraging, and a shortened period of functioning as a forager.

Difference 14: Colonies are not exposed vs. are exposed to novel toxins.

The most important novel toxins to which honey bees are exposed are insecticides and fungicides, substances for which the bees have not had time to evolve detoxification mechanisms. Honey bees are now exposed to an ever-increasing range of insecticides and fungicides that can synergize in causing harm to the bees.

Difference 15: Colonies are not treated vs. are treated for diseases.

When we treat our colonies for diseases, we interfere with the host-parasite arms race between *Apis mellifera* and its pathogens and parasites. Specifically, we weaken natural selection for disease resistance. It is no surprise that most *managed* colonies in North America and Europe possess little resistance to *Varroa* mites, and that there are populations of *wild* colonies on both continents that have evolved strong resistance, as discussed in chapter 10. Treating colonies with miticides and antibiotics may also interfere with the microbiomes of a colony's bees.

Difference 16: Colonies are not managed vs. are managed as sources of honey and pollen.

Colonies managed for honey production are housed in large hives, so they are more productive. However, they are also less apt to reproduce by swarming, which means there is less scope for natural selection for healthy colonies—that is, for the healthiest colonies to be most successful in passing on their genes. Also, the immense quantity of brood in large-hive colonies renders them more vulnerable to population explosions of *Varroa* mites and the other agents of honey bee disease that reproduce in brood. Harvesting pollen makes it harder for colonies to acquire a complete diet.

Difference 17: Combs are not moved vs. are moved between colonies.

Moving combs from one colony to another, especially combs that have been or are being used for rearing brood, is an extremely effective way to spread diseases between colonies, and yet it is commonly done. Sometimes it is done to equalize colony strength, other times to give colonies additional combs in which to store honey. Whatever the purpose, it greatly increases the spread of diseases between colonies.

Difference 18: Honey cappings are recycled by bees vs. are harvested by beekeepers.

Removing beeswax from a colony when uncapping honey combs for harvesting honey imposes a serious energetic burden. The weight-to-weight efficiency of beeswax synthesis from sugar is at best about 0.20, so every kilogram (ca. 2 pounds) of wax taken from a colony costs it some 5 kilograms (ca. 10 pounds) of honey that is not available for other purposes, such as brood rearing and winter survival. Besides the energetic cost of beeswax synthesis, there are the wear-and-tear costs on the worker bees involved in the wax synthesis; the bees expend a portion of their lifetime work capacity in producing wax rather than performing other tasks. This situation is analogous to a human factory in which the total cost of making something involves wear-and-tear (depreciation) costs on the machinery, not just the energy costs of powering the machinery. The way of harvesting honey that is the most energetically burdensome to the bees is removal of entire combs filled with honey (cut comb honey or crushed comb honey). It is less burdensome to produce extracted honey, since this removes just the top layer of wax cappings.

Difference 19: Colonies are allowed vs. are not allowed to choose the larvae for rearing queens.

When we humans graft one-day-old larvae into artificial queen cups for rearing queens, we prevent the bees from choosing which larvae will develop into queens. One study has found that in emergency queen rearing, the bees do not choose larvae at random and instead favor those of certain patrilines. It may be the case, too, that the bees exercise choice that is based on larval health, which is a sign of the overall vigor of the developing bee.

Difference 20: Drones are allowed vs. are not allowed to compete fiercely for matings.

In bee-breeding programs that use artificial insemination, the drones that provide the sperm do not have to prove their vigor by competing among dozens or hundreds of other drones to mount a flying queen and inject his sperm into her oviducts. This lack of drone-drone competition weakens the sexual selection for drones possessing genes for excellent health and flight ability.

Difference 21: Drone brood is not removed vs. is removed from colonies for mite control.

The practice of removing drone brood from colonies partially castrates them and so interferes with natural selection for colonies that are healthy enough to invest heavily in rearing and supporting drones.

SUGGESTIONS FOR DARWINIAN BEEKEEPING

Beekeeping looks different when viewed from an evolutionary perspective. We have seen that honey bees have lived independently of human beings for millions of years and that throughout this immense span of time their biology was tuned by natural selection to favor two things: colony survival and colony reproduction. We have also seen that ever since humans started keeping bees in hives several thousand years ago, we have been disrupting the close fit that previously existed between these bees and their environment. We have done so in two general ways: 1) by moving honey bee colonies to geographic locations to which they are not well adapted, and 2) by manipulating their lives in ways that boost their production of things that we value: honey, beeswax, pollen, royal jelly, and pollination.

What can a beekeeper do now to help his or her honey bees live with a better fit to their environment, hence with less stress and better health? The answer depends on the individual beekeeper's aims. A backyard beekeeper who enjoys supervising a few colonies mainly for the pleasures of bee watching, and who is happy to put the bees' needs before his or her own, has many ways to pursue bee-friendly beekeeping. In contrast, a commercial beekeeper who manages hundreds or thousands of colonies to earn a living by producing honey crops and fulfilling pollination contracts has fewer options for pursuing a kinder and gentler approach to the craft. What follows is a list of suggestions. You can think of the items on it as ingredients for devising a personal recipe of Darwinian beekeeping, one that is realistic given your aims and opportunities as a beekeeper.

1. *Work with bees that are adapted to your location.* For example, if you live in the northeastern United States, then either rear queens from your hardiest survivor colonies or buy queens (or nucleus colonies) produced from stock that has proven itself by thriving in this region despite its long, harsh winters. If you do not want to rear your own queens and you do not have a local queen producer, but you do have wild colonies living in your area, then you can easily get locally adapted stock by capturing swarms pro-

duced by the wild colonies. The most efficient way to do so is to set out bait hives. This approach will work best if you live in a place that is not crowded with fellow beekeepers, some of whom might be purchasing queens shipped in from far away.

2. *Space your hives as widely as possible.* Where I live in central New York State, the wild colonies living in the forests are spaced roughly 800 meters (0.5 miles) apart. We have seen in chapter 10 that wild colonies benefit from this wide separation, but of course such a wide spacing of colonies is impossible for most beekeepers. Fortunately, we have also seen in chapter 10 (Fig. 10.6) that spacing colonies just 30–50 meters (100–160 feet) apart greatly reduces the likelihood of drifting of drones—and probably also workers—among colonies and thus the spreading of disease.

3. *House your colonies in small hives.* Consider providing just one deep hive body for a colony's brood nest and then providing it with just one medium-depth honey super over a queen excluder for securing a modest honey crop. Your honey harvest will be smaller than it would be if you were to supersize the colony by giving it two deep hive bodies for its brood nest and then putting a tall stack of honey supers on top. You will, however, reduce the colony's problems with parasites and pathogens, especially *Varroa destructor* and the viruses it vectors, as discussed in chapter 10 (Fig. 10.8). This is especially true if you allow your colonies to swarm. Swarming removes many mites from a colony, and it creates a break in the colony's brood rearing that deprives the mites of the cells of capped brood that they need for their reproduction.

4. *Roughen the inner wall surfaces of your hives, or build them of rough-sawn lumber.* This will stimulate your colonies to cover the walls inside your hives with propolis and so build antimicrobial shrouds around their combs.

5. *Use hives whose walls provide good insulation.* These might be hives built of thick lumber, or they might be hives made of plastic foam. An important subject for future research is how much insulation is best for colonies living in different climates and how best to provide it.

6. *Position hives high off the ground.* This is not always feasible but may be doable if you have a porch or a flat roof where you can keep your hives.

Another important topic for research is how exactly it is that entrance height affects colony success in different settings and climates. Is it the case that colonies living where winters are snowy benefit greatly from having their entrances high off the ground because this helps their workers avoid crashing onto the snow when they fly out to make wintertime cleansing flights?

7. *Allow colonies to maintain 10–20 percent of the comb in their hives as drone comb.* Doing so will give your colonies the opportunity to rear plentiful drones, and this can help improve the genetics in your area (if you are not treating your colonies). Drones are costly, so it is only the healthiest and strongest colonies that can produce legions of drones. Please note: plentiful drone brood in a colony fosters the reproduction of the *Varroa* mites by providing them with their ideal host, so providing plentiful drone comb also requires diligent monitoring of the mite levels in your colonies and taking appropriate actions for each colony in which the mite level gets dangerously high (see suggestion 14, below).

8. *Minimize disruptions of nest structure, so the functional organization of each colony's nest is maintained.* In practice, this means replacing each frame of comb in its original position and orientation in the hive after removing it for inspection. It also means refraining from inserting frames of empty comb between frames filled with brood in order to inhibit colony swarming.

9. *Minimize the moving of colonies.* Move colonies as rarely as possible, because doing so disrupts many aspects of a colony's functioning, including brood care, nest thermoregulation, and food collection. Besides the stress of the move itself, there is the need for the foragers to memorize the new landmarks around their hive so each bee can find her way home, and the need for them to learn from scratch the locations of good food sources and handy water sources.

10. *Locate your colonies as far as possible from flowers that are contaminated with insecticides and fungicides.* The greater the separation of colonies from these sources of harmful chemicals, the less often the foragers from your

colonies will be exposed to them and will bring them home in the nectar, pollen, and water they collect.

11. *Locate your colonies in places that are surrounded as much as possible by natural areas: wetlands, forests, abandoned fields, moorlands, and the like.* This will help ensure that your colonies have access to diverse sources of pollen and nectar that are not contaminated with insecticides and fungicides, as well as good sources of clean water and propolis.

12. *When you need additional colonies, acquire them by capturing swarms with bait hives or by making "splits" from strong colonies and letting them conduct emergency queen rearing and natural queen mating.* These two ways of acquiring additional colonies will provide you with colonies headed by queens that were reared from larvae chosen by the bees and mated by drones that competed fiercely for mating success.

13. *Minimize pollen trapping and honey harvesting from your colonies.* Both activities consist in robbing resources that the bees have worked hard to collect for their own needs. Taking these things from a colony will, directly or indirectly, lower its success as a living system by reducing its survival or its reproduction, or both.

14. *Refrain from treating colonies for* Varroa. This will help your bees acquire, through natural selection, resistance to the mites. It is now clear that this will eventually happen, probably within five years, if you live where most of the colonies around you are either wild or are being managed by beekeepers who have agreed to refrain from treating for *Varroa* and from importing queens of mite-susceptible stock. The study from Gotland, Sweden, described in chapter 10 shows us that at first there will be heavy colony losses but that a small percentage of the colonies will have natural resistance to the mites and will survive. I strongly advise, however, that you adopt this suggestion to stop treating colonies for *Varroa* only if you can do so as part of a program of extremely diligent beekeeping. If you pursue treatment-free beekeeping without paying close attention to the mite levels in your colonies, then you will create the situation in your apiary in which natural selection is likely to favor virulent *Varroa* mites, not

Varroa-resistant bees. To help natural selection favor *Varroa*-resistant bees, you need to monitor the mite levels in all your colonies and kill those whose mite populations are skyrocketing long before these colonies collapse from heavy infections of viruses spread by the mites.

By preemptively killing your *Varroa*-susceptible colonies, you accomplish two things that are important. First, you eliminate your colonies that lack *Varroa* resistance. Second, you prevent the "mite bomb" phenomenon of mites spreading en masse to your other colonies—and any other colonies in the area—when foragers from neighboring colonies rob honey from the collapsing colonies and bring home their *Varroa* mites. If you don't perform these preemptive killings, then even the most resistant colonies in and around your apiary can become overrun with mites and die, in which case there will be no natural selection for mite resistance among the colonies in your apiary. If you are unwilling to kill your colonies with dangerously high mite loads, then you will need to give them a thorough treatment with a miticide *and* replace their queens with queens of a mite-resistant stock.

CLOSING THOUGHTS

I hope you have enjoyed this review of what we know about how honey bees live in nature. We have seen that *Apis mellifera* remains an untamed creature and that the step from a beekeeper's hive to a tree's hollow remains a short one for these small beings. We have also seen that a honey bee colony is a marvelously integrated living system that has been shaped by natural selection to meet the challenges of getting rooted in a carefully chosen homesite and then surviving and reproducing there for several years. In looking at how a wild colony builds its nest, acquires its food, keeps itself warm, rears its young, defends itself from intruders, and passes on its genes by casting swarms and rearing drones, we have learned that a colony of honey bees presents us with countless mysteries. How does it control the type of comb it builds—at first just worker comb but eventually also drone comb? How does it control the emptying and filling of its drone-comb cells with honey in relation to the seasons? How does it know when to switch on its brood rearing in midwinter and then to switch it off in early autumn? How does it decide when to swarm? And then, when it decides to do so, how does it control the proportion of bees that leave in the swarm, the swarm fraction? And why does a colony seal up its nest cavity with propolis so tightly at the end of summer? Is it so that moisture will condense on its nest cavity's walls and thereby provide its members with drinking water all winter long? These and countless other questions about the lives of colonies living in the wild remind us that the behavior and social life of honey bees still holds many secrets.

If you are a beekeeper, then I hope, too, that this tour of the astonishing natural history of honey bees has inspired you to consider pursuing beekeeping in a way that focuses less on treating a honey bee colony as a honey factory or a pollination unit and more on admiring it as an amazing form of life. More than any other insect, the honey bee has the power to capture our hearts and connect us emotionally with the wonders and mysteries of nature. We love these beautifully social bees, we want them in our backyards, and many of us cannot bear the idea of living without them.

All of us who admire honey bees are seeking ways to improve their lives. This is of vital importance because, as the human population approaches 8 billion, we need the pollination services provided by honey bees more than ever before. A recent, and authoritative, study of the crop production values of different species of bees has concluded that the honey bee provides nearly half of all crop pollination services worldwide. This means that *Apis mellifera* contributes to agriculture almost as much as the hundreds of other crop-pollinating bee species combined. It also means that the honey bee deserves special care. One way we can conserve *Apis mellifera* is to protect forestlands, for these provide habitat for wild colonies. The persistence of honey bee colonies living in woodlands in the Americas, Africa, and Europe, despite the spread of the deadly mite *Varroa destructor*, shows us that honey bees are remarkably resilient. It also shows us that if we conserve forests and other wild places, then we can be confident that wild colonies of honey bees will thrive and provide an important reservoir of this species' genetic diversity.

A second way that we can improve the lives of honey bees is to revise our treatment of the millions of colonies that live not in the wild but in our hives. This is the goal of what I have called Darwinian beekeeping and others have called natural beekeeping, apicentric beekeeping, and bee-friendly beekeeping. Whatever the name, the aim is the same: to put the needs of the bees before those of the beekeeper. This happens when a beekeeper's manipulations of the bees are done with bee-friendly intentions and in ways that harmonize with the bees' natural history. Conventional beekeeping, however, continues to develop along a trajectory that disrupts and endangers the lives of honey bee colonies. Therefore, to truly help the bees, we must do more than just keep the world healthy for them; we must also build a new relationship between human beings and honey bees, one that promotes the health of the millions of managed colonies that we depend on to produce our food. Darwinian beekeeping, which combines respecting the bees and using them for practical purposes, seems to me to be a good way for us to be responsible stewards of the honey bee, *Apis mellifera*, our greatest friend among the insects.

Notes

PREFACE

Page vii: The figure of nearly 4,000 books on honey bees published in the United States comes from the annotated bibliography of American books on bees and beekeeping by Mason (2016).

Page viii: The story of Karl von Frisch's discovery of the meaning of the communication dances performed by honey bees is best told in the book by Munz (2016).

Page viii: The figure of 40% annual mortality of colonies kept by beekeepers comes from the Bee Informed Partnership, which receives reports from more than 5,000 beekeepers each year in the United States of their levels of colony mortality in summer and winter. See, Colony loss 2014–2015: Preliminary results, Bee Informed Partnership, 13 May 2015, https://beeinformed.org/results/colony-loss-2014-2015-preliminary-results/ (accessed 17 March 2017).

Page x: The rule "know thy animal in its natural world" is based on the title of the book of Tinbergen (1974).

CHAPTER 1. INTRODUCTION

Page 1: The Wendell Berry quotation is from his essay "Preserving Wildness." See Berry (1987), p. 147.

Page 2: The studies mentioned here that show that it is only the elderly bees (the foragers) within a colony that get most of their sleep at night, and in relatively long bouts, are those of Klein, Olzsowy et al. (2008) and Klein, Stiegler et al. (2014). The study that used sleep-deprivation methods to explore the functions of sleep for honey bees is that of Klein, Klein et al. (2010).

Page 3: The figure of 40% annual mortality of colonies kept by beekeepers comes from the Bee Informed Partnership, which receives reports from more than 5,000 beekeepers each year in the United States of their levels of colony mortality in summer and winter. See, Colony loss 2014–2015: Preliminary results, Bee Informed Partnership, 13 May 2015, https://beeinformed.org/results/colony-loss-2014-2015-preliminary-results/ (accessed 17 March 2017).

Page 3: The evidence that crowding honey bee colonies fosters the spread of diseases is found in Seeley and Smith (2015). The evidence that housing them in huge hives boosts their honey production but also their vulnerability to parasites is found in Loftus et al. (2016). These matters are discussed in greater detail in chapter 10, "Colony Defense."

Pages 5–6: For a fuller description of the characteristics of the dark European honey bee, see Ruttner (1987) and Ruttner et al. (1990).

Page 5: The archaeological studies that have found the chemical fingerprint of pure beeswax in organic residues preserved on fragments of pottery vessels from sites in Germany and Austria that date to 5200–5500 BCE are reported in Roffet-Salque et al. (2015).

Page 5–6: See Han et al. (2012) and Wallberg et al. (2014) for detailed information, based on genetic analyses, about the evolutionary and demographic history of *Apis mellifera* since it split off from the other members of the genus *Apis*, evidently in western Asia.

Page 7: The best source of detailed information about tree beekeeping in medieval Russia is Galton

294   NOTES TO CHAPTER 2

(1971). Additional information about tree beekeeping in the South Ural area of Bashkortostan is found in Ilyasov et al. (2015).

Pages 7: The spread of the honey bee in North America following its introduction in the 1600s is described in Kritsky (1991). The report on how New Englanders find where "bees hive in the woods" is that of Dudley (1720). The quotation of William Clark about honey bees living west of the Mississippi River in 1804 is from Moulton (2002).

Pages 7–9: A detailed review of the introduction of subspecies of *Apis mellifera* to North America is provided by Sheppard (1989). Also, Schiff et al. (1994) report an analysis of the mitochondrial DNA in bees in collected from 692 wild colonies in the southern United States (from North Carolina to Arizona) before the arrival of Africanized bees; it reveals that most of the wild colonies living in these southern states had mitochondrial DNA haplotypes representing European subspecies: 61.6% had the haplotype common in *A. m. carnica* and *A. m. ligustica*, 36.7% had the haplotype common in *A. m. mellifera*, and 1.7% had the haplotype common in *A. m. lamarckii*.

Page 9: The Africanization of the wild honey bees in southern Texas between 1991 and 2013 is beautifully documented through genetic analyses reported in Pinto et al. (2004, 2005) and Rangel, Giresi, et al. (2016).

Pages 10–11: For a full description of the whole-genome sequencing analysis of the ancestry of the wild colonies of honey bees living in the countryside around Ithaca, New York, see Mikheyev et al. (2015).

Page 14: The rapid evolution of gentle behavior in the Africanized honey bees following their introduction to Puerto Rico in 1994 is described in Rivera-Marchand et al. (2012) and analyzed genetically in Avalos et al. (2017). See Zuk et al. (2006) to learn more about the adaptive disappearance in less than five years of the calling song of male field crickets on the Hawaiian island of Kauai. The genetics underlying this rapid evolution in the male crickets' behavior is described by Tinghitella (2008).

Page 14: For more examples of rapid evolution in the physiological and behavioral traits of animals in natural populations, see Able and Belthoff (1998) on changes over 40 years in migratory behavior of house finches (*Haemorhous mexicanus*) in eastern North America; and Campbell-Stanton et al. (2017) on a striking, genetically based change over the winter of 2013–2014 in the low-temperature tolerance in lizards living in southern Texas. See also Grant and Grant (2014) for a detailed look at evolutionary changes over 40 years in beak size and shape in finch populations on an island in the Galapagos archipelago.

CHAPTER 2. BEES IN THE FOREST, STILL

Page 17: The Mark Twain quotation is the substance of his cable to the Associated Press in response to the news item that he had died. It was printed in the 2 June 1897 issue of the *New York Journal* newspaper.

Page 17: The geological history of the Finger Lakes region is described in von Engeln (1961), and the climate of Ithaca is documented in Dethier and Pack (1963).

Pages 18–21: The social and environmental histories of Ithaca and the surrounding lands are described by Kammen (1985) and Allmon et al. (2017). Smith, Marks et al. (1993) and Thompson et al. (2013) report studies of the return to a predominantly forested landscape around Ithaca, in particular, and the northeastern United States in general.

Pages 25: My studies of the nest-site preferences of honey bees are reported in Seeley and Morse (1978a).

Pages 26–28: The history of the Arnot Forest is described in Hamilton and Fischer (1970) and Odell et al. (1980).

Page 28: The fascinating craft of bee hunting is described in detail in Seeley (2016). See also Edgell (1949).

Pages 31–33: The results of the bee hunting in the Arnot Forest in 1978 are reported in Visscher and Seeley (1982).

Pages 33–34: The study of the wild colonies in Oswego, New York, is reported in Morse et al. (1990).

Pages 34–35: The exceptional study of the wild colonies in the Welder Wildlife Refuge in southern Texas is reported in Pinto et al. (2004).

Pages 36: The study of wild colonies living in trees lining rural roads in northern Poland is reported by Oleksa et al. (2013). The information about the density of managed colonies in the same region of Poland comes from Semkiw and Skubida (2010).

Pages 36–39: The study of the density of colonies in two national parks, and a third rural area, in Germany, done with an indirect, genetic method, is that of Moritz, Kraus et al. (2007). The more recent study of colony density in two areas of natural beech forests in Germany, done by means of bee hunting (bee lining) and inspecting the nest cavities of black woodpeckers, is that of Kohl and Rutschmann (2018).

Page 39: The study for three national parks in Australia is that of Hinson et al. (2015).

Page 41: The survey of the wild colonies in part of the Shindagin Hollow State Forest, in New York State, by means of bee hunting, is reported in Radcliffe and Seeley (2018). For a study of the density of honey bee colonies in various locations across the natural range of *Apis mellifera* (Europe, Africa, and central Asia), see Jaffé et al. (2009). It should be noted, however, that the estimates of colony density reported in Jaffé et al. (2009) include both managed and wild colonies.

Pages 42–44: The biology of *Varroa destructor* is reviewed by De Jong (1997). See also, Varroa mite, Featured Creatures, University of Florida, http://entnemdept.ufl.edu/creatures/misc/bees/varroa_mite.htm (accessed 10 July 2017). For detailed information on this mite's native distribution on *Apis cerana* in eastern Asia, the history of its host shift to *Apis mellifera* in eastern Russia, and its subsequent dispersal with the aid of beekeepers to Europe, Africa, and South America, see De Jong et al. (1982). Anderson and Trueman (2000) found that the *Varroa* mites infesting *Apis cerana* consist of two species—*V. jacobsoni* and *V. destructor*—and that the mites infesting *Apis mellifera* nearly worldwide are all members of the species *V. destructor*. This original host species of *V. destructor* was *A. cerana* on mainland Asia; the host species for *V. jacobsoni* remains *A. cerana* on the islands of the Malaysia–Indonesia–New Guinea region.

Page 44: The introduction of *Varroa destructor* to North America and its rapid spread across the continent are reviewed by Wenner and Bushing (1996) and Sanford (2001). The information on Africanized bee swarms found in Florida on eight ships from Central and South America between 1983 and 1989 comes from the records "Florida Africanized Bee Interceptions," provided to me by Dave Westervelt, of the Bureau of Plant and Apiary Inspection for the state of Florida.

Page 46: The paper by Bernhard Kraus and Robert E. Page Jr. that reported the extremely grim news about California's population of wild colonies is Kraus and Page (1995).

Pages 46–47: Gerald Loper describes his studies of a population of wild colonies living north of Tucson in Loper (1995, 1997, and 2002). He also provides a follow-up report in Loper et al. (2006).

Pages 48–56: Full reports on the mapping of the wild honey bee colonies in the Arnot Forest in the autumn of 2002, and on the testing of these colonies for infestations with *Varroa destructor* in 2003 and 2004, are found in Seeley (2007). The earliest report of this work is found in Seeley (2003).

CHAPTER 3. LEAVING THE WILD

Page 57: The Euell Gibbons quotation is from his book *Stalking the Wild Asparagus*; see Gibbons (1962), p. 235.

Page 57: The view that our prehuman ancestors (other members of the genus *Homo* and their direct ancestors, including Australopithecines) consumed and enjoyed honey is supported by the fact that several species of nonhuman primates—including baboons, macaques, chimpanzees, gorillas, and orangutans—are also successful honey hunters. This topic is reviewed by Crittenden (2011). Chimpanzees in Gabon are so fond of honey that they will prepare a three-piece tool kit for raiding the nest of a honey bee colony. It includes a pounder (a heavy stick for breaking open the nest entrance), an enlarger (to enlarge the opening into the nest), and a collector (a stick with a frayed end to poke into the honey combs and then into the mouth, slurping the honey off). See Boesch et al. (2009)

Page 57: The first correct assignment of a fossil bee to the honey bee genus *Apis* is that of Cockerell (1907). For more information about fossil honey bees, see the monograph on the fossil bees of the world by Zeuner and Manning (1976), the recent review of fossil honey bees by Engel (1998), and the definitive work on the evolution of insects by Grimaldi and Engel (2005).

Page 58: The oldest well-dated evidence of *Homo sapiens* is a fossil skull found during mining operations in the Jebel Irhoud massif near the west coast of Morocco. Its age is 315,000 years, plus or minus 34,000 years, as determined by thermoluminescence dating. For a recent analysis of this skull, see Hublin et al. (2017).

Page 58: The complex story of the origins of *Homo sapiens* in Africa and subsequent migrations into Asia and Europe, and eventually the entire world, is reviewed by Wenke (1999). Gibbons (2017) and Hublin et al. (2017) provide updates.

Page 58: For analyses of the caloric content of honey, see White et al. (1962) or Murray et al. (2001). The latter reference provides results for honey collected by the Hadza of Tanzania, thus for honey as it would have been eaten by early humans: fortified with proteins and fats from crushed bee larvae and pupae.

Pages 58–59: For detailed reports on honey hunting by the Hadza of Tanzania, see Marlowe et al. (2014) and Wood et al. (2014). A recent paper (Smits et al. 2017) describes how the gut microbiome of the Hadza hunter-gatherers changes between the wet and dry seasons so they maintain a healthy gut as they shift from heavy consumption of meat in the dry season to consumption of honey, berries, and other fruits in the wet season. For detailed reports on honey hunting by the Efe (and their close relatives, the Mbuti) of the Ituri Forest in the Democratic Republic of the Congo, see Turnbull (1976), Ichikawa (1981), and Terashima (1998). To enjoy a short novel on honey hunting and rock art painting by early people of southern Africa, see Dixon (2015).

Page 60–61: The evidence from rock art of humans gathering honey from wild colonies of honey bees is thoroughly reviewed in Crane (1999). The monograph by Hernández-Pacheco (1924) is the definitive reference for the prehistoric paintings found in 1917 in the Araña cave complex in Spain. Dams and Dams (1977) report their discovery in 1976 of additional rock art depicting honey gathering during the Mesolithic in eastern Spain.

Page 62: The stone bas-relief in pharaoh Nyuserre's temple to the sun god Re is discussed in further detail in chapter 20 of Crane (1999) and chapter 2 of Kritsky (2015). Another rich source of well-preserved scenes of beekeeping and other activities of daily life in ancient Egypt is the lavishly decorated tomb (at Thebes) of Rekhmire, the vizier for two pharaohs from about 1470 to 1445 BCE; see Garis Davies (1944).

Page 62–63: For more information on the Iron Age apiary found in Tel Rehov in northern Israel, see Mazar and Panitz-Cohen (2007) and Bloch et al. (2010).

Page 64: The final chapter of Kritsky (2015), cleverly titled "The Afterlife of Ancient Egyptian Beekeeping," contains excellent photos and detailed descriptions of the tools and techniques of traditional beekeeping in modern-day Egypt. These probably resemble those used by beekeepers during Pharaonic times, some 2,000 years ago.

Page 64–66: For a translation of Columella's writings on beekeeping in book 9 of his work *De re rustica* (On agriculture), see Columella (1968).

Pages 66–69: For general information about tree beekeeping in northern Europe, see chapter 16 in Crane (1999). The best source of detailed information about tree beekeeping in Russia is Galton (1971). The estimate by Prokopovich of the honeycomb per bee tree harvested by tree beekeepers is reported in Galton (1971), p. 27.

Page 69: Additional information about tree beekeeping in the South Ural area of Bashkortostan is found in Ilyasov et al. (2015).

Pages 69–72: For excellent reviews of traditional beekeeping in northwestern Europe, where the most widespread traditional hive was an inverted basket, or skep, see chapter 27 in Crane (1999) and chapters 3 and 4 in Kritsky (2010). The woodcut of two skeps in a shelter shown in Fig. 3.6 is from Münster (1628), p. 1415.

Pages 72–74: The story of Lorenzo L. Langstroth's discovery of the bee space and his use of it when designing the first fully functional movable-frame hive is told with admirable precision in the biography of Langstroth by Naile (1976). For more information about the context of Langstroth's discovery, especially the many parallel attempts in Europe to develop a movable-frame hive, see Kritský (2010).

Pages 74–76: For the 30 October 1851 entry from Langstroth's journal, see Naile (1976), p. 75. The Langstroth quotation, "will give the apiarian perfect control over his bees," is from the 26 November 1851 entry of his journal; see Naile (1976), p. 79.

Pages 76–78: For reviews of the impact of Langstroth's movable-frame hive on beekeeping worldwide, and of the follow-on inventions of tools and methods that made beekeeping with movable-frame hives so productive, see chapters 41 and 43 in Crane (1999).

CHAPTER 4. ARE HONEY BEES DOMESTICATED?

Page 79: The Lorenzo L. Langstroth quotation is from the title of chapter 2 of his book *Langstroth on the Hive and the Honey-Bee*; see Langstroth (1853).

Page 79: To learn more about domestication in general, see Roberts (2017) and DeMello (2012).

Page 79: The fascinating story of the domestication of the industrial yeasts (*Saccharomyces cerevisiae*) used to brew beers, wine, spirits, sake, and bioethanol is reported by Gallone et al. (2016).

Page 80: For more information about the evidence of the exploitation of honey bees by early farmers of the Middle East, see Roffet-Salque et al. (2015).

Page 81: The biblical quotation "a land flowing with milk and honey" is from the Book of Exodus, chapter 3, verses 8 and 17.

Page 81: The evidence that wild colonies of honey bees of European ancestry preferentially occupy rather modest-size nesting cavities, in the range of 20–40 liters (5.3–10.6 gallons), is found in papers by Seeley and Morse (1976, 1978a), Jaycox and Parise (1980, 1981), and Rinderer, Tucker et al. (1982).

Page 81: To the best of my knowledge, Eva Crane is the first person to hypothesize that hive beekeeping began with swarms of honey bees occupying empty pots and baskets of Neolithic farmers; see Crane (1999), p. 161.

Page 82: The engorgement of worker honey bees before swarming is described in detail by Combs (1972). Free (1968) describes engorgement by worker honey bees whose colony has been smoked.

Page 82: That colonies started by larger swarms have a higher probability of surviving the first winter is shown in Rangel and Seeley (2012).

Page 82: The reduced defensiveness in response to smoke probably arises mainly through its effects on the central nervous system of bees, but the effects of smoke on the bee's sensory (olfactory) system also play a role. Worker honey bees have reduced sensitivity to the smell of their alarm pheromones when smoked. Smoke probably interferes with the bees' olfaction generally, not just their detection of alarm pheromones. See Visscher, Vetter et al. (1995).

Page 83–84: For more information on how wild colonies of honey bees living in the Cape Point Nature Reserve in South Africa survived a wildfire there, see Tribe et al. (2017).

Page 85: The archaeological evidence regarding when humans learned to control fire and make use of it at will is described by Gowlett (2016).

Page 85: For more information on industrial dairy farming in New York State, see Kurlansky (2014).

Page 86: Biologists use the concept of heritability to describe the degree to which a particular trait of an organism (or, for honey bees, a colony) is genetically influenced. It varies between 0 and 1 and is used to assess the reliability of using a certain trait as a guide to judging the breeding value of an individual. Table 1 in Collins (1986) lists heritability estimates for traits of both individual honey bees (e.g., worker longevity) and whole colonies of honey bees (e.g., honey production). See also Bienefeld and Pirchner (1990) and Oxley and Oldroyd (2010) for estimates of heritability for several colony traits (growth rate in spring, wax production, calmness, etc.); the estimates of heritability of honey production, for example, range from 0.15 to 0.54.

Page 87: For the original description of instrumental insemination, see Watson (1928). The subsequent improvements needed to make instrumental insemination a reliable technique are described in Laidlaw (1944). An up-to-date description is that of Harbo (1986). A video showing how instrumental insemination is done is available; see, Instrumental insemination of honey bee queens – Susan Cobey, YouTube video, 5:23, posted by "tlawrence53," on 6 January 2009, https://www.youtube.com/watch?v=Csjy02ofpyI (accessed 22 December 2017).

Page 87: American foulbrood (AFB) is the only pathogen-based disease of honey bees that is virulent— i.e., that can rapidly overcome a colony's defensive mechanisms and kill it. Fries and Camazine (2001) explain how this virulence has evolved because the spores of AFB are easily transmitted from one colony to an unrelated colony (horizontal transmission) when a weakened, infected colony is robbed by bees from other colonies or when a swarm occupies a nest site where the previous colony succumbed to AFB. Ewald (1994) provides a clear evolutionary explanation of why some pathogens—including those that cause malaria, smallpox, tuberculosis, AIDS in humans, and AFB in bees—are extremely deadly, while others are not.

Page 87–89: Rothenbuhler (1958) provides a detailed and well-referenced review of the successful breeding program for resistance to American foulbrood conducted by O. Wallace Park and colleagues in the 1930s and 1940s, as well as briefer descriptions of other breeding programs, such as that by Brother Adam in England for resistance to acarine disease (presumed causative agent, *Acarapis woodi*). For another excellent, and more recent, review of the breeding for resistance to AFB, see Spivak and Gilliam (1998a).

Page 89–90: Spivak and Gilliam (1998b) provide a detailed and well-referenced review of the excellent studies on hygienic behavior, mainly as a mechanism of defense against chalkbrood, European foulbrood, and *Varroa*, since the eras of O. Wallace Park and Walter C. Rothenbuhler.

Page 90–91: To learn more about the remarkable story of the alfalfa bee, see Mackensen and Nye (1966) and Nye and Mackensen (1968, 1970). The breeding of these bees by commercial producers of alfalfa seed is reported by Cale (1971).

Page 91: Oxley and Oldroyd (2010) and Oldroyd (2012) discuss in greater detail the absence of distinct breeds of honey bees and the fact that honey bees have never been truly domesticated.

Page 91–92: Roberts (2017) reviews the deep history of the domestication of dogs and cattle, along with eight other animals (chickens, horses, and humans) and plants (wheat, apples, potatoes, rice, and maize) that used to live wild.

Page 92: Some will say that humans have made a fundamental change in the genetics of *Apis mellifera*, because the honey bee populations that have been established in North America, New Zealand, and Australia are mixtures of most of the honey bee subspecies of Europe (as shown for Canada by Harpur et al. 2012) and because the migratory activities of beekeepers and the shipping of queens between countries are starting to homogenize the honey bee subspecies within Europe (De la Rúa et al. 2009). However, I am not treating these genetic changes as fundamental, because they have not produced distinct subpopulations (breeds) of honey bees that are inbred.

Page 93: Box 5.2 in DeMello (2012) contains a list of all 18 domesticated animals, besides dogs, along with an estimate of the time of domestication for each. The list comprises sheep, cats, goats, pigs, cattle, chickens, guinea pigs, donkeys, ducks, horses, llamas, Bactrian camels, dromedary camels, water buffalo, yaks, alpacas, and turkeys—and, also, honey bees.

Page 94: For more information about the movement of honey bee colonies to the heather moorlands in Scotland, see Manley (1985) and Badger (2016).

Page 94–95: For a general overview of the migratory beekeeping that is required to meet the demand for pollination during California's almond bloom, see Ferris Jabr, The mind-boggling math of migratory beekeeping, *Scientific American*, 1 September 2013, https://www.scientificamerican.com/article/migratory-beekeeping-mind-boggling-math/ (accessed 23 December 2017). Jacobsen (2008) and Nordhaus (2011) provide more detailed portraits of the trucking of millions of colonies each year to farmers across the United States who need pollinators.

Page 95–98: For more information about the tools and methods of modern beekeeping, see Flottum (2014) and Sammataro and Avitabile (2011).

CHAPTER 5. THE NEST

Page 99: The Charles Darwin quotation is from his book *On the Origin of Species*; see Darwin (1964), p. 224.

Page 99: The concept of looking at the nests of animals as part of their survival equipment is best described by Richard Dawkins when he discusses how the structures built by organisms are part of their "extended phenotypes"; see the final chapter in either Dawkins (1982) or Dawkins (1989).

Page 100: The full report of the study of the natural nests of honey bees living in the woods around Ithaca is found in Seeley and Morse (1976). It describes the methods of nest dissection, including how we measured the nest cavities by filling them with sand after removing the combs. Avitabile et al. (1978) report a related study that was conducted in Connecticut and collected information about nest entrance height, size, and orientation for 108 colonies living in trees. These authors also report finding mostly nest entrances that were low (< 5 m/16.4 ft.), small (< 60 cm^2/9.3 sq. in.), and with a southerly orientation.

Page 101: To make a detailed dissection of each nest and conduct an accurate census of the bees within it, I put to death each colony on the day on which its tree was felled. In doing so, I worked to minimize the bees' suffering. My procedure was to arrive at the bee tree in early morning, when it was still cool and just starting to get light, so the bees were not yet flying out. I then climbed up the tree, plugged all but one of the nest's entrance openings with rags, spooned several tablespoons of Cyanogas powder (calcium cyanide) in the one entrance still open, and then plugged it too. Each time, I heard a rapid rise of buzzing sounds coming from inside the sealed-up nest, but within two minutes there was silence. It was grim work, and I doubt that I could do it today, but I believe that the benefits of the information gained from this study justify the deaths of the 21 colonies studied.

Page 101: Analysis of the distribution of compass directions of the entrances of the bee trees using circular statistics yielded the following results: mean vector bearing, 192° (ca. SSW); mean orientation vector, 0.39. This shows that their distribution is biased toward the south and is unquestionably nonrandom; $p < 0.01$, Rayleigh test for nonrandomness. See Batschelet (1981).

Page 103: The three surveys of wild colonies in the Arnot Forest are reported in Visscher and Seeley (1982), Seeley (2007), and Seeley, Tarpy et al. (2015). The survey of the wild colonies in the Shindagin Hollow State Forest is reported in Radcliffe and Seeley (2018).

Page 105: One of the bee-tree nests was odd in having both an immense (204 cm^2/31.6 sq. in.) nest entrance and a huge (448 liter/118.3 gal.) nest cavity. The entrance was a gaping opening at the base of a large beech tree (*Fagus grandifolia*), and the cavity was a ca. 5 m (16.4 ft.) tall space inside it. The colony had built its nest near the top of this cavity, so even though both entrance and cavity were unusually large, the colony occupied a good, snug nesting site.

Page 108: I regret that we did not take more systematic measurements of the thickness of the wooden walls of these tree cavities, for this turns out to be of great importance for understanding how colony thermoregulation works in these natural cavities. See Mitchell (2016).

Page 108: The high percentage of drone comb that we found in these nests (10%–24%, average 17%) has been confirmed in another study that described the nests built by colonies whose comb building was not manipulated. The range reported in this second study is 11%–23%, with an average of 20%. See Smith, Ostwald et al. (2016).

Page 108: My study of small-cell combs as a means of controlling infestations of *Varroa* mites, which includes data on the cells built on small-cell foundation, is reported in Seeley and Griffin (2011).

Page 109: Edgell (1949) reports an estimated average of 8.5 kg (ca. 19 lb.) of honey collected from 56 bee-tree colonies in central New Hampshire. Their nests were taken up at various times across the summer.

Page 111: The seasonal pattern of swarm emergence for Ithaca was measured over six years, when 126 swarms were collected; see Fell et al. (1977).

Page 111: Rangel, Griffin et al. (2010) document how nest-site scouts can start searching for potential homesites even before their colony casts a swarm. This paper reports scouts inspecting potential nest sites starting two to three days before their colony swarms.

Page 111: For a detailed description of the behavior of a nest-site scout as she inspects a potential nesting cavity to assess its size and other properties, see Seeley (1977). For detailed information about how the scout bees in a swarm work together in a process of collective decision-making to choose a new homesite, see Lindauer (1955) and Seeley (2010).

Page 112: The article written by a French beekeeper on how to build bait hives that are attractive to honey bee swarms is Marchand (1967). The long history of the quest by beekeepers for a perfect hive is described beautifully by Kritsky (2010).

Page 112: For details of the methods and results of my study of the nest-site preferences of the wild honey bees around Ithaca, see Seeley and Morse (1978a).

Page 116: The paper by Tibor I. Szabo on the benefits of a south-facing nest entrance for bees living where winters are cold and snowy is Szabo (1983a).

Page 116: The paper by Derek M. Mitchell on the benefits to the bees of *not* having a top entrance is Mitchell (2017).

Pages 116–119: My studies of how scout bees measure the volume of a prospective nest cavity and what they prefer regarding cavity volume are reported in Seeley (1977). The related studies done by Elbert R. Jaycox and Stephen G. Parise are reported in Jaycox and Parise (1980, 1981). The paper that reports the large-scale investigation by Thomas E. Rinderer and colleagues is Rinderer, Tucker et al. (1982). The report of the related work by Justin O. Schmidt is Schmidt and Hurley (1995).

Page 119: For the full report of the study by Tibor I. Szabo on the benefits of installing swarms in hives whose frames are filled with comb, see Szabo (1983b). That tree beekeepers in Russia placed a high value on tree cavities that have already been occupied by colonies is described in Galton (1971). Three good references on bait hives for honey bees are Marchand (1967), Guy (1971), and Seeley (2017a).

Page 119–120: The dimensions of the entrance openings used in the test of vertical slit vs. circle were 1 × 7 cm (0.4 × 2.75 in.) for the slit and 3 cm (1.2 in.) in diameter for the circle. In the test of cavity dryness, each nest box received its 2 liters (0.5 gal.) of sawdust (wet or dry) just before it was mounted on its tree or power-line pole. The wet sawdust was created by mixing 2 liters of sawdust with 1 liter (0.26 gal.) of water. Another liter of water was poured into each box in the wet-sawdust treatment group each time the nest boxes were inspected, which was about every 10 days. For more details on the methods used in the study of the bees' real-estate preferences, see Seeley and Morse (1978a).

Pages 121–122: For a review of how the wax-gland epithelium (and thus the wax production) changes with age in worker honey bees, see chapter 4 in Hepburn (1986). This chapter also reviews the studies that have shown that the elderly bees in a colony, which are normally functioning as foragers and have only a thin wax-gland epithelium, can rejuvenate their wax glands when they are members of a swarm.

Page 122: The formula for calculating the area of a hexagon is $a^2 \cdot 6 \tan(30°)$, or $a^2 \cdot 0.8655$, where a is the wall-to-wall dimension (the apothem) of the hexagon. Using this formula, one can calculate that there are approximately 82,051 worker cells in 1.92 m² (20.7 sq. ft.) of worker comb (apothem = 5.20 mm/0.205 in., hence the area of a worker cell is 23.40 mm²/0.0363 sq. in.) and approximately 13,125 drone cells in 0.48 m² (5.2 sq. ft.) of drone comb (apothem = 6.50 mm/0.256 in., hence the area of a drone cell is 36.57 mm²/0.0567 sq. in.). These areas of worker and drone comb were calculated as 80% and 20%, respectively, of 2.4 m² (25.8 sq. ft.) of total comb surface (i.e., counting both surfaces on a comb). This is the average amount of total comb surface that I have found in nests of wild colonies.

Page 122: Wojciech Skowronek (cited by Hepburn 1986, p. 39) found that a worker honey bee can produce approximately 20 mg (0.0007 oz.) of beeswax; 60,000 bees × 20 mg of beeswax/bee = 1,200,000 mg, or 1.2 kg (2.6 lb.) of beeswax. The estimate that bees need about 1 g (0.04 oz.) of beeswax to build 20 cm² (3.1 sq. in.) of comb surface (i.e., 10 cm²/1.6 sq. in. of double-sided comb) comes from Wojciech Skowronek (cited by Hepburn 1986, p. 64).

Page 122: The figures regarding amount of sugar carried per swarm bee come from Combs (1972). The mean number of worker bees in a swarm comes from Fell et al. (1977). The estimate of efficiency of beeswax synthesis from sugar comes from Horstmann (1965) and Weiss (1965).

Page 123: For detailed information on the likelihood of winter survival by colonies that were started by swarms the previous summer (founder colonies), see Seeley (1978) and Seeley (2017b).

Page 125: For information on the circular cross-section shape of the cells in the nests of solitary bees and social bees other than honey bees (e.g., bumble bees and stingless bees), see Michener (1974) or Michener (2000). In the nests of all these non-*Apis* species, the cells serve only as brood cells. The social bees that are not honey bees build special honey pots for holding their honey stores.

Page 126: For more information on how worker bees use their antennae to judge the thickness of a cell's walls during cell construction, see Martin and Lindauer (1966).

Pages 127–128: For a full report on the patterns of nest construction and colony population dynamics after a swarm moves into an empty nest cavity, see Smith, Ostwald et al. (2016). This paper also reports how the colony's comb area increases, its populations of workers and drones fluctuate, and its honey and pollen stores rise and fall over the 14 months following the colony's occupation of a new nesting cavity. The dimensions of the large observation hives used in this study are 4.3 cm deep, 88 cm wide, and 100 cm tall (1.7 in. × 34.6 in. × 39.7 in.).

Page 127: One measure of the importance of the initial burst of comb building is the high percentage of a colony's first-year comb building that occurs within the first four to six weeks of living in a new nest cavity. Smith, Ostwald et al. (2016) report 57% within the first four weeks. Lee and Winston (1985) report 90% within the first six weeks.

Page 127: The tanking up on honey by the worker bees in swarms is described in Combs (1972).

Pages 129–131: The theoretical and experimental studies by Stephen C. Pratt on optimal timing of comb construction are found in Pratt (1999) and are reviewed in Pratt (2004).

Pages 131: Rösch (1927) and Seeley (1982) have reported the match in the age ranges of the bees responsible for comb building and the bees responsible for unloading nectar from incoming foragers and storing it in comb cells.

Pages 131: See Pratt (1998a) for the study with paint-marked bees that showed that the worker bees engaged in nectar receiving do not form a large percentage of the bees engaged in comb building.

Pages 132: That swarms newly settled in their homesites build only worker cells for the first several weeks of construction has been reported by Free (1967), Taber and Owens (1970), and Lee and Winston (1985), as well as Smith, Ostwald et al. (2016).

Pages 132–135: See Pratt (1998b) for the study of the information pathways used by bees to decide the kind of comb (worker or drone) to build. See also Pratt (2004) for a review of work on this subject and on the matter of when to build comb.

Pages 135–136: See table 12.5 in Crane (1990) for an excellent summary of the plants reported to be sources of propolis collected by the honey bee *Apis mellifera*. See also Simone-Finstrom and Spivak (2010) for references to more recent papers on this subject.

Page 136: See Meyer (1954) and Meyer (1956) for detailed descriptions of the process whereby a resin collector packs her corbiculae with resin.

Pages 136–138: See Nakamura and Seeley (2006) for detailed information on the close observations of resin collectors and resin users made by Jun Nakamura.

Pages 139: See Simone-Finstrom, Gardner et al. (2010) for the study that discovered better associative learning of tactile stimuli in resin collectors than pollen foragers.

CHAPTER 6. ANNUAL CYCLE

Page 140: The Robert Frost quotation is from his poem "A Prayer in Spring"; see Latham (1969).

Page 142: See Avitabile (1978) for information about the weights of honey bee colonies living in central Connecticut from late autumn to early spring. The climate in central Connecticut is a close match to that of central New York State.

Page 142: See Southwick (1982) for information about the metabolic rate of a typical overwintering colony as a function of ambient temperatures.

Page 143: See McLellan (1977) for a critical evaluation of using change in a honey bee colony's weight as an indicator of its acquisition (or loss) of energy.

Page 143: See Milum (1956), Mitchener (1955), and Koch (1967) for reviews of records of weight changes of honey bee colonies across a summer or throughout a year for the United States, Canada, and Germany, respectively.

Page 143: See Seeley and Visscher (1985) for more information about the study in which the weekly weight changes of two unmanaged colonies (simulating wild colonies) were recorded from November 1980 to June 1983.

Page 144: See Farrar (1936) for the details of his study of the influence of pollen reserves on brood rearing and weight losses of honey bee colonies in winter.

Page 148: See Nolan (1925), Allen and Jeffree (1956), Jeffree (1956), and Winston (1981) for examples of describing colony growth patterns by measuring the number of cells of brood in the nest. See

Jeffree (1955) and Loftus et al. (2016) for examples of doing this by repeatedly censusing a colony's population of adult bees.

Page 146: See Simpson (1957a) and Gary and Morse (1962) for information on the occurrence of false starts in queen rearing in preparation for swarming.

Page 148: In Fig. 6.4, the data on low levels of brood rearing from mid-November to the end of February come from Avitabile (1978). The swarm dates data are an extension of the results reported in Fell et al. (1977). Air temperature data come from Brumbach (1965).

Page 148: The evidence that colonies adjust the start of brood rearing in the winter in response to increasing day length comes from a study by Kefuss (1978).

Page 148: For more information on the annual patterns of brood rearing in temperate climates see Nolan (1925) and Jeffree (1955, 1956).

Page 148: See Louveaux (1973) and Strange et al. (2007) for more information about the colony transplant experiments conducted in France that have shown that the annual cycles of brood rearing are partly under genetic control.

Page 150: For more information about the timing of drone production in colonies at the Rothamsted Experimental Station, see Free and Williams (1975).

Page 151: For more reports on the seasonal timing of swarming see Mitchener (1948) for Manitoba; Murray and Jeffree (1955) for Scotland; Simpson (1957b) for southern England; Fell et al. (1977) and Caron (1980) for northeastern United States; and Page (1982) for central California.

Page 151: The probabilities presented here of colony survival over summer and over winter, and for new and established colonies, come from two long-term studies of wild colony survival that I conducted in the 1970s and in the 2010s. See Seeley (1978) and Seeley (2017b).

Page 151: To postpone the onset of brood rearing from midwinter to mid-spring in 6 of the 12 colonies used in this study, we caged each colony's queen between two sheets of plastic queen excluder material, spaced 8 mm (0.3 in.) apart by strips of wood and inserted between the upper and lower stories of each colony's hive. This arrangement enabled each queen to move laterally during the winter and thereby maintain contact with the winter cluster but also prevented her from gaining access to combs in which to lay eggs. In each of the other six colonies in the study, we did not restrict the queen's movements, thus we did not limit the colony's winter brood rearing.

Pages 152: The studies reviewed here, on the critical timing of colony growth and reproduction, are reported in Seeley and Visscher (1985).

Page 153: For more information on the life cycle of bumble bee colonies, see Heinrich (1979) and Goulson (2010).

Page 153: For more information on how insects overwinter in temperate, subarctic, and even arctic regions, where they are exposed to subfreezing temperatures for some part of the year, see Chapman (1998), pp. 518–520.

Page 153: For more information on the tundra-dwelling bumble bee, *Bombus polaris*, see Richards (1973).

Page 153: Michener (1974) provides excellent summaries of the natural history and biogeography of both stingless bees and honey bees.

Page 154: The studies of latitudinal differences in rates of ant predation on undefended wasp larvae are reported in Jeanne (1979).

CHAPTER 7. COLONY REPRODUCTION

Page 155: The George C. Williams quotation is from his book *Adaptation and Natural Selection*. See the opening sentence of chapter 6 ("Reproductive Physiology and Behavior") in Williams (1966).

Page 155: A sexual organism (plant or animal) is a simultaneous hermaphrodite if it produces both female and male gametes (eggs and sperm) in each breeding season. See chapter 2 in Charnov

(1982) for a full discussion of this matter. The gametes of a honey bee colony are its virgin queens and unmated drones.

Page 157: Detailed information on development times for honey bee queens and drones (16 and 24 days), and on the sexual maturation times for queens and drones following emergence from their cells (minimum values: 6 and 10 days), is found in chapters 4 and 5 of Koeniger, Koeniger, Ellis et al. (2014) and Koeniger, Koeniger, and Tiesler (2014).

Page 159: To convert an area of capped drone comb (in Page 1981), or an area of brood-filled drone comb (in Smith, Ostwald et al. 2016), into the number of cells of drone brood, I multiplied the reported values of drone comb (in cm^2) by 2.73 cells/cm^2. This is the density of hexagonal cells that have an apothem (wall-to-wall dimension) of 6.5 mm (0.24 in.), which is the average dimension of drone cells for European honey bees. To calculate the total number of drone brood cell-days for a colony, I calculated the areas under the curves reported in Page (1981) and Smith, Ostwald et al. (2016) of the number of capped (or brood-filled) drone cells in a colony's nest over the entire summer. To calculate the total number of drones reared by a colony across the summer, I divided the total number of drone cell-days for a colony by the number of drone cell-days per drone (either 14 days for capped cells of drone brood or 21 days for all cells with drone brood). This assumes every cell that contains a developing drone yields a viable drone.

Page 159: For more information about the study that tested whether honey bee colonies seasonally switch which type of comb (drone or worker) they use preferentially for honey storage, so their drone comb is empty and ready for brood rearing in the spring, see Smith, Ostwald et al. (2015).

Page 161: For more information about the behavior of worker bees shaking the queen and how it functions to (among other things) prepare queens to fly out of the hive, either with a swarm or on a mating flight, see Allen (1956, 1958, 1959a), Hammann (1957), and Schneider (1989, 1991).

Page 161: The figure of a 25% reduction in the queen's weight in preparation for swarming comes from Fell et al. (1977).

Page 161: The evidence that only about a quarter of a colony's workers are left behind when the swarm containing the mother queen departs is discussed in detail later in the chapter.

Page 161: One wonders how a worker bee in a swarming colony decides whether to stay at home and support a young (sister) queen or leave in the prime swarm and support the old (mother) queen. Does she prefer the former option if the new queen could be a full sister but the latter option if all the young queens are only half sisters? A study (Rangel, Mattila et al. 2009) has checked whether the workers are more inclined to stay at home if some of the immature queens are their full sisters, but no evidence of this was found. Thus, it seems clear that there is no intracolonial nepotism during swarming in *Apis mellifera*.

Page 161: The wondrous process whereby the scout bees in a swarm select their new homesite is described in detail in Seeley (2010).

Page 162: The information about the large dispersal distances of honey bee swarms is found in Seeley and Morse (1978b) and Kohl and Rutschmann (2018).

Page 162: The value of 0.87 for the probability of swarming by an unmanaged colony during a summer comes from my study with simulated wild colonies living in the woods around Ithaca, reported in Seeley (2017b).

Page 163: For an impressively detailed description of the mechanisms by which all but one of the unmated queens produced during the swarming process are eliminated from the parental nest, see Gilley and Tarpy (2005). See also Winston (1980) for detailed information on the occurrence of swarming and afterswarming in unmanaged honey bee colonies living in Lawrence, Kansas.

Page 164: The probabilities of a colony producing a first afterswarm (0.70) and a second afterswarm (0.60) were calculated based on what Winston (1980) and Gilley and Tarpy (2005) have reported. Each study describes the pattern of swarming and afterswarming for five colonies.

Page 164: The work by M. Delia Allen mentioned here is found in Allen (1956).
Page 164: The probability of survival to the following summer of the mother queen that departs in a prime swarm ($p = 0.23$) and of the daughter queen that inherits the original nest ($p = 0.81$) comes from Seeley (2017b, see fig. 5).
Page 165: The methods and results of the two long-term studies of wild-colony demography that are mentioned here are described in detail in Seeley (1978) and Seeley (2017b).
Page 165: I describe how I use bait hives to capture swarms in Seeley (2012, 2017a).
Page 166: The bimodal seasonal pattern of swarming in the Ithaca area is described in Fell et al. (1977).
Page 167: Afterswarms are delayed, relative to prime swarms, in building their nests because they leave the parental nest well after the prime swarm has departed. Detailed information about the delayed departures of afterswarms is found in table I in Gilley and Tarpy (2005). They report first afterswarms departing 5–7 days after the prime swarm and second afterswarms departing 12–18 days after the prime swarm.
Page 169: The formula for calculating how long, on average, a site will be continuously occupied by a honey bee colony is as follows, where A represents site age in years:

$$0.5 + \sum_{A=0}^{20} A[(0.23)(0.81)^{A-1}][0.19] = \text{age in years}$$

Page 171: For more information about the evolution of parental allocation of resources to male and female offspring, see Charnov (1982).
Page 172: The values used here for the average dry weights of a drone and a worker come from the study by Henderson (1992).
Page 174: How large a food reserve is carried off by the worker bees in a swarm when they leave home? Combs (1972) reports that, on average, a worker bee in a swarm holds 37 mg (0.001 oz.) of a 67% sugar solution in her crop (honey stomach). (For comparison, on average, a worker bee in a non-swarming colony holds only 10 mg/0.0003 oz. of a 39% sugar solution.) Therefore, an average swarm with 12,000 workers will be stocked with some 300 g (10.6 oz.) of sugar. (12,000 bees × 37 mg sugar solution per bee × 67% sugar = 297.5 grams of sugar.)
Page 174: I have studied the cost to a colony of producing and maintaining a population of drones (Seeley 2002). To do so, I compared the honey yields of colonies that were managed for honey production, and were either with or without drone comb, and found that colonies with full-scale drone production produced, on average, some 20 kg (44 lb.) less honey than colonies with severely limited drone production.
Page 174: How many mating flights does a drone make across his life? A drone lives for about 20 days after reaching sexual maturity, and he makes two to four mating flights on a day of good weather (Winston 1987, pp. 56 and 202). If we assume that half the days in summer have weather good enough for drones and queens to conduct mating flights, then we can estimate that a typical drone makes 20–40 mating flights in his brief (and probably sexually unfulfilled) life.
Page 175: The two-part investigation that is mentioned here is Rangel, Reeve et al. (2013).
Page 176: For more information on how we measured the winter survival probabilities for mother-queen and sister-queen colonies as a function of swarm fraction, see Rangel and Seeley (2012).
Page 176: The three studies that report measurements of the swarm fraction are Martin (1963), Getz et al. (1982), and Rangel and Seeley (2012).
Pages 178–182: The fascinating biology of mating by honey bees is now thoroughly reviewed in two books: Koeniger, Koeniger, and Tiesler (2014), in German, and Koeniger, Koeniger, Ellis et al. (2014), in English.

Page 178: The discovery that 9-oxo-2-decenoic acid is the sex attractant pheromone of honey bees is reported in Gary (1962).

Page 180: The figures for the average distance to a queen's mating site (2–3 km/1.2–1.9 mi.) and for the maximum distances to which drones fly in search of a mating opportunity (5–7 km/3.0–4.2 mi.) come from Ruttner and Ruttner (1972).

Pages 180: The original report of the impressive mark-and-recapture study of the distance of drone mating flights is Ruttner and Ruttner (1966).

Page 181: The elegant experimental study of the maximum mating range of honey bees, conducted in Ontario, Canada, is reported in Peer (1957).

Page 182: The evolution of polyandry (multiple mating by queens) in the genus *Apis* is reviewed by Palmer and Oldroyd (2000). For a report of the average paternity numbers of all eight species of *Apis*, see Tarpy et al. (2004). For *Apis mellifera*, they report 12.0 ± 6.3 (mean ± one standard deviation) for the observed insemination number, and 12.1 ± 8.6 or 11.6 ± 7.9 for the effective paternity frequency. There are two estimates of the effective paternity frequency because there are two ways to estimate this number.

Page 182: The multistage process whereby drones inject their sperm into the queen's oviducts and then a small portion of it is transferred to the queen's spermatheca for long-term storage is described well in chapter 10 in Koeniger, Koeniger, Ellis et al. (2014).

Page 182: There are many studies that show how a honey bee colony benefits from the queen acquiring sperm from numerous drones and then producing a genetically diverse workforce. For information about improved disease resistance, see Tarpy (2003), Seeley and Tarpy (2007), and Simone-Finstrom, Walz et al. (2016); about greater nest microclimate stability, see Jones et al. (2004); and about increased productivity through enhanced acquisition of food resources, see Mattila and Seeley (2007) and Mattila et al. (2008).

Page 183: Heather Mattila's discovery that queen promiscuity helps ensure that a colony possesses a critical minority of workers who are social facilitators of foraging-related activities is reported in Mattila and Seeley (2010).

Pages 183–186: For more information on the study that compared the mating frequencies of queens living in the wild, where colonies are widely dispersed, and queens living in apiaries, where colonies are tightly bunched, see Tarpy, Delaney et al. (2015).

CHAPTER 8. FOOD COLLECTION

Page 187: The Thomas Smibert quotation is from his poem "The Wild Earth-Bee"; see Smibert (1851).

Page 187: The evidence that worker bees will travel 14 km (8.7 mi.) to collect food comes from a study conducted in a semidesert region of Wyoming. Colonies were placed at various distances (up to 14 km/8.7 mi.) from irrigated fields of alfalfa and yellow sweet clover, and it was found that even the most distant colonies collected nectar and pollen from these fields. See Eckert (1933).

Page 187: The statement that a colony will field about a third of its members as foragers is based on the results reported in Thom et al. (2000).

Page 189: That pollen foragers preferentially unload pollen near cells containing brood (esp. cells containing eggs and larvae) was shown by Dreller and Tarpy (2000). For broader analyses of the behavioral rules of honey bees that create the consistent pattern of brood, pollen, and honey in the combs of their nests—brood at the bottom, surrounded by pollen, and honey above—see Camazine (1991), Johnson (2009), and Montovan et al. (2013).

Page 190: Even during a strong honey flow—a time of intense nectar collection—water collectors constitute a small percentage of the bees coming into a hive bearing a load of liquid cargo. See, for example, fig. 2 in Seeley (1986).

Page 190: For more information about the age distributions of the nectar receivers and water receivers in a colony, see Seeley (1989) and Kühnholz and Seeley (1997).
Page 192: The details of the study of these undisturbed colonies living in Connecticut are described in Seeley and Visscher (1985).
Page 192: There are approximately 7,700 bees per kg (3,500 bees/lb.); see Otis (1982).
Page 193: The figure of 130 mg (0.004 oz.) of pollen to produce a worker bee comes from Haydak (1935).
Page 193: The statement that, on average, a worker bee lives about one month in summer is based on Sekiguchi and Sakagami (1966) and Sakagami and Fukuda (1968).
Page 193: The estimates of total brood rearing and food consumption by colonies managed for honey production come from Brünnich (1923) and Nolan (1925), regarding brood production; Eckert (1942), Hirschfelder (1951), and Louveaux (1958), regarding pollen consumption; and Weipple (1928) and Rosov (1944), regarding honey consumption.
Page 193–194: The average weight of a pollen load comes from Parker (1926) and Fukuda et al. (1969). The average flight distance (= twice the average flower-patch distance) comes from Visscher and Seeley (1982). The estimate of flight cost comes from Scholze et al. (1964) and Heinrich (1980). The energy value for pollen comes from Southwick and Pimentel (1981). The average sugar concentration of nectar comes from Park (1949), Southwick et al. (1981), and Seeley (1986). The average sugar concentration of honey comes from White (1975). The average weight of a nectar load comes from Park (1949) and Wells and Giacchino (1968).
Page 195: The figure of 6 km (3.7 mi.) for the foraging range of a colony is based on the work reported in Visscher and Seeley (1982), which described the spatial patterns of foraging by a full-size colony living in the Arnot Forest. It found that a circle large enough to enclose 95% of the sites indicated by the waggle dances of this colony's foragers had a radius of 6 kilometers.
Page 195: The indicated flight speed of a forager—30 km/hr (18.6 mph)—is the approximate average of the cruising flight speed of nectar foragers for their empty, outbound flights (34.2 km/hr, 21.3 mph) and their laden, homebound flights (24.2 km/hr, 15.0 mph). For details on how these flight speeds were measured, see Seeley (1986) or Biology Box 5 in Seeley (2016).
Pages 195–196: For examples of the foraging-range studies using the standard mark-and-recapture method, see Berlepsch (1860, p. 176), Levin et al. (1960), Levin (1961), and Robinson (1966). The study conducted in the semidesert region of Wyoming is Eckert (1933). The method of magnetic retrieval of steel labels in a capture-recapture system for honey bees is described in Gary (1971). A good example of a study in which the method of magnetic retrieval of steel labels was used to determine the distribution of a colony's foragers is Gary et al. (1978).
Page 197: The pioneering work by Herta Knaffl on the spatial range of a honey bee colony's foraging operation based on reading the bees' waggle dances is described in detail in Knaffl (1953).
Pages 197–200: For a detailed description of the study conducted with Kirk Visscher of the spatial and temporal patterns in the foraging work of a full-size colony living in the Arnot Forest, see Visscher and Seeley (1982).
Page 201: The remarkable study of long-range foraging by bees flying to the heather in the high moors outside of Sheffield, England, is that of Beekman and Ratnieks (2000). For more remarkable studies conducted in England that have used the technique of spying on waggle dances to explore the effects of colony size and season on foraging range and dynamics, see Beekman et al. (2004) and Couvillon et al. (2014, 2015).
Pages 201–204: The treasure hunt study of the abilities of colonies to conduct reconnaissance for profitable food sources far from their nests is described in Seeley (1987).
Page 207: For more examples of maps that depict the day-by-day dynamics of the recruitment targets of the colony living in the Arnot Forest, see fig. 3 in Visscher and Seeley (1982).

Pages 206–209: The experimental study of the ability of a colony to choose among a changing array of potential food sources by skillfully adjusting the number of foragers engaged at these sources is described in detail in Seeley, Camazine et al. (1991). For a review of all the studies on this topic, see chapters 3 and 5 in Seeley (1995).

Page 210: For an example of the distribution of sugar concentrations in the loads of liquid collected by worker bees in Connecticut, see fig. 2.12 in Seeley (1995). The range of sugar concentrations shown in this figure extends from about 15% to 65% and averages around 40%. Park (1949) and Southwick et al. (1981) report the same average concentration of sugar in nectar collected by worker bees in Iowa and New York, respectively.

Page 210: For examples of studies that have looked at what stimulates robbing among colonies in apiaries, see Butler and Free (1952) and Ribbands (1954).

Pages 211–212: The study of the speed and occurrence of robbing among widely separated wild colonies is reported in Peck and Seeley (forthcoming).

Page 212: The bait hive on the shed attached to my barn is a small, five-frame Langstroth hive. To make my bait hives conspicuous and attractive to scout bees, I insert five frames filled with dark, aromatic comb. See Seeley (2012) and Seeley (2017a).

Page 213: The paper that describes the spectacular skill of *Varroa* mites at climbing onto worker bees is Peck et al. (2016). The paper that shows that when *Varroa* mites are in colonies that are weakened by mite-transmitted diseases, they no longer discriminate against climbing onto foragers (including robbers) is Cervo et al. (2014).

Page 214: I believe that all the colonies in the Arnot Forest are now infested with *Varroa* mites because every swarm that I capture in this forest is infested with mites, as discussed in chapter 2.

CHAPTER 9. TEMPERATURE CONTROL

Page 215: The Thomas Hood quotation is from his poem "November"; see Hood (1873), p. 332.

Page 215: There are several good sources of information about the temperatures in broodless winter clusters. See Hess (1926), Owens (1971), Fahrenholz et al. (1989), and Stabentheiner et al. (2003). For detailed information about the brood-nest temperatures of honey bee colonies, see Himmer (1927), Owens (1971), Levin and Collison (1990), and Kraus et al. (1998).

Pages 216: Two studies that have shown that the proper behavioral performance of worker bees depends on their being kept at $34.5°$–$35.5°$C ($94°$–$96°$F) throughout their pupal development are Tautz et al. (2003) and Groh et al. (2004). Their findings show that this narrow range of temperatures is typical for the brood nests of honey bee colonies.

Page 217: At present, we lack good data on the thickness of the walls of natural tree cavities occupied by honey bees, and we do not know if wall thickness is assessed by scout bees when they are inspecting prospective nest cavities. Fig. 5.1 shows the width of the walls of one natural nest cavity; the walls vary in width from ca. 8 cm to 13 cm (ca. 3 to 5 in.).

Page 217: The isotherm lines for the hive depicted in Fig. 9.1 are based on readings from 192 thermocouples that were mounted in 12 horizontal rows in the central plane of the hive, between the two centermost frames of comb. The thickness of the hive's wooden walls was 19 mm (0.75 in.).

Page 218: That the flight muscles of insects have some of the highest known levels of metabolic activity is discussed in Bartholomew (1981). The weight-specific rates of power output for a honey bee come from Heinrich (1980) and for an Olympic rower come from Neville (1965). The efficiency of the insect flight apparatus in converting fuel to mechanical power is discussed in Kammer and Heinrich (1978).

Page 219: For more information on how strongly bees can elevate their thorax temperature above ambient temperature, see Esch (1960) and Heinrich (1979b). The need of worker honey bees to maintain thoracic temperature above about 27°C (81°F) is reported by Esch (1976) and Heinrich (1979b), and is explained by Josephson (1981) and Heinrich (1977). The ability of worker honey bees to warm their flight muscles through isometric contractions of these muscles is analyzed by Esch (1964).

Page 219: That honey bees use the same mechanism for warming their flight muscles and heating their nests is shown in Esch (1960). Bujok et al. (2002) and Kleinhenz et al. (2003) describe how a nurse bee can warm a pupa by pressing her thorax to a cell capping or by entering an empty cell adjacent to one containing a pupa and then producing heat with her flight muscles.

Page 221: That sustained brood-nest temperatures over 37°C (99°F) disrupt larval metamorphosis was shown by Himmer (1927). Chadwick (1931) reports that honey-laden combs will start to collapse at 40°C (104°F). The upper lethal temperature for adult honey bees is reported by Allen (1959b) and Free and Spencer-Booth (1962). That honey bees can survive several days at 15°C (59°F) was shown by Free and Spencer-Booth (1960).

Pages 221–222: Vern G. Milum's study of the effect of brood-nest temperature on worker development time is reported in his paper Milum (1930). Anna Maurizio's study that found that a reduction to 30°C (86°F) for just a few hours is sufficient for a successful infection of larvae by chalkbrood fungus is reported in her landmark paper Maurizio (1934).

Page 222: The study that reports an elevated brood-nest temperature (fever) response to infection by the chalkbrood fungus, *Ascosphaera apis*, is reported in Starks et al. (2000).

Page 223: The work that found that bees chilled below about 18°C (64°F) cannot activate their flight muscles is Allen (1959b) and Esch and Bastian (1968). The study that showed worker bees enter a chill coma when cooled below about 10°C (50°F) is that of Free and Spencer-Booth (1960).

Page 224: Conduction is the transfer of heat through a substance that is motionless. Convection is the transfer of heat through a substance by means of motion of the substance; it requires a flow of air or water. Evaporation of water takes heat away, because water absorbs considerable heat when it changes from a liquid to a gas. Thermal radiation is heat transfer that occurs when an object emits electromagnetic radiation, such as infrared radiation.

Page 226: Detailed information about the temperature at which clustering starts is reported in Free and Spencer-Booth (1958), Kronenberg and Heller (1982), and Southwick (1982, 1985). Charles D. Owens's studies on the structure-temperature relations of colonies in winter are reported in Owens (1971). The statement that the volume of a colony's cluster shrinks fivefold as the temperature falls from 14°C to −10°C (57° and 14°F)—where the bees reach their lower limit of cluster contraction—is based on fig. 22 in Owens (1971).

Page 227: The measurement of heat conductance from a 17,000 bee (ca. 2.2 kg/4.9 lb.) winter cluster of honey bees living in a Langstroth hive is reported in Southwick and Mugaas (1971). The similarity in heat conductance for an overwintering colony of bees and birds and mammals is shown in Fig. 5 of this paper.

Pages 227–229: The pioneering studies on the differences in heat conductance between the walls of natural tree cavities and various man-made hives and on the consequences of these differences are reported in Mitchell (2016, 2017).

Page 230: The two cavities have the same dimensions for width, depth, and height: 24 cm × 24 cm × 87 cm (9.5 in. × 9.5 in. × 34 in.). The tree cavity was made by cutting into a sugar maple tree whose diameter at the height of the cavity is 96 cm (37.8 in.), removing slices of wood, and smoothing the inner surfaces with an adze. The walls around the cavity differ in thickness. The thinnest is the removable front wall (with the entrance): 15 cm (6 in.). The thickest is the back wall: 57 cm (22.4 in.). Side walls are 36 cm (14.2 in.) thick. Temperature data are collected by an array of Raspberry Pi microcontroller temperature sensor/recorder units mounted in each box and powered by a solar panel in the tree.

Page 232: The figures for the pooled, resting metabolism of brood and adult bees at 35°C (95°F) come from Allen (1959b), Cahill and Lustick (1976), and Kronenberg and Heller (1982). The figure of 500 watts/kg (230 watts/lb.) for the maximum metabolic rate for honey bee flight muscle comes from Jongbloed and Wiersma (1934), Bastian and Esch (1970), and Heinrich (1980).

Page 232: The study that revealed that a nurse bee will incubate pupal brood by heating her thorax and pressing it to the capping of a brood cell is Bujok et al. (2002).

Page 232: The strong increase in metabolic rate of small groups of bees—from ca. 30 watts/kg (13.5 watts/lb.) at 36°C (97°F) to ca. 300 watts/kg (135 watts/lb.) at 5°C (41°F)—to resist chilling was reported by Cahill and Lustick (1976).

Page 233: The plot of colony metabolic rate as a function of ambient temperature and cluster formation (Fig. 9.7) is from Southwick (1982).

Page 234: The experiment by Lindauer in Italy is described in his 1954 paper on the temperature regulation and water economy of honey bee colonies; see Lindauer (1954).

Page 235: Two early reports that bees start fanning (for cooling) when the temperature inside the brood nest reaches 36°C (97°F) are Hess (1926) and Wohlgemuth (1957).

Page 235–236: The work by Jacob Peters and colleagues is reported in Peters et al. (2017). The reports by Engel H. Hazelhoff of honey bee nest ventilation for temperature control and carbon dioxide removal are Hazelhoff (1941) and Hazelhoff (1954). An experimental study of nest ventilation in response to high levels of carbon dioxide is Seeley (1974).

Page 236–237: The natural experiment in southern California that demonstrated the importance of water for cooling honey bee colonies is described in Chadwick (1931).

Page 237: The evidence that some foraging-age bees specialize in water collection for days, if not weeks, comes from Lindauer (1954), Robinson et al. (1984), Kühnholz and Seeley (1997), and Ostwald et al. (2016). The analysis of how water collectors fuel their flights home is found in Visscher, Crailsheim et al. (1996). The various purposes for which honey bees collect water are described in Park (1949), Nicolson (2009), and Human et al. (2006).

Page 238–239: The observations of water collection in winter by bees in northern Scotland are reported in Chilcott and Seeley (2018). The analysis, using an infrared camera, of the thermoregulation techniques of water collectors in winter is described in Kovac et al. (2010). For a detailed report on thermoregulation by water collectors—also investigated using an infrared camera—see Schmaranzer (2000).

Page 239: To read Derek Mitchell's analysis of the effects of top entrances on the temperature and humidity of colonies living in well-insulated hives, see Mitchell (2017).

Page 240: The studies of the activation and deactivation of the water collectors in a honey bee colony as it experiences a temporary heat stress are reported in Ostwald et al. (2016) and Kühnholz and Seeley (1997).

Page 242: The reports by beekeepers in Australia and South Africa of water storage in the combs are Rayment (1923) and Eksteen and Johannsmeier (1991). The report by O. Wallace Park of clusters of reservoir bees filled with water is Park (1923).

CHAPTER 10. COLONY DEFENSE

Page 243: The Henry David Thoreau quotation is from his essay "Walking"; see Thoreau (1862), p. 665.

Page 243: For an extensive review of the several hundred organisms that will consume part or all of a honey bee colony if they can penetrate the bees' defenses, see Morse and Flottum (1997).

Page 244: The view that honey bees have a long evolutionary history with most of their infectious diseases, and that beekeepers sometimes seriously interfere with the bees' natural mechanisms for controlling the agents of their diseases, is discussed in the authoritative books on honey bee pathology by Bailey (1963, 1981) and Bailey and Ball (1991).

Page 244–245: The history of the transport of colonies of European *Apis mellifera* from Ukraine to the Far East region of Russia is summarized in Crane (1999, pp. 366–367).

Page 245: High virulence is expected to evolve in pathogens and parasites that spread easily between genetically unrelated hosts (horizontally) rather than from parent to offspring (vertically). This is because the horizontal transmission of pathogens/parasites favors strains that reproduce vigorously, and this high reproduction generally harms the host. In contrast, vertical transmission favors pathogens/parasites that reproduce slowly enough to leave the host healthy enough to produce the offspring that the pathogen or parasite needs for its next hosts. For a more detailed explanation of how the evolution of virulence depends on the ecology of the pathogen/parasite, see Ewald (1994, 1995). In honey bees, horizontal transmission (between unrelated colonies) of deformed wing virus (DWV) can occur in two ways: 1) when workers infested with *Varroa* mites carrying DWV drift into uninfested colonies, and 2) when workers rob a colony with *Varroa* mites carrying DWV and then bring these mites home.

Page 245: I have not been able to find reliable information on when *Varroa* mites began to infest colonies of *Apis mellifera* in the Primorsky region of Russia, but a report by Crane (1978) suggests that it occurred soon after peasants from Ukraine, who brought with them colonies of *Apis mellifera*, started settling in the region in 1883 (see Ihor Samokysh, Ukrainians in Zeleny Klyn, *Day Kyiv*, 17 November 2011, https://day.kyiv.ua/en/article/day-after-day/ukrainians-zeleny-klyn; accessed 28 June 2018).

Page 245: The studies of Martin (1998), Martin (2001), and Martin et al. (2012) have shown that the primary cause of the widespread mortality of honey bee (*Apis mellifera*) colonies is coinfections of the bees by *Varroa destructor* and viruses, especially the deformed wing virus.

Page 246–247: The mechanisms of resistance to *Varroa destructor* by honey bees from far-eastern Russia were elucidated by a team of researchers working at the Honey Bee Breeding, Genetics, and Physiology Research Laboratory of the U.S. Department of Agriculture, in Louisiana. Their detailed and multifaceted studies are reviewed in Rinderer, Harris et al. (2010). The detailed report of the controlled field study that demonstrated conclusively the genetically based resistance to *Varroa destructor* of the bees imported from far-eastern Russia is Rinderer, Guzman et al. (2001).

Pages 247–252: For detailed information about the design and results of the experiment that has tracked an isolated, *Varroa*-infested population of honey bees living on the island of Gotland, Sweden, see Fries, Hansen et al. (2003), Fries, Imdorf et al. (2006), and Locke (2016). For information about the mechanisms of *Varroa*-mite resistance of the Gotland bees, see Fries and Bommarco (2007), Locke and Fries (2011), Locke (2015, 2016), and Oddie, Büchler et al. (2018).

Page 252: The amazing skill of *Varroa* mites in climbing onto worker bees while they are foraging on flowers is described in Peck et al. (2016). This paper includes a link to a lovely video that shows the incredible nimbleness of these mites.

Page 253: The methods of bee hunting alluded to here are described in full in Seeley (2016).

Pages 253–254: For more information about the investigation of whether the honey bees living in the Arnot Forest are genetically distinct from the bees living in the apiaries nearest to this forest, see Seeley, Tarpy et al. (2015).

Page 254–257: The genetic study of the old (museum) and new (modern) samples of wild-colony bees, using whole-genome sequencing, is reported in Mikheyev et al. (2015).

Page 257: The reports of other collapses (but not extinctions) of populations of wild or abandoned honey bee colonies after the arrival of *Varroa destructor* come from Texas, Arizona, Louisiana, Sweden, Norway, and France. See, Texas: Pinto et al. (2004); Arizona: Loper et al. (2006); Louisiana: Villa et al. (2008); Sweden: Fries, Imdorf et al. (2006), Norway: Oddie, Dahle et al. (2017), and France: Le Conte et al. (2007) and Kefuss et al. (2016).

Page 257–258: The work of David Peck is not yet published but will appear shortly in a paper titled "Multiple mechanisms of behavioral resistance to an introduced parasite, *Varroa destructor*, in a survivor population of European honey bees."

Page 258: The evidence that simply uncapping and recapping mite-infested cells of worker brood is an effective way for worker bees to reduce *Varroa* mite reproductive success (and does so without killing their worker brood) is reported in Oddie, Büchler et al. (2018).

Page 258–259: The history of the shift from bee hunting to beekeeping is described most fully by Crane (1999). For information on how colonies grouped in apiaries, relative to ones living in widely dispersed nests, experience greater competition for forage, see Crane (1990, p. 194); experience higher risk of having honey stolen during nectar dearths, see Free (1954) and Downs and Ratnieks (2000); experience more problems in reproduction (especially queen loss), see Crane (1990, p. 196); and experience greater risk of acquiring disease, see Free (1958) and Goodwin and Van Eaton (1999).

Pages 260: The figure of 40% or more bees drifting to a non-natal colony comes from Jay (1965, 1966a, 1966b). Studies of the effectiveness of various ways to reduce bee drift among colonies in apiaries are reported in Jay (1965, 1966a, 1966b) and Pfeiffer and Crailsheim (1998).

Pages 260–263: The experimental study that is described here, on the effects of crowding honey bee colonies in apiaries, is Seeley and Smith (2015). A related study, Frey and Rosenkranz (2014), has looked at how differences in colony spacing on the landscape scale (i.e., in regions with low vs. high colony densities) can also strongly affect the rate at which colonies acquire *Varroa destructor* mites from their neighbors in autumn.

Pages 263–268: The experimental study of the importance of small nests and frequent swarming in helping wild colonies survive despite being infested with *Varroa destructor* is Loftus et al. (2016).

Page 264–265: A study that shows that colonies living in small hives swarm more often than ones living in large hives is Simpson and Riedel (1963). A study that shows that approximately 50% of the mites in a colony are on the adult bees and 50% are in the sealed brood is Fuchs (1990).

Page 268: Donzé and Guerin (1997) provide a superb description of where and how immature *Varroa destructor* mites spend their time in the capped brood cells of honey bees. For a comprehensive description of the life cycle of *Varroa destructor*, see Rosenkranz et al. (2010). Among the first to suggest that smaller comb cells might cause higher mortality of immature *Varroa* mites were Erickson et al. (1990) and Medina and Martin (1999).

Page 269: The three previous tests of the idea of giving colonies of European honey bees combs of small cells to reduce their vulnerability to *Varroa* mites are Ellis et al. (2009), Berry et al. (2010), and Coffey et al. (2010). The test of this idea that I made with Sean Griffin is Seeley and Griffin (2011).

Page 270: The study by John McMullan and Mark J. F. Brown of the fill factor of brood combs with large and small cells is McMullan and Brown (2006).

Page 270: The study that shows the benefits to colonies of south-facing entrances is Szabo (1983a).

Page 273: My reference book on the mammals of eastern North America is Whitaker and Hamilton (1998).

Page 273: Several important in vitro studies of the inhibitory effects of propolis on the growth of various bacterial and fungal pathogens of honey bees are Antúnez et al. (2008), Bastos et al. (2008), Bilikova et al. (2013), Lindenfelser (1968), and Wilson et al. (2015).

Pages 274–276: The work from the laboratory of Marla Spivak that is summarized here is reported in two papers: Simone et al. (2009) and Borba et al. (2015). Another study that presents strong evidence of the health benefits of propolis in the nests of honey bees—a strong correlation between the intensity of a colony's propolis collection and its life span and brood viability—is Nicodemo et al. (2014).

CHAPTER 11. DARWINIAN BEEKEEPING

Page 277: The Leslie Bailey quotation is from his book *Honey Bee Pathology*; see Bailey (1981), p. 7.

Page 277: The concept of Darwinian beekeeping is an application of the ideas of Darwinian medicine, as discussed by Williams and Nesse (1991) and Nesse and Williams (1994), to the subject of beekeeping. The fundamental insight of both Darwinian medicine and Darwinian beekeeping is that living systems experience differences between their modern environment (current environment) and the environment that they evolved to live in (environment of evolutionary adaptation) and that these differences cause many problems because living systems are often poorly equipped to deal with the novelties of their modern environments.

Page 278: The experiments that have shown that the unusual annual brood cycle of colonies in the Landes region of southwestern Frances is an adaptive, genetically based trait are reviewed in Louveaux (1973) and Strange et al. (2007). Hatjina et al. (2014) describe a large-scale study that investigated locally adaptive differences in the timing of colony development. They describe an experimental analysis that used five European subspecies of *Apis mellifera*: *carnica*, *ligustica*, *macedonica*, *mellifera*, and *siciliana*.

Page 280: A study that has looked specifically at the effects of crowding colonies in apiaries on the problems of colony reproduction and disease transmission is Seeley and Smith (2015). Brosi et al. (2017) present a model of infectious disease epidemiology that shows the key role of hive/nest density on the spread of infectious diseases.

Page 280: A study that has looked explicitly at the effects of hive size on both the production of honey and the problems of brood diseases is Loftus et al. (2016).

Page 280: An experimental investigation of the effects on the immune systems of worker bees of having a propolis coating on the walls of their colony's nest cavity (or hive) is Borba et al. (2015).

Page 281: The best source of information about the differences in insulation value between walls of tree cavities and standard wooden hives and about the effects of these differences on the energetics of colony thermoregulation is Mitchell (2016).

Page 281: To the best of my knowledge, there are no published studies of the effect of nest entrance height on the riskiness of taking cleansing flights in winter when there is snow on the ground. I have, however, performed a pilot study in which I placed two colonies on the gently sloping roof of the storage shed at my laboratory and placed two more colonies nearby on hive stands at ground level. When ca. 20 cm (8 in.) of snow covered the ground, the bees exiting the higher hives started out approximately 200 cm (ca. 6.5 ft.) above the snow, whereas those exiting the lower hives started out only a few cm (1–2 in.) above the snow. On three sunny days in winter, when the air warmed enough for the bees to make cleansing flights, I counted the number of bees that had flown from and then crashed in the snow outside each of these four hives. The average for each of the two high hives was 8 bees per warm day, while for the low hives it was 113 bees.

Page 281: The evidence that inhibiting drone production boosts a colony's honey production is reported and reviewed in Seeley (2002) and that it slows reproduction by *Varroa destructor* is described in Martin (1998).

Page 281: Several investigators have explored the behavioral rules of worker bees that produce the cell-allocation pattern in honey bee nests. The pioneering studies are reported in Camazine (1991) and Camazine et al. (1990), which show that bees can produce these patterns by following simple rules that do not require the bees to have global knowledge (a "blueprint") of their nest's final layout. Johnson (2009) added a gravity-based rule that biases the movement of the nectar storers toward the top of the nest, to produce the pattern of nectar storage primarily in the top of the nest. The most recent work, by Montovan et al. (2013), adds two more behavioral rules: 1) the consumption of nectar and pollen by the workers is brood-density dependent (strongest near the brood) and 2) the movements by the queen are biased toward the center of the comb by responding to temperature gradients. The richness of mechanisms for building and maintaining the cell allocation pattern in honey bee nests is a strong indicator of the adaptive value of this pattern to the bees.

Pages 282: The evidence that moving a colony overnight to a new location can reduce a colony's weight gain for the following week is reported in Moeller (1975).

Page 282: The study that measured the effects of colony disturbance on colony weight gain (honey production) for the day is Taber (1963).

Page 283: The statement that *Varroa destructor* (and the viruses it spreads) has killed millions of honey bee colonies comes from Martin et al. (2012).

Page 283: The study of the effects of pollen diversity in the diets of nurse bees is Di Pasquale et al. (2013).

Pages 283: A study that compares the effectiveness of various pollen substitutes to that of real pollen is Oliver (2014). See also Randy Oliver, A comparative test of the pollen subs, ScientificBeekeeping.com (n.d.), http://scientificbeekeeping.com/a-comparative-test-of-the-pollen-sub/. Oliver found that "natural pollen still reigns supreme." The study that has demonstrated that workers reared in pollen-stressed colonies become poor foragers is Scofield and Mattila (2015).

Page 283: One study that has documented the high levels of agrochemicals in honey bee colonies in North America is Mullin et al. (2010). Traynor et al. (2016) have reported that colonies incurred higher risks of brood being poisoned by insecticides and fungicides during development when the colonies were performing pollination services (for apples, blueberries, cranberries, citrus fruits, and cucumbers) than when they were producing honey or were sitting in holding yards. Other studies have shown that commercial pollination environments almost invariably expose managed honey bee colonies to higher levels of pesticide residues because drift of pesticides onto noncultivated plants nearby creates a summer-long route of pesticide exposure (Botías et al. 2015).

Page 283: For a review of the known populations of wild honey bee colonies that are surviving without treatments for *Varroa* mites, see Locke (2016). For a paper that examines how treating honey bee colonies with miticides and antibiotics can alter their microbiomes, see Engel et al. (2016).

Page 284: A study by Loftus et al. (2016) compared colonies living in large hives to ones living in small hives in terms of their vulnerability to population explosions of *Varroa* mites and agents of disease that reproduce in cells containing brood, such as chalkbrood (causative agent the fungus *Ascosphaera apis*) and American foulbrood (causative agent the bacterium *Paenibacillus larvae*).

Page 284: The figure of 5 unit weights of honey consumed to produce 1 unit weight of beeswax comes from data of Weiss (1965) as analyzed by Hepburn (1986).

Page 285: The study that found that workers can favor larvae of certain patrilines when they rear queens is that of Moritz, Lattorff et al. (2005). Sometimes these are even subfamilies that are poorly represented among the workers.

Page 286: The construction and use of bait hives is described in papers by Seeley (2012) and Seeley (2017a) and in the book by Magnini (2015).

Page 287: The evidence of the effectiveness of spacing colonies for reducing the spread of parasites and pathogens between colonies is presented in Seeley and Smith (2015) and the references cited in it.

Page 287: The effectiveness of reducing hive size in lowering a colony's disease load is described in Loftus et al. (2016).

Page 287: The evidence that having a thick coating of propolis on the interior wall surfaces of hives is reviewed in Simone-Finstrom and Spivak (2010).

Page 287: The influence of thick insulation of a hive's ceiling and walls on the cost of colony thermoregulation is discussed in Mitchell (2016, 2017).

Page 289: The behaviors of the bees that create the "mite bomb" phenomenon have recently been analyzed (Peck and Seeley, forthcoming). The main mechanism by which colonies around a collapsing colony suddenly acquire higher infestations of *Varroa* mites is robbing of honey from the dying colony by foragers from the healthy colonies nearby. The mites are quite skilled at climbing onto worker bees when they are standing still, filling themselves with a dying colony's honey (Peck and Seeley, forthcoming).

Page 291: The preeminent importance of *Apis mellifera* as a crop pollinator is reported in the detailed study by Kleijn et al. (2015) of crop pollination services worldwide. In this study, the crop production values of different species of bees were calculated based on data from 90 studies conducted on 1,394 crop fields distributed across five continents. In each study, the investigators measured the abundance and density of the bees visiting the flowers of a crop that depends on bee pollination for maximum yield. Twenty different crops were examined. On average, honey bees contributed $2,913 per ha ($1,179 per ac.) to the production of the crops, while the community of "wild bees" (= non-*Apis* bees) contributed $3,251 per ha ($1,316 per ac.). This indicates that honey bees contributed nearly as much value to the production of these crops as all the other kinds of bees combined.

Page 292: For a discussion of the various names that have been used to describe a type of beekeeping that differs from conventional forms of honey bee management, see Phipps (2016). He points out that despite their different names, they all refer to the methods used by beekeepers who want to let their bees live in homes made of natural materials, build their combs freely, swarm as they see fit, and handle diseases on their own.

Page 292: A book by Heaf (2010) provided the first detailed discussion of the health consequences of conventional beekeeping and of the range of attitudes of beekeepers to their bees. Neumann and Blacquière (2016) and Seeley (2017c) were the first to review systematically the various ways in which the practices of conventional beekeeping—such as treatments against disease, artificial selection against propolis usage, and crowding colonies in apiaries—interfere with natural selection for healthy honey bee colonies and to suggest that lasting solutions to the problems of beekeeping are most likely to come by making full use of the power of natural selection. Blacquière and Panziera (2018) make an explicit plea to let natural selection, rather than artificial selection, be the main way forward to acquire bees that have natural resistance to the mite *Varroa destructor* and other environmental threats.

References

Able, K. P., and J. R. Belthoff. 1998. Rapid 'evolution' of migratory behavior in the introduced house finch of eastern North America. *Proceedings of the Royal Society of London B* 265: 2063–2071.

Allen, M. D. 1956. The behaviour of honeybees preparing to swarm. *Animal Behaviour* 4: 14–22.

———. 1958. Shaking of honeybee queens prior to flight. *Nature* 181: 68.

———. 1959a. The occurrence and possible significance of the 'shaking' of honeybee queens by the workers. *Animal Behaviour* 7: 66–69.

———. 1959b. Respiration rates of worker honeybees at different ages and at different temperatures. *Journal of Experimental Biology* 36: 92–101.

Allen, M. D., and E. P. Jeffree. 1956. The influence of stored pollen and of colony size on the brood rearing of honeybees. *Annals of Applied Biology* 44: 649–656.

Allmon, W. D., M. P. Pritts, P. L. Marks, B. P. Epstein, D. A. Bullis, and K. A. Jordan. 2017. *Smith Woods: The Environmental History of an Old Growth Forest Remnant in Central New York State.* Paleontological Research Institution, Ithaca, New York.

Anderson, D. L., and J.W.H. Trueman. 2000. *Varroa jacobsoni* (Acari: Varroidae) is more than one species. *Experimental and Applied Acarology* 24: 165–189.

Antúnez, K., J. Harriet, L. Gende, M. Maggi, M. Eguaras, and P. Zunino. 2008. Efficacy of natural propolis extract in the control of American Foulbrood. *Veterinary Microbiology* 131: 324–331.

Avalos, A., H. Pan, C. Li, J. P. Acevedo-Gonzalez, G. Rendon, C. J. Fields, P. J. Brown, T. Giray, G. E. Robinson, M. E. Hudson, and G. Zhang. 2017. A soft selective sweep during rapid evolution of gentle behaviour in an Africanized honeybee. *Nature Communications* 8: 1550, doi: 10.1038/s41467-017-01800-0.

Avitabile, A. 1978. Brood rearing in honey bee colonies from late autumn to early spring. *Journal of Apicultural Research* 17: 69–73.

Avitabile, A., D. P. Stafstrom, and K. J. Donovan. 1978. Natural nest sites of honeybee colonies in trees in Connecticut, USA. *Journal of Apicultural Research* 17: 222–226.

Badger, M. 2016. *Heather Honey: A Comprehensive Guide.* Beecraft, Stoneleigh, England.

Bailey, L. 1963. *Infectious Diseases of the Honey-Bee.* Land Books, London.

———. 1981. *Honey Bee Pathology.* Academic Press, London.

Bailey, L., and B. Ball. 1991. *Honey Bee Pathology.* 2nd ed. Academic Press, London.

Bartholomew, G. A. 1981. A matter of size: an examination of endothermy in insects and terrestrial vertebrates. In: *Insect Thermoregulation*, B. Heinrich, ed., pp. 45–78. Wiley, New York.

Bastian, J., and H. Esch. 1970. The nervous control of the indirect flight muscles of the honey bee. *Zeitschrift für Vergleichende Physiologie* 67: 307–324.

Bastos, E.M.A.F., M. Simone, D. M. Jorge, A.E.E. Soares, and M. Spivak. 2008. *In vitro* study of the antimicrobial activity of Brazilian propolis against *Paenibacillus larvae*. *Journal of Invertebrate Pathology* 97: 273–281.

Batschelet, E. 1981. *Circular Statistics in Biology.* Academic Press, London and New York.

Beekman, M., and F.L.W. Ratnieks. 2000. Long-range foraging by the honey-bee, *Apis mellifera* L. *Functional Ecology* 14: 490–496.

Beekman, M., D.J.T. Sumpter, N. Seraphides, and F.L.W. Ratnieks. 2004. Comparing foraging behaviour of small and large honey-bee colonies by decoding waggle dances made by foragers. *Functional Ecology* 18: 829–835.

Berlepsch, A. von. 1860. *Die Biene und die Bienenzucht in honigarmen Gegenden.* Heinrichshofen, Mühlhausen, Germany.

Berry, J. A., W. B. Owens, and K. S. Delaplane. 2010. Small-cell comb foundation does not impede Varroa mite population growth in honey bee colonies. *Apidologie* 41: 40–44.

Berry, W. 1987. *Home Economics.* North Point Press, New York.

Bienefeld, K., and F. Pirchner. 1990. Heritabilities for several colony traits in the honeybee (*Apis mellifera carnica*). *Apidologie* 21: 175–183.

Bilikova, K., M. Popova, B. Trusheva, and V. Bankova. 2013. New anti-*Paenibacillus larvae* substances purified from propolis. *Apidologie* 44: 278–285.

Blacquière, T., and D. Panziera. 2018. A plea for use of honey bee's natural resilience in beekeeping. *Bee World* 95: 34–38.

Bloch, G., T. M. Francoy, I. Wachtel, N. Panitz-Cohen, S. Fuchs, and A. Mazar. 2010. Industrial apiculture in the Jordan valley during Biblical times with Anatolian honeybees. *Proceedings of the National Academy of Sciences (USA)* 107: 11240–11244.

Boesch, C., J. Head, and M. M. Robbins. 2009. Complex tool sets for honey extraction among chimpanzees in Loango National Park, Gabon. *Journal of Human Evolution* 56: 560–590.

Borba, R. S., K. K. Klyczek, K. L. Mogen, and M. Spivak. 2015. Seasonal benefits of a natural propolis envelope to honey bee immunity and colony health. *Journal of Experimental Biology* 218: 3689–3699.

Botías, C., A. David, J. Horwood, A. Abdul-Sada, E. Nicholls, E. Hill, and D. Goulson. 2015. Neonicotinoid residues in wildflowers: A potential route of chronic exposure to bees. *Environmental Science and Technology* 49: 12731–12740.

Brosi, B. J., K. Delaplane, M. Boots, and J. C. deRoode. 2017. Ecological and evolutionary approaches to managing honey bee disease. *Nature Ecology and Evolution* 1: 1250–1262.

Brumbach, J. J. 1965. The climate of Connecticut. *Bulletin of the Connecticut Geological and Natural History Survey* 99: 1–215.

Brünnich, K. 1923. A graphic representation of the oviposition of a queen bee. *Bee World* 4: 208–210, 223–224.

Bujok, B., M. Kleinhenz, S. Fuchs, and J. Tautz. 2002. Hot spots in the bee hive. *Naturwissenschaften* 89: 299–301.

Butler, C. G., and J. B. Free. 1952. The behaviour of worker honeybees at the hive entrance. *Behaviour* 4: 263–292.

Cahill, K., and S. Lustick. 1976. Oxygen consumption and thermoregulation in *Apis mellifera* workers and drones. *Comparative Biochemistry and Physiology, Part A: Physiology* 55: 355–357.

Cale, G. H., Jr. 1971. The Hy-Queen story. Pt. 1: Breeding bees for alfalfa pollination. *American Bee Journal* 111: 48–49.

Camazine, S. 1991. Self-organizing pattern formation on the combs of honey bee colonies. *Behavioral Ecology and Sociobiology* 28: 61–76.

Camazine, S., J. Sneyd, M. J. Jenkins, and J. D. Murray. 1990. A mathematical model of self-organized pattern formation on the combs of honeybee colonies. *Journal of Theoretical Biology* 147: 553–571.

Campbell-Stanton, S. C., Z. A. Cheviron, N. Rochette, J. Catchen, J. B. Losos, and S. V. Edwards. 2017. Winter storms drive rapid phenotypic, regulatory, and genomic shifts in the green anole lizard. *Science* 357: 495–498.

Caron, D. M. 1980. Swarm emergence date and cluster location in honeybees. *American Bee Journal* 119: 24–25.

Cervo, R., C. Bruschini, F. Cappa, S. Meconcelli, G. Pieraccini, D. Pradella, and S. Turillazzi. 2014. High *Varroa* mite abundance influences chemical profiles of worker bees and mite-host preferences. *Journal of Experimental Biology* 217: 2998–3001.

Chadwick, P. C. 1931. Ventilation of the hive. *Gleanings in Bee Culture* 59: 356–358.

Chapman, R. F. 1998. *The Insects. Structure and Function.* Cambridge University Press, Cambridge.

Charnov, E. L. 1982. *The Theory of Sex Allocation.* Princeton University Press, Princeton, New Jersey.

Cheshire, F. R. 1888. *Bees and Bee-Keeping; Scientific and Practical.* Vol. 2: *Practical.* L. Upcott Gill, London.

Chilcott, A. B., and T. D. Seeley. 2018. Cold flying foragers: honey bees in Scotland seek water in winter. *American Bee Journal* 158: 75–77.

Cockerell, T.D.A. 1907. A fossil honey-bee. *Entomologist* 40: 227–229.

Coffey, M. F., J. Breen, M.J.F. Brown, and J. B. McMullan. 2010. Brood-cell size has no influence on the population dynamics of *Varroa destructor* mites in the native western honey bee, *Apis mellifera mellifera*. *Apidologie* 41: 522–530.

Collins, A. M. 1986. Quantitative genetics. In: *Bee Genetics and Breeding*, T. E. Rinderer, ed., 283–303. Academic Press, Orlando, Florida.

Columella, L.J.M. 1954. *On Agriculture.* Translated by Edward H. Heffner. Vol. 2, bks. 5–9. Harvard University Press, Cambridge, Massachusetts.

Combs, G. F. 1972. The engorgement of swarming worker honeybees. *Journal of Apicultural Research* 11: 121–128.

Couvillon, M. J., F. C. Riddell Pearce, C. Accleton, K. A. Fensome, S.K.L. Quah, E. L. Taylor, and F.L.W. Ratnieks. 2015. Honey bee foraging distance depends on month and forage type. *Apidologie* 46: 61–70.

Couvillon, M. J., R. Schürch and F.L.W. Ratnieks. 2014. Waggle dance distances as integrative indicators of seasonal foraging challenges. *PLoS ONE* 9 (4): e93495.

Crane, E. 1978. The Varroa mite. *Bee World* 59: 164–167

———. 1990. *Bees and Beekeeping: Science, Practice and World Resources.* Cornell University Press, Ithaca, New York.

———. 1999. *The World History of Beekeeping and Honey Hunting.* Routledge, New York.

Crittenden, A. N. 2011. The importance of honey consumption in human evolution. *Food and Foodways* 219: 257–273.

Dams, M., and L. Dams. 1977. Spanish rock art depicting honey gathering during the Mesolithic. *Nature* 268: 228–230.

Darwin, C. R. 1964. *On the Origin of Species: A Facsimile of the First Edition.* Harvard University Press, Cambridge, Massachusetts.

Dawkins, R. 1982. *The Extended Phenotype.* W. H. Freeman, Oxford.

———. 1989. *The Selfish Gene.* New ed. Oxford University Press, Oxford.

De Jong, D. 1997. Mites: *Varroa* and other parasites of brood. In: *Honey Bee Pests, Predators, and Diseases*, R. A. Morse and K. Flottum, eds., 279–327. A. I. Root, Medina, Ohio.

De Jong, D., R. A. Morse, and G. C. Eickwort. 1982. Mite pests of honey bees. *Annual Review of Entomology* 27: 229–252.

De la Rúa, P., R. Jaffé, R. Dall'Olio, I. Muñoz, and J. Serrano. 2009. Biodiversity, conservation and current threats to European honeybees. *Apidologie* 40: 263–284.

DeMello, M. 2012. *Animals and Society: An Introduction to Human-Animal Studies.* Columbia University Press, New York.

Dethier, B. E., and A. Boyd Pack. 1963. *The Climate of Ithaca, New York.* New York State College of Agriculture, Ithaca, New York.

Di Pasquale, G., M. Salignon, Y. Le Conte, L. P. Belzunces, A. Decourtye, A. Kretzschmar, S. Suchail,

J.-L. Brunet, and C. Alaux. 2013. Influence of pollen nutrition on honey bee health: Do pollen quality and diversity matter? *PLoS ONE* 8: e72016.

Dixon, L. 2015. *A Time There Was: A Story of Rock Art, Bees and Bushmen*. Northern Bee Books, Hebden Bridge, England.

Donzé, G., and P. M. Guerin. 1997. Time-activity budgets and space structuring by the different life stages of *Varroa jacobsoni* in capped brood of the honey bee, *Apis mellifera*. *Journal of Insect Behavior* 10: 371–393.

Doolittle, G. M. 1889. *Scientific Queen-Rearing as Practically Applied; Being A Method by Which the Best of Queen-Bees Are Reared in Perfect Accord with Nature's Ways. For the Amateur and Veteran in Bee-Keeping*. Newman and Son, Chicago.

Downs, S. G., and F. L. W. Ratnieks. 2000. Adaptive shifts in honey bee (*Apis mellifera* L.) guarding behavior support predictions of the acceptance threshold model. *Behavioral Ecology* 11: 233–240.

Dreller, C., and D. R. Tarpy. 2000. Perception of the pollen need by foragers in a honeybee colony. *Animal Behaviour* 59: 91–96.

Dudley, P. 1720. An account of a method lately found out in New-England, for discovering where the bees hive in the woods, in order to get their honey. *Philosophical Transactions of the Royal Society of London* 31: 148–150.

Eckert, J. E. 1933. The flight range of the honey bee. *Journal of Agricultural Research* 47: 257–285.

———. 1942. The pollen required by a colony of honeybees. *Journal of Economic Entomology* 35: 309–311.

Edgell, G. H. 1949. *The Bee Hunter*. Harvard University Press, Cambridge, Massachusetts.

Eksteen, J. K., and M. F. Johannsmeier. 1991. Oor bye en byeplante van die Noord-Kaap. *South African Bee Journal* 63: 128–136.

Ellis, A. M., G. W. Hayes, and J. D. Ellis. 2009. The efficacy of small cell foundation as a varroa mite (*Varroa destructor*) control. *Experimental and Applied Acarology* 47: 311–316.

Engel, M. S. 1998. Fossil honey bees and evolution in the genus *Apis* (Hymenoptera: Apidae). *Apidologie* 29: 265–281.

Engel, P., W. K. Kwong, Q. McFrederick, K. E. Anderson, S. M Barribeau, J. A. Chandler, R. S. Cornman, J. Dainat, J. R. de Miranda, V. Doublet, O. Emery, J. D. Evans, and 21 more authors. 2016. The bee microbiome: impact on bee health and model for evolution and ecology of host-microbe interactions. *mBio* 7: e02164-15.

Erickson, E. H., D. A. Lusby, G. D. Hoffman, and E. W. Lusby. 1990. On the size of cells: Speculations on foundation as a colony management tool. *Gleanings in Bee Culture* 118: 98–101, 173–174.

Esch, H. 1960. Über die Körpertemperaturen und den Wärmehaushalt von *Apis mellifica*. *Zeitschrift für Vergleichende Physiologie* 43: 305–335.

———. 1964. Über den Zusammenhang zwischen Temperatur, Aktionspotentialen, und Thoraxbewegungen bei der Honigbiene (*Apis mellifica* L.). *Zeitschrift für Vergleichende Physiologie* 48: 547–551.

———. 1976. Body temperature and flight performance of honey bees in a servo-mechanically controlled wind tunnel. *Journal of Comparative Physiology* 109: 265–277.

Esch, H., and J. A. Bastian. 1968. Mechanical and electrical activity in the indirect flight muscles of the honey bee. *Zeitschrift für Vergleichende Physiologie* 58: 429–440.

Ewald, P. W. 1994. *Evolution of Infectious Disease*. Oxford University Press, New York.

———. 1995. The evolution of virulence: A unifying link between parasitology and ecology. *Journal of Parasitology* 81: 659–669.

Fahrenholz, L., I. Lamprecht, and B. Schricker. 1989. Thermal investigations of a honey bee colony: Thermoregulation of the hive during summer and winter and heat production of members of different bee castes. *Journal of Comparative Physiology B* 159: 551–560.

Farrar, C. L. 1936. Influence of pollen reserves on the surviving population of overwintered colonies. *American Bee Journal* 76: 452–454.

Fell, R. D., J. T. Ambrose, D. M. Burgett, D. De Jong, R. A. Morse, and T. D. Seeley. 1977. The seasonal cycle of swarming in honeybees. *Journal of Apicultural Research* 16: 170–173.

Flottum, K. 2014. *The Backyard Beekeeper*. Quarry Books, Beverly, Massachusetts.

Free, J. B. 1954. The behaviour of robber honeybees. *Behaviour* 7: 233–240.

———. 1958. The drifting of honey-bees. *Journal of Agricultural Science* 51: 294–306.

———. 1967. The production of drone comb by honeybee colonies. *Journal of Apicultural Research* 6: 29–36.

———. 1968. Engorging of honey by worker honeybees when their colony is smoked. *Journal of Apicultural Research* 7: 135–138.

Free, J. B., and Y. Spencer-Booth. 1958. Observations on the temperature regulation and food consumption of honeybees (*Apis mellifera*). *Journal of Experimental Biology* 35: 930–937.

———. 1960. Chill-coma and cold death temperatures of *Apis mellifera*. *Entomologia Experimentalis et Applicata* 3: 222–230.

———. 1962. The upper lethal temperatures of honeybees. *Entomologia Experimentalis et Applicata* 5: 249–254.

Free, J. B., and I. H. Williams. 1975. Factors determining the rearing and rejection of drones by the honeybee colony. *Animal Behaviour* 23: 650–675.

Frey, E., and P. Rosenkranz. 2014. Autumn invasion rates of *Varroa destructor* (Mesostigmata: Varroidae) into honey bee (Hymenoptera: Apidae) colonies and the resulting increase in mite populations. *Journal of Economic Entomology* 107: 508–515.

Fries, I., and R. Bommarco. 2007. Possible host-parasite adaptations in honey bees infested by *Varroa destructor* mites. *Apidologie* 38: 525–533.

Fries, I., and S. Camazine. 2001. Implications of horizontal and vertical pathogen transmission for honey bee epidemiology. *Apidologie* 32: 199–214.

Fries, I., H. Hansen, A. Imdorf, and P. Rosenkranz. 2003. Swarming in honey bees (*Apis mellifera*) and *Varroa destructor* population development in Sweden. *Apidologie* 34: 389–397.

Fries, I., A. Imdorf, and P. Rosenkranz. 2006. Survival of mite infested (*Varroa destructor*) honey bee (*Apis mellifera*) colonies in a Nordic climate. *Apidologie* 37: 564–570.

Frost, R. 1969. *The Poetry of Robert Frost*. Henry Holt, New York.

Fuchs, S. 1990. Preference for drone brood cells by *Varroa jacobsoni* Oud in colonies of *Apis mellifera carnica*. *Apidologie* 21:193–199.

Fukuda, H., K. Moriya, and K. Sekiguchi. 1969. The weight of crop contents in foraging honeybee workers. *Annotationes Zoologicae Japonenses* 42: 80–90.

Gallone, B., J. Steensels, T. Prahl, L. Soriaga, and 15 more authors. 2016. Domestication and divergence of *Saccharomyces cerevisiae* beer yeasts. *Cell* 166: 1397–1410.

Galton, D. 1971. *Survey of a Thousand Years of Beekeeping in Russia*. Bee Research Association, London.

Garis Davies, N. de. 1944. *The Tomb of Rekh-mi-Rē' at Thebes*. Metropolitan Museum of Art, New York.

Gary, N. E. 1962. Chemical mating attractants in the queen honey bee. *Science* 136: 773–774.

———. 1971. Magnetic retrieval of ferrous labels in a capture-recapture system for honey bees and other insects. *Journal of Economic Entomology* 64: 961–965.

Gary, N. E., and R. A. Morse. 1962. The events following queen cell construction in honeybee colonies. *Journal of Apicultural Research* 1: 3–5.

Gary, N. E., P. C. Witherell, and K. Lorenzen. 1978. The distribution and foraging activities of common Italian and "Hy-Queen" honey bees during alfalfa pollination. *Environmental Entomology* 7: 228–232.

Getz, W. M., D. Brückner, and T. R. Parisian. 1982. Kin structure and the swarming behavior of the honey bee *Apis mellifera*. *Behavioral Ecology and Sociobiology* 10: 265–270.

Gibbons, A. 2017. Oldest members of our species discovered in Morocco. *Science* 356: 993–994.

Gibbons, E. 1962. *Stalking the Wild Asparagus*. David McKay Co., New York.

Gilley, D. G., and D. R. Tarpy. 2005. Three mechanisms of queen elimination in swarming honey bee colonies. *Apidologie* 36: 461–474.

Goodwin, M., and C. Van Eaton. 1999. *Elimination of American Foulbrood Without the Use of Drugs*. National Beekeepers' Association of New Zealand, Napier.

Goulson, D. 2010. *Bumblebees: Behaviour, Ecology, and Conservation*. Oxford University Press, London.

Gowlett, J.A.J. 2016. The discovery of fire by humans: A long and convoluted process. *Philosophical Transactions of the Royal Society B* 371: 20150164, doi:10.1098/rstb.2015.0164.

Grant, P. R., and B. R. Grant. 2014. *40 Years of Evolution: Darwin's Finches on Daphne Major Island*. Princeton University Press, Princeton, New Jersey.

Grimaldi, D., and M. S. Engel. 2005. *Evolution of the Insects*. Cambridge University Press, New York.

Groh, C., J. Tautz, and W. Rössler. 2004. Synaptic organization in the adult honey bee brain is influenced by brood-temperature control during pupal development. *Proceedings of the National Academy of Sciences (USA)* 101: 4268–4273.

Guy, R. D. 1971. A commercial beekeeper's approach to the use of primitive hives. *Bee World* 52: 18–24.

Hamilton, L. S., and M. M. Fischer. 1970. *The Arnot Forest: A Natural Resources Research and Teaching Area*. Extension Bulletin 1207. New York State College of Agriculture, Cornell University, Ithaca, New York. https://cpb-us-e1.wpmucdn.com/blogs.cornell.edu/dist/3/6154/files/2015/07/history-of-the-Arnot-1970-2agzopw.pdf (accessed 9 January 2019).

Hammann, E. 1957. Wer hat die Initiative bei den Ausflügen der Jungkönigin, die Königin oder die Arbeitsbienen? *Insectes Sociaux* 4: 91–106.

Han, F., A. Wallberg, and M. T. Webster. 2012. From where did the Western honeybee (Apis mellifera) originate? *Ecology and Evolution* 2: 1949–1957.

Harbo, J. R. 1986. Propagation and instrumental insemination. In: *Bee Genetics and Breeding*, T. E. Rinderer, ed., 361–389. Academic Press, Orlando, Florida.

Harpur, B. A., S. Minaei, C. F. Kent, and A. Zayed. 2012. Management increases genetic diversity of honey bees via admixture. *Molecular Ecology* 21: 4414–4421.

Hatjina, F., C. Costa, R. Büchler, A. Uzunov, M. Drazic, J. Filipi, L. Charistos, L. Ruottinen, S. Andonov, M. D. Meixner, M. Bienkowska, G. Dariusz, and 13 more authors. 2014. Population dynamics of European honey bee genotypes under different environmental conditions. *Journal of Apicultural Research* 53: 233–247.

Haydak, M. H. 1935. Brood rearing by honeybees confined to a pure carbohydrate diet. *Journal of Economic Entomology* 28: 657–660.

Hazelhoff, E. H. 1941. De luchtverversching van een bijenkast gedurende den zomer. *Maandscrift voor Bijenteelt* 44: 10–14, 27–30, 45–48, 65–68.

———. 1954. Ventilation in a bee-hive during summer. *Physiologia Comparata et Oecologia* 3: 343–364.

Heaf, D. 2010. *The Bee-Friendly Beekeeper: A Sustainable Approach*. Northern Bee Books, Mytholmroyd, England.

Heinrich, B. 1977. Why have some animals evolved to regulate a high body temperature? *American Naturalist* 111: 623–640.

———. 1979a. *Bumblebee Economics*. Harvard University Press, Cambridge, Massachusetts.

———. 1979b. Thermoregulation of African and European honeybees during foraging, attack, and hive exits and returns. *Journal of Experimental Biology* 80: 217–229.

———. 1980. Mechanisms of body-temperature regulation in honeybees, *Apis mellifera*. *Journal of Experimental Biology* 85: 73–87.

Henderson, C. E. 1992. Variability in the size of emerging drones and of drone and worker eggs in honey bee (*Apis mellifera* L.) colonies. *Journal of Apicultural Research* 31: 114–118.

Hepburn, H. R. 1986. *Honeybees and Wax*. Springer Verlag, Berlin.

Hernández-Pacheco, E. 1924. *Las Pinturas Prehistóricas de Las Cuevas de la Araña (Valencia)*. Museo Nacional de Ciencas Naturales, Madrid.

Hess, W. R. 1926. Die Temperaturregulierung im Bienenvolk. *Zeitschrift für Vergleichende Physiologie* 4: 465–487.

Himmer, A. 1927. Ein Beitrag zur Kenntnis des Wärmeshaushalts im Nestbau sozialer Hautflügler. *Zeitschrift für Vergleichende Physiologie* 5: 375–389.

Hinson, E. M., M. Duncan, J. Lim, J. Arundel, and B. P. Oldroyd. 2015. The density of feral honey bee (*Apis mellifera*) colonies in South East Australia is greater in undisturbed than in disturbed habitats. *Apidologie* 46: 403–413.

Hirschfelder, H. 1951. Quantitative Untersuchungen zum Polleneintragen der Bienenvölker. *Zeitschrift für Bienenforschung* 1: 67–77.

Hood, Thomas. 1873. *The Complete Poetical Works*. Vol. 1. G. P. Putnam's Sons, New York.

Horstmann, H.-J. 1965. Einige biochemischen Überlegungen zur Bildung von Bienenwachs aus Zucker. *Zeitschrift für Bienenforschung* 8: 125–128.

Hublin, J.-J., A. Ben-Ncer, S. E. Bailey, S. E. Freidline, S. Neubauer, M. M. Skinner, I. Bergmann, A. Le Cabec, S. Benazzi, K. Harvati, and P. Gunz. 2017. New fossils from Jebel Irhoud, Morocco and the pan-African origin of *Homo sapiens*. *Nature* 546: 289–292.

Human, H., S. W. Nicolson, and V. Dietemann. 2006. Do honeybees, *Apis mellifera scutellata*, regulate humidity in their nest? *Naturwissenschaften* 93: 397–401.

Ichikawa, M. 1981. Ecological and sociological importance of honey to the Mbuti net hunters, Eastern Zaire. *African Study Monographs* 1: 55–68.

Ilyasov, R. A., M. N. Kosarev, A. Neal, and F. G. Yumaguzhin. 2015. Burzyan wild-hive honeybee *A.m. mellifera* in South Ural. *Bee World* 92: 7–11.

Jacobsen, R. 2008. *Fruitless Fall: The Collapse of the Honey Bee and the Coming Agricultural Crisis*. Bloomsbury, New York.

Jaffé, R., V. Dietemann, M. H. Allsopp, C. Costa, R. M. Crewe, R. Dall'Olio, P. de la Rúa, M.A.A. El-Niweiri, I. Fries, N. Kezic, M. S. Meusel, R. J. Paxton, and 3 more authors. 2009. Estimating the density of honeybee colonies across their natural range to fill the gap in pollinator decline censuses. *Conservation Biology* 24: 583–593.

Jay, S. C. 1965. Drifting of honeybees in commercial apiaries. Pt. 1: Effect of various environmental factors. *Journal of Apicultural Research* 4: 167–175.

———. 1966a. Drifting of honeybees in commercial apiaries. Pt. 2: Effect of various factors when hives are arranged in rows. *Journal of Apicultural Research* 5: 103–112.

———. 1966b. Drifting of honeybees in commercial apiaries. Pt. 3: Effect of apiary layout. *Journal of Apicultural Research* 5: 137–148.

Jaycox, E. R., and S. G. Parise. 1980. Homesite selection by Italian honey bee swarms, *Apis mellifera ligustica* (Hymenoptera: Apidae). *Journal of the Kansas Entomological Society* 53: 171–178.

———. 1981. Homesite selection by swarms of black-bodied honey bees, *Apis mellifera caucasica* and *A. m carnica* (Hymenoptera: Apidae). *Journal of the Kansas Entomological Society* 54: 697–703.

Jeanne, R. L. 1979. A latitudinal gradient of rates of ant predation. *Ecology* 60: 1211–1224.

Jeffree, E. P. 1955. Observations on the decline and growth of honey bee colonies. *Journal of Economic Entomology* 48: 723–726.

———. 1956. Winter brood and pollen in honey bee colonies. *Insectes Sociaux* 3: 417–422.

Johnson, B. R. 2009. Pattern formation on the combs of honeybees: Increasing fitness by coupling self-organization with templates. *Proceedings of the Royal Society of London B* 276: 255–261.

Jones, J. C., M. R. Myerscough, S. Graham, and B. P. Oldroyd. 2004. Honey bee nest thermoregulation: Diversity promotes stability. *Science* 305: 402–404.

Jongbloed, J., and C.A.G. Wiersma. 1934. Der Stoffwechsel der Honigbiene während des Fliegens. *Zeitschrift für Vergleichende Physiologie* 21: 519–533.

Josephson, R. K. 1981. Temperature and the mechanical performance of insect muscle. In: *Insect Thermoregulation*, B. Heinrich, ed., pp. 20–44. Wiley, New York.

Kammen, C. 1985. *The Peopling of Tompkins County: A Social History*. Heart of the Lakes Publishing, Interlaken, New York.

Kammer, A. E., and B. Heinrich. 1978. Insect flight metabolism. *Advances in Insect Physiology* 13: 133–228.

Kefuss, J. A. 1978. Influence of photoperiod on the behaviour and brood-rearing activities of honeybees in a flight room. *Journal of Apicultural Research* 17: 137–151.

Kefuss, J., J. Vanpoucke, M. Bolt, and C. Kefuss. 2016. Selection for resistance to *Varroa destructor* under commercial beekeeping conditions. *Journal of Apicultural Research* 54: 563–576.

Kleijn, D., R. Winfree, I. Bartomeus, L. G. Carvalheiro, and 56 more authors. 2015. Delivery of crop pollination services is an insufficient argument for wild pollinator conservation. *Nature Communications* 6: 7414, doi:10.1038/ncomms8414.

Klein, B. A., A. Klein, M. K. Wray, U. G. Mueller, and T. D. Seeley. 2010. Sleep deprivation impairs precision of waggle dance signaling in honey bees. *Proceedings of the National Academy of Sciences (USA)* 107: 22705–22709.

Klein, B. A., K. M. Olzsowy, A. Klein, K. M. Saunders, and T. D. Seeley. 2008. Caste-dependent sleep of worker honey bees. *Journal of Experimental Biology* 211: 3028–3040.

Klein, B. A., M. Stiegler, A. Klein, and J. Tautz. 2014. Mapping sleeping bees within their nest: Spatial and temporal analysis of worker honey bee sleep. *PLoS ONE* 9 (7): e102316, doi:10.1371/journal.pone.0102316.

Kleinhenz, M., B. Bujok, S. Fuchs, and J. Tautz. 2003. Hot bees in empty broodnest cells: Heating from within. *Journal of Experimental Biology* 206: 4217–4231.

Knaffl, H. 1953. Über die Flugweite und Entfernungsmeldung der Bienen. *Zeitschrift für Bienenforschung* 2: 131–140.

Koch, H. G. 1967. Der Jahresgang der Nektartracht von Bienenvölkern als Ausdruck der Witterungs-singularitäten und Trachtverhältnisse. *Zeitschrift für Angewandte Meteorologie* 5: 206–216.

Koeniger, G., N. Koeniger, J. Ellis, and L. Connor. 2014. *Mating Biology of Honey Bees (Apis mellifera)*. Wicwas Press, Kalamazoo, Michigan.

Koeniger, G., N. Koeniger, and F.-T. Tiesler. 2014. *Paarungsbiologie und Paarungskontrolle bei der Honigbiene*. Buchshausen Druck und Verlagshaus, Herten, Germany.

Kohl, P. L., and B. Rutschmann. 2018. The neglected bee trees: European beech forests as a home for feral honey bee colonies. *PeerJ* 6: e4602, doi:10.7717/peerj.4602.

Kovac, H., A. Stabentheiner, and S. Schmaranzer. 2010. Thermoregulation of water foraging honeybees—balancing of endothermic activity with radiative heat gain and functional requirements. *Journal of Insect Physiology* 56: 1834–1845.

Kraus, B., and R. E. Page, Jr. 1995. Effect of *Varroa jacobsoni* (Mesostigmata: Varroidae) on feral *Apis mellifera* (Hymenoptera: Apidae) in California. *Environmental Entomology* 24: 1473–1480.

Kraus, B., H.H.W. Velthuis, and S. Tingek. 1998. Temperature profiles of the brood nests of *Apis cerana* and *Apis mellifera* colonies and their relation to varroosis. *Journal of Apicultural Research* 37: 175–181.

Kritsky, G. 1991. Lessons from history: The spread of the honey bee in North America. *American Bee Journal* 131: 367–370.

———. 2010. *The Quest for the Perfect Hive: A History of Innovation in Bee Culture*. Oxford University Press, New York.

———. 2015. *The Tears of Re: Beekeeping in Ancient Egypt*. Oxford University Press, New York.

Kronenberg, F. C., and H. C. Heller. 1982. Colonial thermoregulation in honey bees (*Apis mellifera*). *Journal of Comparative Physiology* 148: 65–76.

Kühnholz, S., and T. D. Seeley. 1997. The control of water collection in honey bee colonies. *Behavioral Ecology and Sociobiology* 41: 407–422.

Kurlansky, M. 2014. Inside the milk machine: How modern dairy works. *Modern Farmer*. https://modernfarmer.com/2014/03/real-talk-milk/ (accessed 15 December 2017).

Laidlaw, H. H. 1944. Artificial insemination of the queen bee (*Apis mellifera* L.): Morphological basis and results. *Journal of Morphology* 74: 429–465.

Langstroth, L. L. 1853. *Langstroth on the Hive and the Honey-Bee: A Bee Keeper's Manual*. Hopkins, Bridgman, and Co., Northampton, Massachusetts.

Latham, E. C. 1969. *The Poetry of Robert Frost*. Henry Holt, New York.

Le Conte, Y., G. de Vaublanc, D. Crauser, F. Jeanne, J.-C. Rousselle, and J. J. Bécard. 2007. Honey bee colonies that have survived *Varroa destructor*. *Apidologie* 38: 566–572.

Lee, P. C., and M. L. Winston. 1985. The effect of swarm size and date of issue on comb construction in newly founded colonies of honeybees (*Apis mellifera* L.). *Canadian Journal of Zoology* 63: 524–527.

Levin, C. G., and C. H. Collison. 1990. Broodnest temperature differences and their possible effect on drone brood production and distribution in honeybee colonies. *Journal of Apicultural Research* 29: 35–45.

Levin, M. D. 1961. Distribution of foragers from honey bee colonies placed in the middle of a large field of alfalfa. *Journal of Economic Entomology* 54: 431–434.

Levin, M. D., G. E. Bohart, and W. P. Nye. 1960. Distance from the apiary as a factor in alfalfa pollination. *Journal of Economic Entomology* 53: 56–60.

Lindauer, M. 1954. Temperaturregulierung und Wasserhaushalt im Bienenstaat. *Zeitschrift für Vergleichende Physiologie* 36: 391–432.

———. 1955. Schwarmbienen auf Wohnungssuche. *Zeitschrift für Vergleichende Physiologie* 37: 263–324.

Lindenfelser, L. A. 1968. In vivo activity of propolis against *Bacillus larvae*. *Journal of Invertebrate Pathology* 12: 129–131.

Locke, B. 2015. Inheritance of reduced *Varroa* mite reproductive success in reciprocal crosses of mite-resistant and mite-susceptible honey bees (*Apis mellifera*). *Apidologie* 47: 583–588.

———. 2016. Natural *Varroa* mite-surviving *Apis mellifera* honeybee populations. *Apidologie* 47: 467–482.

Locke, B., and I. Fries. 2011. Characteristics of honey bee colonies (*Apis mellifera*) in Sweden surviving *Varroa destructor* infestation. *Apidologie* 42: 533–542.

Loftus, J. C., M. L. Smith, and T. D. Seeley. 2016. How honey bee colonies survive in the wild: Testing the importance of small nests and frequent swarming. *PLoS ONE* 11 (3): e0150362, doi:10.1371/journal.pone.0150362.

Loper, G. M. 1995. A documented loss of feral bees due to mite infestations in S. Arizona. *American Bee Journal* 135: 823–824.

———. 1997. Over-winter losses of feral honey bee colonies in southern Arizona, 1992–1997. *American Bee Journal* 137: 446.

———. 2002. Nesting sites, characterization and longevity of feral honey bee colonies in the Sonoran

Desert of Arizona: 1991–2000. In: *Proceedings of the 2nd International Conference on Africanized Honey Bees and Bee Mites*, E. H. Erickson, Jr., Robert E. Page, Jr., and A. A. Hanna, eds., 86–96. A. I. Root, Medina, Ohio.

Loper, G. M., D. Sammataro, J. Finley, and J. Cole. 2006. Feral honey bees in southern Arizona 10 years after *Varroa* infestation. *American Bee Journal* 134: 521–524.

Louveaux, J. 1958. Recherches sur la récolte du pollen par les abeilles (*Apis mellifica* L.). *Annales de l'Abeille* 1: 113–188, 197–221.

———. 1973. The acclimatization of bees to a heather region. *Bee World* 54: 105–111.

Mackensen, O., and W. P. Nye. 1966. Selecting and breeding honeybees for collecting alfalfa pollen. *Journal of Apicultural Research* 5: 79–86.

Magnini, R. M. 2015. *Swarm Traps: Principles and Design*. Sweet Clover, Scotch Lake, Nova Scotia.

Manley, R.O.B. 1985. *Honey Farming*. Northern Bee Books, Hebden Bridge, England.

Marchand, C. 1967. Préparons le piégeage des essaims. *L'Abeille de France* 46 (490): 59–61.

Marlowe, F. W., J. C. Berbesque, B. Wood, A. Crittenden, C. Porter, and A. Mabulla. 2014. Honey, Hadza, hunter-gatherers, and human evolution. *Journal of Human Evolution* 71: 119–128.

Martin, H., and M. Lindauer. 1966. Sinnesphysiologische Leistungen beim Wabenbau der Honigbiene. *Zeitschrift für Vergleichende Physiologie* 53: 372–404.

Martin, P. 1963. Die Steuerung der Volksteilung beim Schwärmen der Bienen. Zugleich ein Beitrage zum Problem der Wanderschwärme. *Insectes Sociaux* 10: 13–42.

Martin, S. 1998. A population model for the ectoparasitic mite *Varroa jacobsoni* in honey bee (*Apis mellifera*) colonies. *Ecological Modelling* 109: 267–281.

Martin, S. J. 2001. The role of *Varroa* and viral pathogens in the collapse of honeybee colonies: A modelling approach. *Journal of Applied Ecology* 38: 1082–1093.

Martin, S. J., A. C. Highfield, L. Brettell, E. M. Villalobos, G. E. Budge, M. Powell, S. Nikaido, and D. C. Schroeder. 2012. Global honey bee viral landscape altered by a parasitic mite. *Science* 336: 1304–1306.

Mason, P. A. 2016. *American Bee Books: An Annotated Bibliography of Books on Bees and Beekeeping 1492–2010*. Club of Odd Volumes, Boston.

Mattila, H. R., and T. D. Seeley. 2007. Genetic diversity in honey bee colonies enhances productivity and fitness. *Science* 317: 362–364.

———. 2010. Promiscuous honeybee queens generate colonies with a critical minority of waggle-dancing foragers. *Behavioral Ecology and Sociobiology* 64: 875–889.

Mattila, H. R., K. M. Burke, and T. D. Seeley. 2008. Genetic diversity within honeybee colonies increases signal production by waggle-dancing foragers. *Proceedings of the Royal Society of London B* 275: 809–816.

Maurizio, A. 1934. Über die Kalkbrut (Perisystis-Mykose) der Bienen. *Archiv für Bienenkunde* 15: 165–193.

Mazar, A., and N. Panitz-Cohen. 2007. It is the land of honey: Beekeeping at Tel Rehov. *Near Eastern Archaeology* 70: 202–219.

McLellan, A. R. 1977. Honeybee colony weight as an index of honey production and nectar flow: A critical evaluation. *Journal of Applied Ecology* 14: 401–408.

McMullan, J. B., and M.J.F. Brown. 2006. The influence of small-cell brood combs on the morphometry of honeybees (*Apis mellifera*). *Apidologie* 37: 665–672.

Medina, L. M., and S. J. Martin. 1999. A comparative study of *Varroa jacobsoni* reproduction in worker cells of honey bees (*Apis mellifera*) in England and Africanized bees in Yucatan, Mexico. *Experimental and Applied Acarology* 23: 659–667.

Meyer, W. 1954. Die "Kittharzbienen" und ihre Tätigkeiten. *Zeitschrift für Bienenforschung* 2: 185–200.

———. 1956. "Propolis bees" and their activities. *Bee World* 37: 25–36.

Michener, C. D. 1974. *The Social Behavior of the Bees.* Harvard University Press, Cambridge, Massachusetts.

———. 2000. *The Bees of the World.* Johns Hopkins University Press, Baltimore, Maryland.

Mikheyev, A. S., M.M.Y. Tin, J. Arora, and T. D. Seeley. 2015. Museum samples reveal rapid evolution by wild honey bees exposed to a novel parasite. *Nature Communications* 6: 7991, doi:10.1038/ncomms8991.

Milum, V. G. 1930. Variations in time of development of the honey bee. *Journal of Economic Entomology* 23: 441–446.

———. 1956. An analysis of twenty years of honey bee colony weight changes. *Journal of Economic Entomology* 49: 735–738.

Mitchell, D. 2016. Ratios of colony mass to thermal conductance of tree and man-made nest enclosures of *Apis mellifera*: Implications for survival, clustering, humidity regulation and *Varroa destructor*. *International Journal of Biometeorology* 60: 629–638.

———. 2017. Honey bee engineering: Top ventilation and top entrances. *American Bee Journal* 157: 887–889.

Mitchener, A. V. 1948. The swarming season for honey bees in Manitoba. *Journal of Economic Entomology* 41: 646.

———. 1955. Manitoba nectar flows 1924–1954, with particular reference to 1947–1954. *Journal of Economic Entomology* 48: 514–518.

Moeller, F. E. 1975. Effect of moving honeybee colonies on their subsequent production and consumption of honey. *Journal of Apicultural Research* 14: 127–130.

Montovan, K. J., N. Karst, L. E. Jones, and T. D. Seeley. 2013. Local behavioral rules sustain the cell allocation pattern in the combs of honey bee colonies (*Apis mellifera*). *Journal of Theoretical Biology* 336: 75–86.

Moritz, R.F.A., F. B. Kraus, P. Kryger, and R. M. Crewe. 2007. The size of wild honeybee populations (*Apis mellifera*) and its implications for the conservation of honeybees. *Journal of Insect Conservation* 11: 391–397.

Moritz, R.F.A., H.M.G. Lattorff, P. Neumann, F. B. Kraus, S. E. Radloff, and H. R. Hepburn. 2005. Rare royal families in honeybees, *Apis mellifera*. *Naturwissenschaften* 92: 488–491.

Morse, R. A., S. Camazine, M. Ferracane, P. Minacci, R. Nowogrodzki, F.L.W. Ratnieks, J. Spielholz, and B. A. Underwood. 1990. The population density of feral colonies of honey bees (Hymenoptera: Apidae) in a city in upstate New York. *Journal of Economic Entomology* 83: 81–83.

Morse, R. A., and K. Flottum. 1997. *Honey Bee Pests, Predators, and Diseases.* 3rd ed. A. I. Root Company, Medina, Ohio.

Moulton, G. E. 2002. *The Definitive Journals of Lewis and Clark.* University of Nebraska Press, Lincoln, Nebraska.

Mullin, C. A., M. Frazier, J. L. Frazier, S. Ashcraft, R. Simonds, D. vanEnglesdorp, and J. S. Pettis. 2010. High levels of miticides and agrochemicals in North American apiaries: Implications for honey bee health. *PLoS ONE* 5: e9754, doi.org/10.1371/journal.pone.0009754.

Münster, S. 1628. *Cosmographia.* Heinrich Petri, Basel.

Munz, T. 2016. *The Dancing Bees. Karl von Frisch and the Discovery of the Honeybee Language.* University of Chicago Press, Chicago, Illinois.

Murray, L., and E. P. Jeffree. 1955. Swarming in Scotland. *Scottish Beekeeper* 31: 96–98.

Murray, S. S., M. J. Schoeninger, H. T. Bunn, T. R. Pickering, and J. A. Marlett. 2001. Nutritional composition of some wild plant foods and honey used by Hadza foragers of Tanzania. *Journal of Food Composition and Analysis* 14: 3–13.

Naile, F. 1976. *America's Master of Bee Culture: The Life of L. L. Langstroth.* Cornell University Press, Ithaca, New York.

Nakamura, J., and T. D. Seeley. 2006. The functional organization of resin work in honeybee colonies. *Behavioral Ecology and Sociobiology* 60: 339–349.

Nesse, R. M., and G. C. Williams. 1994. *Why We Get Sick: The New Science of Darwinian Medicine*. Times Books, New York.

Neumann, P., and T. Blacquière. 2016. The Darwin cure for apiculture? Natural selection and managed honeybee health. *Evolutionary Applications* 2016: 1–5, doi:10.1111/eva.12448.

Neville, A. C. 1965. Energy economy in insect flight. *Science Progress* 53: 203–219.

Nicodemo, D., E. B. Malheiros, D. De Jong, and R.H.N. Couto. 2014. Increased brood viability and longer lifespan of honeybees selected for propolis production. *Apidologie* 45: 269–275.

Nicolson, S. W. 2009. Water homeostasis in bees, with the emphasis on sociality. *Journal of Experimental Biology* 212: 429–434.

Nolan, W. J. 1925. The brood-rearing cycle of the honeybee. *Bulletin of the United States Department of Agriculture* 1349: 1–56.

Nordhaus, H. 2011. *The Beekeeper's Lament: How One Man and Half a Billion Honey Bees Help Feed America*. HarperCollins, New York.

Nye, W. P., and O. Mackensen. 1968. Selective breeding of honeybees for alfalfa pollen: Fifth generation and backcrosses. *Journal of Apicultural Research* 7: 21–27.

———. 1970. Selective breeding of honeybees for alfalfa pollen collection: With tests in high and low alfalfa pollen collection regions. *Journal of Apicultural Research* 9: 61–64.

Oddie, M., R. Büchler, B. Dahle, M. Kovacic, Y. LeConte, B. Locke, J. R. de Miranda, F. Mondet, and P. Neumann. 2018. Rapid parallel evolution overcomes global honey bee parasite. *Scientific Reports* 8: 7704, doi:10.1038/s41598-018-26001-7.

Oddie, M.A.Y., B. Dahle, and P. Neumann. 2017. Norwegian honey bees surviving *Varroa destructor* mite infestations by means of natural selection. *PeerJ* 5: e3956, doi:10.7717/peerj.3956.

Odell, A. L., J. P. Lassoie, and R. R. Morrow. 1980. *A History of Cornell University's Arnot Forest*. Dept. of Natural Resources Research and Extension Ser. no. 14. Cornell University, Ithaca, New York. https://blogs.cornell.edu/arnotforest/files/2015/07/history-of-the-Arnot-1980-yoa9tl.pdf (accessed 9 January 2019).

Oldroyd, B. P. 2012. Domestication of honey bees was associated with expansion of genetic diversity. *Molecular Ecology* 21: 4409–4411.

Oleksa, A., R. Gawroński, and A. Tofilski. 2013. Rural avenues as a refuge for feral honey bee population. *Journal of Insect Conservation* 17: 465–472.

Oliver, R. 2014. A comparative test of the pollen subs. *American Bee Journal* 154: 795–801, 869–874, 1021–1025.

Ostwald, M. M., M. L. Smith, and T. D. Seeley. 2016. The behavioral regulation of thirst, water collection and water storage in honey bee colonies. *Journal of Experimental Biology* 219: 2156–2165.

Otis, G. 1982. Weights of worker honeybees in swarms. *Journal of Apicultural Research* 21: 88–92.

Owens, C. D. 1971. The thermology of wintering honey bee colonies. *Technical Bulletin, United States Department of Agriculture* 1429: 1–32.

Oxley, P. R., and B. P. Oldroyd. 2010. The genetic architecture of honeybee breeding. *Advances in Insect Physiology* 39: 83–118.

Page, R. E., Jr. 1981. Protandrous reproduction in honey bees. *Environmental Entomology* 10: 359–362.

———. 1982. The seasonal occurrence of honey bee swarms in north-central California. *American Bee Journal* 121: 266–272.

Palmer, K. A., and B. P. Oldroyd. 2000. Evolution of multiple mating in the genus *Apis*. *Apidologie* 31: 235–248.

Park, O. W. 1923. Water stored by bees. *American Bee Journal* 63: 348–349.

———. 1949. Activities of honey bees. In: *The Hive and the Honey Bee*, R. A. Grout, ed., pp. 79–152. Dadant and Sons, Hamilton, Illinois.

Parker, R. L. 1926. The collection and utilization of pollen by the honeybee. *Cornell University Agricultural Experiment Station Memoir* 98: 1–55.

Peck, D. T., and T. D. Seeley. Forthcoming. Mite bombs or robber lures? The roles of drifting and robbing in *Varroa destructor* transmission from collapsing colonies to their neighbors.

———. Forthcoming. Multiple mechanisms of behavioral resistance to an introduced parasite, *Varroa destructor*, in a survivor population of European honey bees.

———. Forthcoming. Robbing by honey bees (*Apis mellifera*): assessing its importance for disease spread in the wild.

Peck, D. T., M. L. Smith, and T. D. Seeley. 2016. *Varroa destructor* mites can nimbly climb from flowers onto foraging honey bees. *PLoS ONE* 11:e0167798, doi.org/10.1371/journal.pone.0167798.

Peer, D. F. 1957. Further studies on the mating range of the honey bee, *Apis melliera* L. *Canadian Entomologist* 89: 108–110.

Peters, J. M., O. Peleg, and L. Mahadevan. 2017. Fluid-mediated self-organization of ventilation in honeybee nests. *BioRxiv*. Preprint, posted 31 October 2017, doi:http://dx.doi.org/10.1101/212100.

Pfeiffer, K. J., and J. Crailsheim. 1998. Drifting of honeybees. *Insectes Sociaux* 45: 151–167.

Phillips, M. G. 1956. *The Makers of Honey.* Crowell, New York.

Phipps, J. 2016. Editorial. *Natural Bee Husbandry* 1: 3.

Pinto, M. A., W. L. Rubink, R. N. Coulson, J. C. Patton, and J. S. Johnston. 2004. Temporal pattern of Africanization in a feral honeybee population from Texas inferred from mitochondrial DNA. *Evolution* 58: 1047–1055.

Pinto, M. A., W. L. Rubink, J. C. Patton, R. N. Coulson, and J. S. Johnston. 2005. Africanization in the United States: Replacement of feral European honeybees (*Apis mellifera* L.) by an African hybrid swarm. *Genetics* 170: 1653–1665.

Pratt, S. C. 1998a. Condition-dependent timing of comb construction by honeybee colonies: How do workers know when to start building? *Animal Behaviour* 56: 603–610.

———. 1998b. Decentralized control of drone comb construction in honey bee colonies. *Behavioral Ecology and Sociobiology* 42: 193–205.

———. 1999. Optimal timing of comb construction by honeybee (*Apis mellifera*) colonies: A dynamic programming model and experimental tests. *Behavioral Ecology and Sociobiology* 46: 30–42.

———. 2004. Collective control of the timing and type of comb construction by honey bees (*Apis mellifera*). *Apidologie* 35: 193–205.

Radcliffe, R. W., and T. D. Seeley. 2018. Deep forest bee hunting: A novel method for finding wild colonies of honey bees in old-growth forests. *American Bee Journal* 158: 871–877.

Rangel, J., M. Giresi, M. A. Pinto, K. A. Baum, W. L. Rubink, R. N. Coulson, and J. S. Johnston. 2016. Africanization of a feral honey bee (*Apis mellifera*) population in South Texas: Does a decade make a difference? *Ecology and Evolution* 6: 2158–2169.

Rangel, J., S. R. Griffin, and T. D. Seeley. 2010. An oligarchy of nest-site scouts triggers a honeybee swarm's departure from the hive. *Behavioral Ecology and Sociobiology* 64: 979–987.

Rangel, J., H. R. Mattila, and T. D. Seeley. 2009. No intracolonial nepotism during colony fissioning in honey bees. *Proceedings of the Royal Society of London B* 276: 3895–3900.

Rangel, J., H. K. Reeve, and T. D. Seeley. 2013. Optimal colony fissioning in social insects: Testing an inclusive fitness model with honey bees. *Insectes Sociaux* 60: 445–452.

Rangel, J., and T. D. Seeley. 2012. Colony fissioning in honey bees: Size and significance of the swarm fraction. *Insectes Sociaux* 59: 453–462.

Rayment, T. 1923. Through Australian eyes: Water in cells. *American Bee Journal* 63: 135–136.

Ribbands, C. R. 1954. The defence of the honeybee community. *Proceedings of the Royal Society of London B* 142: 514–524.

Richards, K. W. 1973. Biology of *Bombus polaris* Curtis and *B. hyperboreus* Schönherr at Lake Hazen, Northwest Territories (Hymenoptera: Bombini). *Quaestiones Entomologicae* 9: 115–157.

Rinderer, T. E., L. I. de Guzman, G. T. Delatte, J. A. Stelzer, V. A. Lancaster, V. Kuznetsov, L. Beaman, R. Watts, and J. W. Harris. 2001. Resistance to the parasitic mite *Varroa destructor* in honey bees from far-eastern Russia. *Apidologie* 32: 381–394.

Rinderer, T. E., J. W. Harris, G. J. Hunt, and L. I. de Guzman. 2010. Breeding for resistance to *Varroa destructor* in North America. *Apidologie* 41: 409–424.

Rinderer, T. E., K. W. Tucker, and A. M. Collins. 1982. Nest cavity selection by swarms of European and Africanized honeybees. *Journal of Apicultural Research* 21: 98–103.

Rivera-Marchand, B., D. Oskay, and T. Giray. 2012. Gentle Africanized bees on an oceanic island. *Evolutionary Applications* 5: 746–756.

Roberts, A. 2017. *Tamed*. Hutchinson, London.

Robinson, F. A. 1966. Foraging range of honey bees in citrus groves. *Florida Entomologist* 49: 219–223.

Robinson, G. E., B. A. Underwood, and C. E. Henderson. 1984. A highly specialized water-collecting honey bee. *Apidologie* 15: 355–358.

Roffet-Salque, M., M. Regert, R. P. Evershed, A. K. Outram, L. J. E. Cramp, O. Decavallas, J. Dunne, P. Gerbault, S. Mileto, S. Mirabaud, M. Pääkkönen, J. Smyth, and 53 more authors. 2015. Widespread exploitation of the honeybee by early Neolithic farmers. *Nature* 527: 226–231.

Rösch, G. A. 1927. Über die Bautätigkeit im Bienenvolk und das Alter der Baubienen. Weiterer Beitrag zur Frage nach der Arbeitsteilung im Bienenstaat. *Zeitschrift für Vergleichende Physiologie* 6: 264–298.

Rosenkranz, P., P. Aumeier, and B. Ziegelmann. 2010. Biology and control of *Varroa destructor*. *Journal of Invertebrate Pathology* 103: S96–S119.

Rosov, S. A. 1944. Food consumption by bees. *Bee World* 25: 94–95.

Rothenbuhler, W. C. 1958. Genetics and breeding of the honey bee. *Annual Review of Entomology* 3: 161–180.

Ruttner, F. 1987. *Biogeography and Taxonomy of Honeybees*. Springer Verlag, Berlin.

Ruttner, F., E. Milner, and J. E. Dews. 1990. *The Dark European Honeybee: Apis mellifera mellifera Linnaeus 1758*. British Isles Bee Breeders Association / Beard and Son, Brighton.

Ruttner, F., and H. Ruttner. 1966. Untersuchungen über die Flugaktivität und das Paarungsverhalten der Drohnen. 3. Flugweite and Flugrichtung der Drohnen. *Zeitschrift für Bienenforschung* 8: 332–354.

———. 1972. Untersuchungen über die Flugaktivität und das Paarungsverhalten der Drohnen. V. Drohnensammelplätze und Paarungsdistanz. *Apidologie* 3: 203–232.

Sakagami, S. F., and H. Fukuda. 1968. Life tables for worker honeybees. *Researches on Population Ecology* 10: 127–139.

Sammataro, D., and A. Avitabile. 2011. *The Beekeeper's Handbook*. Cornell University Press, Ithaca, New York.

Sanford, M. T. 2001. Introduction, spread and economic impact of *Varroa* mites in North America. In: *Mites of the Honey Bee*, T. C. Webster and K. S. Delaplane, eds., 149–162. Dadant and Sons, Hamilton, Illinois.

Schiff, N. M., W. S. Sheppard, G. M. Loper, and H. Shimanuki. 1994. Genetic diversity of feral honey bee (Hymenoptera: Apidae) populations in the southern United States. *Annals of the Entomological Society of America* 87: 842–848.

Schmaranzer, S. 2000. Thermoregulation of water collecting honey bees (*Apis mellifera*). *Journal of Insect Physiology* 46: 1187–1194.

Schmidt, J. O., and R. Hurley. 1995. Selection of nest cavities by Africanized and European honey bees. *Apidologie* 26: 467–475.

Schneider, S. S. 1989. Queen behavior and worker-queen interactions in absconding and swarming colonies of the African honeybee, *Apis mellifera scutellata* (Hymenoptera: Apidae). *Journal of the Kansas Entomological Society* 63: 179–186.

———. 1991. Modulation of queen activity by the vibration dance in swarming colonies of the African honey bee, *Apis mellifera scutellata* (Hymenoptera: Apidae). *Journal of the Kansas Entomological Society* 64: 269–278.

Scholze, E., H. Pichler, and H. Heran. 1964. Zur Entfernungsschätzung der Bienen nach dem Kraftaufwand. *Naturwissenschaften* 51: 69–90.

Scofield, H. N., and H. R. Mattila. 2015. Honey bee workers that are pollen stressed as larvae become poor foragers and waggle dancers as adults. *PLoS ONE* 10(4):e0121731, doi.org/10.1371/journal.pone.0121731.

Seeley, T. D. 1974. Atmospheric carbon dioxide regulation in honey-bee (*Apis mellifera*) colonies. *Journal of Insect Physiology* 20: 2301–2305.

———. 1977. Measurement of nest cavity volume by the honey bee (*Apis mellifera*). *Behavioral Ecology and Sociobiology* 2: 201–227.

———. 1978. Life history strategy of the honey bee, *Apis mellifera*. *Oecologia* 32: 109–118.

———. 1982. Adaptive significance of the age polyethism schedule in honeybee colonies. *Behavioral Ecology and Sociobiology* 11: 287–293.

———. 1986. Social foraging by honeybees: How colonies allocate foragers among patches of flowers. *Behavioral Ecology and Sociobiology* 19: 343–356.

———. 1987. The effectiveness of information collection about food sources by honey bee colonies. *Animal Behaviour* 35: 1572–1575.

———. 1989. Social foraging in honey bees: how nectar foragers assess their colony's nutritional status. *Behavioral Ecology and Sociobiology* 24: 181–199.

———. 1995. *The Wisdom of the Hive*. Harvard University Press, Cambridge, Massachusetts.

———. 2002. The effect of drone comb on a honey bee colony's production of honey. *Apidologie* 33: 75–86.

———. 2003. Bees in the forest, still. *Bee Culture* 131 (January): 24–27.

———. 2007. Honey bees of the Arnot Forest: A population of feral colonies persisting with *Varroa destructor* in the northeastern United States. *Apidologie* 38: 19–29.

———. 2010. *Honeybee Democracy*. Princeton University Press, Princeton, New Jersey.

———. 2012. Using bait hives. *Bee Culture* 140 (April): 73–75.

———. 2016. *Following the Wild Bees: The Craft and Science of Bee Hunting*. Princeton University Press, Princeton, New Jersey.

———. 2017a. Bait hives: A valuable tool for natural beekeeping. *Natural Bee Husbandry* 2 (February): 15–18.

———. 2017b. Life-history traits of honey bee colonies living in forests around Ithaca, NY, USA. *Apidologie* 48: 743–754.

———. 2017c. Darwinian beekeeping: An evolutionary approach to apiculture. *American Bee Journal* 157: 277–282.

Seeley, T. D., S. Camazine, and J. Sneyd. 1991. Collective decision-making in honey bees: How colonies choose among nectar sources. *Behavioral Ecology and Sociobiology* 28: 277–290.

Seeley, T. D., and S. R. Griffin. 2011. Small-cell comb does not control *Varroa* mites in colonies of honeybees of European origin. *Apidologie* 42: 526–532.

Seeley, T. D., and R. A. Morse. 1976. The nest of the honey bee (*Apis mellifera* L). *Insectes Sociaux* 23: 495–512.

———. 1978a. Nest site selection by the honey bee. *Insectes Sociaux* 25: 323–337.

———. 1978b. Dispersal behavior of honey bee swarms. *Psyche* 84: 199–209.

Seeley, T. D., and M. L. Smith. 2015. Crowding honeybee colonies in apiaries can increase their vulnerability to the deadly ectoparasitic mite *Varroa destructor*. *Apidologie* 46: 716–727.

Seeley, T. D., and D. R. Tarpy. 2007. Queen promiscuity lowers disease within honeybee colonies. *Proceedings of the Royal Society of London B* 274: 67–72.

Seeley, T. D., D. R. Tarpy, S. R. Griffin, A. Carcione, and D. A. Delaney. 2015. A survivor population of wild colonies of European honeybees in the northeastern United States: Investigating its genetic structure. *Apidologie* 46: 654–666.

Seeley, T. D., and P. K. Visscher. 1985. Survival of honey bees in cold climates: The critical timing of colony growth and reproduction. *Ecological Entomology* 10: 81–88.

Sekiguchi, K., and S. F. Sakagami. 1966. Structure of foraging population and related problems in the honeybee, with considerations on the division of labor in bee colonies. *Hokkaido National Agricultural Experiment Station Report* 69: 1–65.

Semkiw, P., and P. Skubida. 2010. Evaluation of the economical aspects of Polish beekeeping. *Journal of Apicultural Science* 54: 5–15.

Sheppard, W. S. 1989. A history of the introduction of honey bee races into the United States. *American Bee Journal* 129: 617–619, 664–666.

Simone, M., J. D. Evans, and M. Spivak. 2009. Resin collection and social immunity in honey bees. *Evolution* 63: 3016–3022.

Simone-Finstrom, M., J. Gardner, and M. Spivak. 2010. Tactile learning in resin foraging honeybees. *Behavioral Ecology and Sociobiology* 64: 1609–1617.

Simone-Finstrom, M., M. Walz, and D. R. Tarpy. 2016. Genetic diversity confers colony-level benefits due to individual immunity. *Biology Letters* 12: 20151007, doi:10.1098/rsbl.2015.1007.

Simone-Finstrom, M., and M. Spivak. 2010. Propolis and bee health: The natural history and significance of resin use by honey bees. *Apidologie* 41: 295–311.

Simpson, J. 1957a. Observations on colonies of honey-bees subjected to treatments designed to induce swarming. *Proceedings of the Royal Entomological Society of London (A)* 32: 185–192.

———. 1957b. The incidence of swarming among colonies of honey-bees in England. *Journal of Agricultural Science* 49: 387–393.

Simpson, J., and I.B.M. Riedel. 1963. The factor that causes swarming in honeybee colonies in small hives. *Journal of Apicultural Research* 2: 50–54.

Smibert, T. 1851. *Io Anche! Poems, Chiefly Lyrical.* James Hogg, Edinburgh.

Smith, B. E., P. L. Marks, and S. Gardescu. 1993. Two hundred years of forest cover changes in Tompkins County, New York. *Bulletin of the Torrey Botanical Club* 120: 229–247.

Smith, M. L., M. M. Ostwald, and T. D. Seeley. 2015. Adaptive tuning of an extended phenotype: Honeybees seasonally shift their honey storage to optimize male production. *Animal Behaviour* 103: 29–33.

———. 2016. Honey bee sociometry: Tracking honey bee colonies and their nest contents from colony founding until death. *Insectes Sociaux* 63: 553–563.

Smits, S. A., J. Leach, E. D. Sonnenburg, C. G. Gonzalez, J. S. Lichtman, G. Reid, R. Knight, A. Manjurano, J. Changalucha, J. E. Elias, M. G. Dominguez-Bello, and J. L. Sonnenburg. 2017. Seasonal cycling in the gut microbiome of the Hadza hunter-gatherers of Tanzania. *Science* 357: 802–806.

Southwick, E. E. 1982. Metabolic energy of intact honey bee colonies. *Comparative Biochemistry and Physiology* 71: 277–281.

———. 1985. Allometric relations, metabolism and heat conductance in clusters of honey bees at cool temperatures. *Journal of Comparative Physiology B* 156: 143–149.

Southwick, E. E., G. M. Loper, and S. E. Sadwick. 1981. Nectar production, composition, energetics and pollinator attractiveness in spring flowers of western New York. *American Journal of Botany* 68: 994–1002.

Southwick, E. E., and J. N. Mugaas. 1971. A hypothetical homeotherm: The honeybee hive. *Comparative Biochemistry and Physiology Part A: Physiology* 40: 935–944.

Southwick, E. E., and D. Pimentel. 1981. Energy efficiency of honey production by bees. *Bioscience* 31: 730–732.

Spivak, M., and M. Gilliam. 1998a. Hygienic behaviour of honey bees and its application for control of brood diseases and varroa. Pt. 1: Hygienic behaviour and resistance to American foulbrood. *Bee World* 79: 124–134.

———. 1998b. Hygienic behaviour of honey bees and its application for control of brood diseases and varroa. Pt. 2: Studies on hygienic behaviour since the Rothenbuhler era. *Bee World* 79: 169–186.

Stabentheiner, A., H. Pressl, T. Papst, N. Hrassnigg, and K. Crailsheim. 2003. Endothermic heat production in honeybee winter clusters. *Journal of Experimental Biology* 206: 353–358.

Starks, P. T., C. A. Blackie, and T. D. Seeley. 2000. Fever in honeybee colonies. *Naturwissenschaften* 87: 229–231.

Strange, J. P., L. Garnery, and W. S. Sheppard. 2007. Persistence of the Landes ecotype of *Apis mellifera mellifera* in southwest France: Confirmation of a locally adaptive annual brood cycle trait. *Apidologie* 38: 259–267.

Szabo, T. I. 1983a. Effects of various entrances and hive direction on outdoor wintering of honey bee colonies. *American Bee Journal* 123: 47–49.

———. 1983b. Effect of various combs on the development and weight gain of honeybee colonies. *Journal of Apicultural Research* 22: 45–48.

Taber, S. 1963. The effect of disturbance on the social behavior of the honey bee colony. *American Bee Journal* 103: 286–288.

Taber, S., and C. D. Owens. 1970. Colony founding and initial nest design of honey bees, *Apis mellifera* L. *Animal Behaviour* 18: 625–632.

Tarpy, D. R. 2003. Genetic diversity within honeybee colonies prevents severe infections and promotes colony growth. *Proceedings of the Royal Society of London B* 270: 99–103.

Tarpy, D. R., D. A. Delaney, and T. D. Seeley. 2015. Mating frequencies of honey bee queens (*Apis mellifera* L.) in a population of feral colonies in the northeastern United States. *PLoS ONE* 10 (3): e0118734, doi:10.1371/journal.pone.0118734.

Tarpy, D. R., R. Nielsen, and D. I. Nielsen. 2004. A scientific note on the revised estimates of effective paternity frequency in *Apis*. *Insectes Sociaux* 51: 203–204.

Tautz, J., S. Maier, C. Groh, W. Rössler, and A. Brockmann. 2003. Behavioral performance in adult honey bees is influenced by the temperature experienced during their pupal development. *Proceedings of the National Academy of Sciences of the USA* 100: 7343–7347.

Terashima, H. 1998. Honey and holidays: The interactions mediated by honey between Efe hunter-gatherers and Lese farmers in the Ituri forest. *African Study Monographs*, supplementary issue 25: 123–134.

Thom, C., T. D. Seeley, and J. Tautz. 2000. A scientific note on the dynamics of labor devoted to nectar foraging in a honey bee colony: Number of foragers versus individual foraging activity. *Apidologie* 31: 737–738.

Thompson, J. R., D. N. Carpenter, C. V. Cogbill, and D. R. Foster. 2013. Four centuries of change in northeastern United States forests. *PLoS ONE* 8(9): e72540, doi:10.1371/journal.pone.0072540.

Thoreau, H. D. 1862. Walking. *Atlantic Monthly* 9: 657–674.

Tinbergen, N. 1974. *The Animal in Its World (Explorations of an Ethologist, 1932–1972)*. Harvard University Press, Cambridge, Massachusetts.

Tinghitella, R. M. 2008. Rapid evolutionary change in a sexual signal: Genetic control of the mutation 'flatwing' that renders male field crickets (*Teleogryllus oceanicus*) mute. *Heredity* 100: 261–267.

Traynor, K. S., J. S. Pettis, D. R. Tarpy, C. A. Mullin, J. L. Frazier, M. Frazier, and D. vanEngelsdorp. 2016. In-hive pesticide exposome: Assessing risks to migratory honey bees from in-hive pesticide contamination in the Eastern United States. *Scientific Reports* 6: 33207.

Tribe, G., J. Tautz, K. Sternberg, and J. Cullinan. 2017. Firewalls in bee nests—survival value of propolis walls of wild Cape honeybee (*Apis mellifera capensis*). *Naturwissenschaften* 104: 29, doi.org/10.1007/s00114-017-1449-5.

Turnbull, C. M. 1976. *Man in Africa*. Anchor Press, Garden City, New Jersey.

Villa, J. D., D. M. Bustamante, J. P. Dunkley, and L. A. Escobar. 2008. Changes in honey bee (Hymenoptera: Apidae) colony swarming and survival pre- and postarrival of *Varroa destructor* (Mesostigmata: Varroidae) in Louisiana. *Annals of the Entomological Society of America* 101: 867–871.

Visscher, P. K., K. Crailsheim, and G. Sherman. 1996. How do honey bees (*Apis mellifera*) fuel their water foraging flights? *Journal of Insect Physiology* 42: 1089–1094.

Visscher, P. K., and T. D. Seeley. 1982. Foraging strategy of honeybee colonies in a temperate deciduous forest. *Ecology* 63: 1790–1801.

Visscher, P. K., R. S. Vetter, and G. E. Robinson. 1995. Alarm pheromone perception in honey bees is decreased by smoke (Hymenoptera: Apidae). *Journal of Insect Behavior* 8: 11–18.

von Engeln, O. D. 1961. *The Finger Lakes Region: Its Origin and Nature*. Cornell University Press, Ithaca, New York.

von Frisch, K. 1967. *The Dance Language and Orientation of Bees*. Harvard University Press, Cambridge, Massachusetts.

Wallberg, A., F. Han, G. Wellhagen, B. Dahle, M. Kawata, N. Haddad, Z. Simões, M. Allsopp, I. Kandemir, P. De la Rúa, C. Pirk, and M. T. Webster. 2014. A worldwide survey of genome sequence variation provides insight into the evolutionary history of the honeybee *Apis mellifera*. *Nature Genetics* 46: 1081–1088.

Watson, L. R. 1928. Controlled mating in honeybees. *Quarterly Review of Biology* 3: 377–390.

Weipple, T. 1928. Futterverbrauch und Arbeitsteilung eines Bienenvolkes im Laufe eines Jahres. *Archiv für Bienenkunde* 9: 70–79.

Weiss, K. 1965. Über den Zuckerverbrauch und die Beanspruchung der Bienen bei der Wachserzeugung. *Zeitschrift für Bienenforschung* 8: 106–124.

Wells, P. H., and J. Giacchino Jr. 1968. Relationship between the volume and the sugar concentration of loads carried by honeybees. *Journal of Apicultural Research* 7: 77–82.

Wenke, R. J. 1999. *Patterns in Prehistory: Humankind's First Three Million Years*. Oxford University Press, New York.

Wenner, A. M., and W. W. Bushing. 1996. *Varroa* mite spread in the United States. *Bee Culture* 124: 341–343.

Whitaker, J. O., Jr., and W. D. Hamilton, Jr. 1998. *Mammals of the Eastern United States*. Cornell University Press, Ithaca, New York.

White, J. W., Jr. 1975. Composition of honey. In: *Honey: A Comprehensive Survey*, E. Crane, ed., pp. 157–206. Heinneman, London.

White, J. W., Jr., M. L. Riethof, M. H. Subers, and I. Kushnir. 1962. *Composition of American Honeys*. U.S. Government Printing Office, Washington, D.C.

Williams, G. C. 1966. *Adaptation and Natural Selection*. Princeton University Press, Princeton, New Jersey.

Williams, G. C., and R. M. Nesse. 1991. The dawn of Darwinian medicine. *Quarterly Review of Biology* 66: 1–22.

Wilson, M. B., D. Brinkman, M. Spivak, G. Gardner, and J. D. Cohen. 2015. Regional variation in composition and antimicrobial activity of US propolis against *Paenibacillus larvae* and *Ascosphaera apis*. *Journal of Invertebrate Pathology* 124: 44–50.

Winston, M. L. 1980. Swarming, afterswarming, and reproductive rate of unmanaged honeybee colonies (*Apis mellifera*). *Insectes Sociaux* 27: 391–398.

———. 1981. Seasonal patterns of brood rearing and worker longevity in colonies of the Africanized honey bee (Hymenoptera: Apidae) in South America. *Journal of the Kansas Entomological Society* 53: 157–165.

———. 1987. *The Biology of the Honey Bee*. Harvard University Press, Cambridge, Massachusetts.

Wohlgemuth, R. 1957. Die Temperaturregulation des Bienenvolkes unter regeltheoretischen Gesichtpunkten. *Zeitschrift für Vergleichende Physiologie* 40: 119–161.

Wood, B. M., H. Pontzer, D. A. Raichlen, and F. W. Marlowe. 2014. Mutualism and manipulation in Hadza-honeyguide interactions. *Evolution and Human Behavior* 35: 540–546.

Zeuner, F. E., and F. J. Manning. 1976. A monograph on fossil bees (Hymenoptera: Apoidea). *Bulletin of the British Museum of Natural History (Geology)* 27: 151–268.

Zuk, M., J. T. Rotenberry, and R. M. Tinghitella. 2006. Silent night: Adaptive disappearance of a sexual signal in a parasitized population of field crickets. *Biology Letters* 2: 521–524.

Acknowledgments

Little by little, over the last 40 years, much has been learned about how colonies of honey bees live in the wild. Much of the knowledge that is summarized in this book comes from studies made by me and by my students, often in collaboration with biologists based at various universities. I thank everyone involved. In temporal succession, my collaborators are Roger A. Morse, Richard D. Fell, John T. Ambrose, D. Michael Burgett, David De Jong, Daniel H. Seeley, P. Kirk Visscher, Paul W. Sherman, H. Kern Reeve, Scott Camazine, Susanne Kühnholz, Anja Weidenmüller, Susanne C. Buhrmann, Philip T. Starks, Caroline A. Blackie, Alexander S. Mikheyev, Stephen C. Pratt, Jürgen Tautz, David C. Gilley, David R. Tarpy, Brian R. Johnson, Adrian M. Reich, Kevin M. Passino, Jun Nakamura, Heather R. Mattila, Katherine M. Burke, Madeleine B. Girard, Barrett A. Klein, Juliana Rangel, Sean R. Griffin, Kathryn J. Montovan, Nathaniel Karst, Laura E. Jones, Michael L. Smith, Madeleine M. Ostwald, J. Carter Loftus, Deborah A. Delaney, Ann B. Chilcott, David T. Peck, Hailey N. Scofield, and Robin W. Radcliffe. Many of these people are continuing their work with the bees, and I am sure they will never run out of material for study.

I also take here the opportunity to express my immense gratitude to the late Professor Roger A. Morse, the first director of the Dyce Laboratory for Honey Bee Studies, at Cornell University, for helping me find my way in life. He hired me to work in his lab every summer when I was a college student, and he provided what I needed—a pickup truck, chain saw, lab space, and help from an ex-logger from Maine, Herb Nelson—when I investigated the natural nests of honey bees. Roger let each of his students find his or her own project; he helped us obtain the wherewithal for our studies, and then he let us work freely. I very much regret that he cannot read this book and see what has come of his support of me.

When it came time to go to graduate school, the two famous "ant men" at Harvard University, Bert Hölldobler and Edward O. Wilson, welcomed me to their program. I am greatly indebted to these gentlemen, because in joining their research group, I interacted with people doing behavioral and evolutionary studies on all sorts of social insects, and this broadened my scope as a biologist. In the fall of 1976, I shared an office in the Museum of Comparative Zoology at Harvard with Bernd Heinrich, who was writing his wonderful book *Bumblebee Economics*, and I thank Bernd too for being another hugely important teacher and role model.

Many others have helped make this book a reality, and I want to acknowledge them as well. Herb Nelson taught me how to operate a chain saw, fell large trees, and drive a pickup truck deep into the woods and get back out . . . all invaluable skills for studying the nests of wild honey bee colonies. Alfred Fontana and Donald Schaufler, along with Professors Aaron Moen and Peter Smallidge, the managers and the directors, respectively, of the Arnot Forest of Cornell University, allowed me to work freely in this magnificent woodland over the past 40 years. Barbara Locke Grandér, at the Swedish University of Agricultural Sciences, has kept me informed about the long-term experiment with honey bee colonies left to fend for themselves on the island of Gotland. Bonnie and Gary Morse, organizers of the Bee Audacious Conference in 2016, spurred me to collect my thoughts on Darwinian beekeeping. Ann Chilcott, David Peck, Leo Sharashkin, Michael Smith, Francis Ratnieks, and Mark Winston read drafts of various chapters and gave me numerous suggestions for improvement.

Others have supported my work in indirect, but critical, ways. I am thankful for institutional support from Yale University and Cornell University, and for financial support from the U.S. National Science Foundation, the U.S. Department of Agriculture, the Alexander von Humboldt Foundation in Germany, the North American Pollinator Protection Campaign, the Eastern Apicultural Society, and, the Honeybee Capital Foundation. Over the years, this support has removed many financial obstacles, and I give warm thanks to these institutions and organizations.

I am also grateful to Margaret C. Nelson for creating all the figures in this book. Margy and I have worked together for more than 30 years, and I rely heavily on her advice about the visual display of quantitative information. I also thank the many individuals who have provided photographs for this book: Renata Borba, Laurie Burnham, Scott Camazine, Ann Chilcott, Linton Chilcott, Jenny Cullinan, Megan Denver, Mary Holland, Zachary Huang, Rustyem Ilyasov, Gene Kritsky, Kenneth Lorenzen, Åke Lyberg, Andrzej Oleksa, Robin Radcliffe, Juliana Rangel, Michael Smith, Armin Spürgin, Jürgen Tautz, Eric Tourneret, and Alexander Wild. Their photos have helped me present a vivid account of the lives of the wild bees.

It is a pleasure to express my warm thanks to Alison Kalett, executive editor of Biology and Earth Sciences at Princeton University Press, for encouraging me to write this book and for providing valuable guidance after I took up the challenge. I am also grateful to Amy Hughes, who helped me with her thoughtful editing, and to Brigitte Pelner, who skillfully shepherded the manuscript through the production process.

Finally, I want to give special thanks to my wife, Robin Hadlock Seeley, a fellow field biologist, who understands my passion for studying the honey bees living in the woods, and to our two daughters, Saren and Maira, for their encouragement and for their help in finding the right title for this book.

Illustration Credits

Fig. 1.1. *Left:* photo by Thomas D. Seeley. *Right:* photo by Felix Remter.

Fig. 1.2. Modified from fig. 2.2 in Ruttner, F., 1992, *Naturgeschichte der Honigbienen*, Ehrenwirth, Munich.

Fig. 1.3. Modified from fig. 1 in Kritsky, G., 1991, Lessons from history: The spread of the honey bee in North America, *American Bee Journal* 131: 367–370.

Fig. 1.4. Modified from fig. 4 in Mikheyev, A. S., M.M.Y. Tin, J. Arora, and T. D. Seeley, 2015, Museum samples reveal rapid evolution by wild honey bees exposed to a novel parasite, *Nature Communications* 6: 7991, doi:10.1038/ncomms8991.

Fig. 1.5. Photo by Thomas D. Seeley.

Fig. 2.1. Aerial photo from Google Earth.

Fig. 2.2. Photo by Thomas D. Seeley.

Fig. 2.3. Photo provided by Rustem A. Ilyasov.

Fig. 2.4. *Top:* aerial photo from Google Earth, with boundary lines added by Michael L. Smith. *Bottom:* photo by Thomas D. Seeley.

Fig. 2.5. Photo by Thomas D. Seeley.

Fig. 2.6. Original drawing by Margaret C. Nelson.

Fig. 2.7. Photo by Juliana Rangel.

Fig. 2.8. Photo by Andrzej Oleksa.

Fig. 2.9. Photos by Thomas D. Seeley.

Fig. 2.10. Photo by Alex Wild.

Fig. 2.11. Original drawing by Margaret C. Nelson, based on data in Loper, G., 1997, Over-winter losses of feral honey bee colonies in southern Arizona, 1992–1997, *American Bee Journal* 137: 446; and Loper, G. M., D. Sammataro, J. Finley, and J. Cole, 2006, Feral honey bees in southern

Arizona 10 years after *Varroa* infestation, *American Bee Journal* 134: 521–524.

Fig. 2.12. Photo by Mary Holland.

Fig. 2.13. Original drawing by Margaret C. Nelson.

Fig. 2.14. Photo by Thomas D. Seeley.

Fig. 2.15. Photo by Thomas D. Seeley.

Fig. 3.1. Photo provided by Laurie Burnham.

Fig. 3.2. Reproductions by Margaret C. Nelson. *Left:* based on drawing in Hernández-Pacheco, E., 1924, *Las Pinturas Prehistóricas de Las Cuevas de la Araña (Valencia)*, Museo Nacional de Cienças Naturales, Madrid. *Right:* based on drawing in Dams, M., and L. Dams, 1977, Spanish rock art depicting honey gathering during the Mesolithic, *Nature* 268: 228–230.

Fig. 3.3. Reproduction by Margaret C. Nelson of fig. 20.3a in Crane, E., 1999, *The World History of Beekeeping and Honey Hunting*, Routledge, New York.

Fig. 3.4. Photo provided by Gene Kritsky.

Fig. 3.5. Photo provided by Rustem A. Ilyasov.

Fig. 3.6. From Münster, S., 1628, *Cosmographia*, Heinrich Petri, Basel, Switzerland.

Fig. 3.7. Photo by Thomas D. Seeley.

Fig. 3.8. From Cheshire, F. R., 1888, *Bees and Bee-Keeping; Scientific and Practical*, vol. 2: *Practical*, L. Upcott Gill, London.

Fig. 4.1. Photo provided by Eric Tourneret.

Fig. 4.2. Photo provided by Jenny Cullinan.

Fig. 4.3. Photo from PhD thesis of Lloyd R. Watson: Watson, L. R., 1928, Controlled mating in honeybees, *Quarterly Review of Biology* 3: 377–390.

Fig. 4.4. Original drawing by Margaret C. Nelson, based on data in Rothenbuhler, W. C., 1958, Genetics and breeding of the honey bee, *Annual Review of Entomology* 3: 161–180.

Fig. 4.5. Photo by Alex Wild.

Fig. 4.6. Photo by Ann B. Chilcott.

Fig. 4.7. Photo by Alex Wild.

Fig. 5.1. Photos by Thomas D. Seeley.

Fig. 5.2. Original drawing by Margaret C. Nelson, based on original data of Thomas D. Seeley and data in Seeley, T. D., and R. A. Morse, 1976, The nest of the honey bee (*Apis mellifera* L), *Insectes Sociaux* 23: 495–512.

Fig. 5.3. Original drawing by Margaret C. Nelson, based on data in Seeley, T. D., and R. A. Morse, 1976, The nest of the honey bee (*Apis mellifera* L), *Insectes Sociaux* 23: 495–512.

Fig. 5.4. Photo by Thomas D. Seeley.

Fig. 5.5. Photo by Scott Camazine.

Fig. 5.6. Photo by Thomas D. Seeley.

Fig. 5.7. Photo by Thomas D. Seeley.

Fig. 5.8. Photo by Alex Wild.

Fig. 5.9. Photo by Armin Spürgin.

Fig. 5.10. Original drawings by Margaret C. Nelson. Drawing on right is based on fig. 11 in Martin, H., and M. Lindauer, 1966, Sinnesphysiologische Leistungen beim Wabenbau der Honigbiene, *Zeitschrift für Vergleichende Physiologie* 53: 372–404.

Fig. 5.11. Original drawing by Margaret C. Nelson, based on data in Smith, M. L., M. M. Ostwald, and T. D. Seeley, 2016, Honey bee sociometry: Tracking honey bee colonies and their nest contents from colony founding until death, *Insectes Sociaux* 63: 553–563.

Fig. 5.12. Original drawing by Margaret C. Nelson, based on data in Pratt, S. C., 1999, Optimal timing of comb construction by honeybee (*Apis mellifera*) colonies: A dynamic programming model and experimental tests, *Behavioral Ecology and Sociobiology* 46: 30–42.

Fig. 5.13. *Top:* photo by Thomas D. Seeley. *Bottom:* original drawing by Margaret C. Nelson, based on data in Seeley, T. D., and R. A. Morse, 1976, The nest of the honey bee (*Apis mellifera* L), *Insectes Sociaux* 23: 495–512; and in Smith, M. L., M. M. Ostwald, and T. D. Seeley, 2016, Honey bee sociometry: Tracking honey bee colonies and their nest contents from colony founding until death, *Insectes Sociaux* 63: 553–563; and on data collected (but not reported) in Seeley, T. D., 2017, Life-history traits of honey bee colonies living in forests around Ithaca, NY, USA, *Apidologie* 48: 743–754.

Fig. 5.14. Original drawing by Margaret C. Nelson, based on figure in Pratt, S. C., 1998, Decentralized control of drone comb construction in honey bee colonies, *Behavioral Ecology and Sociobiology* 42: 193–205.

Fig. 5.15. Photo by Kenneth Lorenzen.

Fig. 5.16. Original drawing by Margaret C. Nelson, based on data in Nakamura, J., and T. D. Seeley, 2006, The functional organization of resin work in honeybee colonies, *Behavioral Ecology and Sociobiology* 60: 339–349.

Fig. 6.1. Photo by Zachary Huang, beetography.com.

Fig. 6.2. Original drawing by Margaret C. Nelson, based on fig. 4.1 in Seeley, T. D., 1985. *Honeybee Ecology*, Princeton University Press, Princeton, New Jersey.

Fig. 6.3. Photo by Kenneth Lorenzen.

Fig. 6.4. Original drawing by Margaret C. Nelson, based on fig. 4.2 in Seeley, T. D., 1985, *Honeybee Ecology*, Princeton University Press, Princeton, New Jersey.

Fig. 6.5. Original drawing by Margaret C. Nelson, based on data in fig. 3 in Smith, M. L., M. M. Ostwald, and T. D. Seeley, 2016, Honey bee sociometry: Tracking honey bee colonies and their nest contents from colony founding until death, *Insectes Sociaux* 63: 553–563.

Fig. 7.1. Photo by Kenneth Lorenzen.

Fig. 7.2. Original drawing by Margaret C. Nelson, based on fig. 2 in Page, R. E., Jr., 1981, Protandrous reproduction in honey bees, *Environmental Entomology* 10: 359–362.

Fig. 7.3. Photos by Michael L. Smith.

Fig. 7.4. Original drawing by Margaret C. Nelson.

Fig. 7.5. Photo by Thomas D. Seeley.

Fig. 7.6. Photo by Megan E. Denver.

Fig. 7.7. Original drawing by Margaret C. Nelson.

Fig. 7.8. Photo by Alex Wild.

Fig. 7.9. Original drawing by Margaret C. Nelson, based on fig. 2 in Rangel,

J., H. K. Reeve, and T. D. Seeley, 2013, Optimal colony fissioning in social insects: Testing an inclusive fitness model with honey bees, *Insectes Sociaux* 60: 445–452.

Fig. 7.10. Original drawing by Margaret C. Nelson, based on fig. 2 in Ruttner, F., and H. Ruttner, 1966, Untersuchungen über die Flugaktivität und das Paarungsverhalten der Drohnen. 3. Flugweite and Flugrichtung der Drohnen, *Zeitschrift für Bienenforschung* 8: 332–354.

Fig. 7.11. Original drawing by Margaret C. Nelson.

Fig. 7.12. Original drawing by Margaret C. Nelson, based on data in Tarpy, D. R., D. A. Delaney, and T. D. Seeley. 2015. Mating frequencies of honey bee queens (*Apis mellifera* L.) in a population of feral colonies in the northeastern United States. *PLoS ONE* 10 (3): e0118734.

Fig. 8.1. Photo by Alex Wild.

Fig. 8.2. Original drawing by Margaret C. Nelson.

Fig. 8.3. Photo by Kenneth Lorenzen.

Fig. 8.4. Photo by Alex Wild.

Fig. 8.5. Original drawing by Margaret C. Nelson.

Fig. 8.6. Original drawings by Margaret C. Nelson. *Top:* based on data in fig. 5 in Visscher, P. K., and T. D. Seeley, 1982, Foraging strategy of honeybee colonies in a temperate deciduous forest, *Ecology* 63: 1790–1801. *Bottom:* based on data in table 13 in von Frisch, K., 1967, *The Dance Language and Orientation of Bees*, Harvard University Press, Cambridge, Massachusetts.

Fig. 8.7. *Top:* photo by Thomas D. Seeley. *Bottom:* original drawing by Margaret C. Nelson.

Fig. 8.8. Original drawing by Margaret C. Nelson, based on data in fig. 3 in Visscher, P. K., and T. D. Seeley, 1982, Foraging strategy of honeybee colonies in a temperate deciduous forest, *Ecology* 63: 1790–1801.

Fig. 8.9. Original drawing by Margaret C. Nelson, based on fig. 1 in Seeley, T. D., S. Camazine, and J. Sneyd, 1991, Collective decision-making in honey bees: How colonies choose among nectar sources, *Behavioral Ecology and Sociobiology* 28: 277–290.

Fig. 8.10. Original drawing by Margaret C. Nelson, based on fig. 3 in Peck,

D. T., and T. D. Seeley, forthcoming, Robbing by honey bees in forest and apiary settings: Implications for horizontal transmission of the mite *Varroa destructor*, *Journal of Insect Behavior*.

Fig. 9.1. Original drawing by Margaret C. Nelson, based on part E of fig. 5 in Owens, C. D., 1971, The thermology of wintering honey bee colonies, *Technical Bulletin, United States Department of Agriculture* 1429: 1–32.

Fig. 9.2. Photo by Jürgen Tautz.

Fig. 9.3. Original drawing by Margaret C. Nelson, based on fig. 3 in Starks, P. T., C. A. Blackie, and T. D. Seeley, 2000, Fever in honeybee colonies, *Naturwissenschaften* 87: 229-231.

Fig. 9.4. Original drawing by Margaret C. Nelson, based on part A of fig. 5 in Owens, C. D., 1971, The thermology of wintering honey bee colonies, *Technical Bulletin, United States Department of Agriculture* 1429: 1–32.

Fig. 9.5. Original drawing by Margaret C. Nelson, based on fig. 3 in Mitchell, D., 2016, Ratios of colony mass to thermal conductance of tree and man-made nest enclosures of *Apis mellifera*: Implications for survival, clustering, humidity regulation and *Varroa destructor*, *International Journal of Biometeorology* 60: 629–638.

Fig. 9.6. *Top:* photos by Robin Radcliffe. *Bottom:* original drawing by Margaret C. Nelson.

Fig. 9.7. Original drawing by Margaret C. Nelson, based on fig. 2 in Southwick, E. E., 1982, Metabolic energy of intact honey bee colonies, *Comparative Biochemistry and Physiology* 71: 277–281.

Fig. 9.8. *Top:* Photo by Thomas D. Seeley. *Bottom:* Original drawing by Margaret C. Nelson, based on fig. 1 in Peters, J. M., O. Peleg, and L. Mahadevan, 2019. Collective ventilation in honeybee nests, *Journal of the Royal Society Interface* 16: 20180561.doi.org/10.1098/rsif.2018.0561.

Fig. 9.9. Photo by Linton Chilcott.

Fig. 9.10. Original drawing by Margaret C. Nelson, based on fig. 3 in Ostwald, M. M., M. L. Smith, and T. D. Seeley, 2016, The behavioral regulation of thirst, water collection and water storage in honey bee colonies, *Journal of Experimental Biology* 219: 2156–2165.

Fig. 10.1. Original drawing by Margaret C. Nelson, based on fig. 1 in Rinderer, T. E., L. I. de Guzman, G. T. Delatte, J. A. Stelzer, and 5 more authors, 2001, Resistance to the parasitic mite *Varroa destructor* in honey bees from far-eastern Russia, *Apidologie* 32: 381–394.

Fig. 10.2. Photo by Åke Lyberg.

Fig. 10.3. Original drawing by Margaret C. Nelson, based on figs. 1, 2, and 3 and on data in table 1 in Fries, I., A. Imdorf, and P. Rosenkranz, 2006, Survival of mite infested (*Varroa destructor*) honey bee (*Apis mellifera*) colonies in a Nordic climate, *Apidologie* 37: 564–570.

Fig. 10.4. Aerial photo from Google Earth, with locations of wild honey bee colonies added by Thomas D. Seeley.

Fig. 10.5. Original drawing by Margaret C. Nelson, based on fig. 3 in Mikheyev, A. S., M. M. Y. Tin, J. Arora, and T. D. Seeley, 2015, Museum samples reveal rapid evolution by wild honey bees exposed to a novel parasite, *Nature Communications* 6: 7991, doi:10.1038/ncomms8991.

Fig. 10.6. Original drawing by Margaret C. Nelson, based on fig. 1 in Seeley, T. D., and M. L. Smith, 2015, Crowding honeybee colonies in apiaries can increase their vulnerability to the deadly ectoparasitic mite *Varroa destructor*, *Apidologie* 46: 716–727.

Fig. 10.7. Photo by Thomas D. Seeley.

Fig. 10.8. Original drawing by Margaret C. Nelson, based on fig. 1 and fig. 3 in Loftus, J. C., M. L. Smith, and T. D. Seeley, 2016, How honey bee colonies survive in the wild: Testing the importance of small nests and frequent swarming, *PLoS ONE* 11 (3): e0150362, doi:10.1371/journal.pone.0150362.

Fig. 10.9. Photos by Thomas D. Seeley.

Fig. 10.10. Photo by Renata S. Borba.

Fig. 10.11. Original drawing by Margaret C. Nelson, based on fig. 3 in Borba, R. S., K. K. Klyczek, K. L. Mogen, and M. Spivak, 2015, Seasonal benefits of a natural propolis envelope to honey bee immunity and colony health, *Journal of Experimental Biology* 218: 3689–3699.

Index

Acarapis woodi. See tracheal mites
Aethina tumida. See small hive beetle
afterswarming, 162–164
American foulbrood, 87–90, 244, 259, 273
ancestry of honey bees, 57
annual cycle: of colony food intake and consumption, 142–146; of colony weight gains and losses, 143–146; of honey bee vs. bumble bee colonies, 152–154; origins and evolution, 152–154; of rearing drones, 157–158; of rearing workers, 141–142, 146–152, 278–280; of swarming, 145, 149–152, 158
antibiotics, 284
Apis cerana, 13, 245–246
Apis henshawi, 57–59
Apis mellifera: capensis, 82–84; *carnica*, 6, 9–12, 35, 117; *caucasica*, 6, 9–11; *cypria*, 9; *intermissa*, 9; *lamarckii*, 9, 35; *ligustica*, 6, 9–12, 35, 117; *macedonica*, 6; *mellifera*, 5–12, 35, 278–280; *scutellata*, 9–11, 33–35; *syriaca*, 9; *yemenetica*, 10–11
Arnot Forest: bee foraging in, 197–200, 205–206; bee hunting in, 48–51, 103; description of, 26–28; history of, 18–20; map of, 27, 32, 51, 184; mechanisms of *Varroa* resistance of bees living in, 252–258; natural selection on bees living in, 255–257; presence of black bears in, 54–55, 271–273; presence of *Varroa* in, 51–54, 205–206
artificial insemination, 16, 85–93, 295
artificial selection. *See* breeding of honey bees
Ascosphaera apis. See chalkbrood

bait hives, 52–53, 55, 112, 289
Bashkortostan, Republic of, 7, 25, 68–69
bee hunting (bee lining), 28–31

bee space, 74–77
beekeeping: in antiquity, 62–63; Egyptian, 64–65; movable-frame hive, 73–78; Roman, 64–66; skep, 69–72; tree (Zeidlerei), 66–69; tools of modern, 98
Beekman, Madeleine, 201
beeswax: ancient, 5, 80; economy of use by bees, 124–131; harvesting of, 284; importance in tree beekeeping, 68–69; production by bees, 121–123, 284–285
Berry, Wendell, 2
black bears, 54–55, 116, 185, 271–273
Bombus polaris, 153
breeding of honey bees, 86–93
brood rearing: annual total amount, 159, 172, 193; importance of early onset, 151–152; seasonal pattern of, 148–152
Brown, Mark J. F., 270

carbon dioxide, stimulus for nest ventilation, 235–236
Chadwick, P. C., 236
chalkbrood, 89–91, 109, 210, 222–223, 244, 273, 282
Chilcott, Ann B., 238
chill coma, 223–224, 270
Cingle de la Ermita del Barranc Fondo, 60–61
clustering colonies in apiaries: causes, 258–259; effects, 259–263, 280
Columella, Lucius J. M., 64–66
comb: building, 96–97, 105–106, 120–135; cell size, 108, 269; cell size effects on *Varroa* mites, 269–270; contents over year, 150–151; for drone rearing, 108, 132–135; foundation, 96–97
Cordovan mutation, 260–262

Cranberry Lake Biological Station, 206–209
Crane, Eva, 64
crowding of colonies. *See* clustering of colonies in apiaries
Cueva de la Araña, 60–61
Cullinan, Jenny, 82

dairy cows, 85–86
Darwin, Charles R., 99
Darwinian beekeeping, 3, 277–292
deformed wing virus, 42, 48, 244–245, 259
Delaney, Deborah A., 183–185, 253–254
density of wild colonies, 24–26, 35–41, 202
diseases of honey bees. *See* American foulbrood; chalkbrood; deformed wing virus; European foulbrood; sacbrood; tracheal mite; *Varroa destructor*
domestication: of *Apis mellifera*, 80–82, 93–97; general process, 79–80
Doolittle, Gilbert M., 87
drifting, of bees between colonies, 259–263, 287
drone comb: cell size, 133; control of building, 132–135; control of use for drone rearing, 159–161; percentage of comb area, 52, 108–109, 122, 132–133, 279–281
drone congregation area, 39, 178–181, 186
drones: cost of feeding, 174; drifting among colonies, 260–263, 287; importance of, 86–87, 155–157, 182–183, 290; killing of, 66; level of production, 158–159, 171–173; mating behavior, 285; mating flight range, 178–182; number in colony, 109, 128; timing of production, 150–151, 157–158; use in estimating colony density, 37–39
Dudley, Paul, 7
Dyce Laboratory, 101, 151

Eckert, John C., 196
Edgell, George H., 28, 48
Ellis Hollow, 13–20
energy budget of colony, 193–195
engorgement response: to smoke, 82–84; when swarming, 82
environment of evolutionary adaptation (EEA), 278

European foulbrood, 109
evaporative cooling. *See* temperature control of broodnest

fanning behavior, 234–237
Farrar, Clayton L., 144
fever response of colony, 222–223
field crickets, 14
flight muscles of worker bees, 217–219, 223
flight range: of queens and drones, 38, 179–182; of worker bees, 187, 195–196, 199–200
food collection by colony: forage types, 188–190; spatial scope of, 195–201; total amount for year, 191–195
food sources: colony's skill in choosing among, 204–209; colony's skill in finding, 201–204; distances to, 187, 195–201
fossil honey bees, 57–59
Fries, Ingemar, 247, 251
Frost, Robert, 140
fungicides, 279, 283, 288–289

Galleria mellonella. *See* wax moth
Galton, Dorothy, 24–25
Gary, Norman E., 178, 196
Gibbons, Euell, 57
Gilley, David C., 163–164
Gotland, 247–252, 289
Griffin, Sean R., 183, 253, 269–270
grooming behavior, for resistance to *Varroa destructor*, 246, 257–258

Hainrich Forest, 38
Harz National Park, 37
Haudenosaunee, 19
Hazelhoof, Engel H., 235–236
heat production: by individual bees, 218–220, 232, 234; by whole colonies, 232–233
hive beekeeping: Egyptian, 62–65; with movable-frame hives, 72–78, 86, 95–98; origins of, 62–63, 81–82; Roman, 64–66; with skeps, 69–72
hive size: consequences of large vs. small, 264–268, 280–287; typical range for beekeeping, 263; with skeps, 70–72
Homo sapiens, ancestry, 58
honey, energy value of, 58

honey hunters: Efe, 59; Hadza, 58–59; prehistoric, 60–62
Hood, Thomas, 215
hygienic behavior, 87–91, 251, 258
hypothermia. *See* chill coma

insecticides, 3, 279, 283, 288–289
instrumental insemination. *See* artificial insemination
insulation: of bee's body, 218; of colony's nest, 217–218
Iroquois, 19
Ithaca, New York, 17–20

Jaycox, Elbert R., 117
Jeanne, Robert L., 154

Knaffl, Herta, 197, 201
Koeniger, Gudrun, 179
Koeniger, Nikolaus, 179
Kohl, Patrick L., 38–39
Kovac, Helmut, 238–239
Kraus, Bernhard, 46
Kühnholz, Susanne, 240

Laidlaw, Harry R., 87
Langstroth, Lorenzo L., 72–78, 79, 82, 86
Langstroth hive, 32, 53, 70, 74–79, 101, 104–105, 108, 117–118, 143, 152, 165, 211, 225, 244, 248, 260, 264
Lewis and Clark Expedition, 7
lifespan of a bee tree colony, 169–171
Lindauer, Martin, 234
Linnaeus, Carl, 5
local adaptation, of bees, 286–287
Locke, Barbara, 251
Loftus, Carter L., 264
longevity of wild colonies, 165–171
Loper, Gerald M., 46–47

Mackensen, Otto, 90
mating behavior: controlling, 88–89, 93; flight range, 38, 179–182; nuptial flights, 178–179; sex attractant pheromone, 178
mating sites, 178–181
Mattila, Heather R., 183
Maurizio, Anna, 222
McMullan, John B., 270

Melissococcus plutonius. See European foulbrood
Mikheyev, Alexander S., 10, 254–257
Milum, Vern G., 221
Mitchell, Derek M., 116, 227–229, 239
"mite bomb" phenomenon, 290
mites. *See Varroa destructor*; *Varroa jacobsoni*
miticides, 15, 45, 56, 245–246, 252, 284
Moritz, Robin, 36–38
Morse, Roger A., 33, 100–101, 105, 112, 228
Moses, 81
Müritz National Park, 37
mushroom bodies, 216

Nakamura, Jun, 136–139
nectar: annual collection amount, 192–195; collection efficiency 194; composition, 190; handling/use, 190–191; load size, 194
Nelson, Herb, 100–101
nest: architecture, 108–109, 281; construction, 96–97, 105–106, 120–139; cooling, 234–242; spacing, 51; ventilation, 234–237
nest entrance opening: direction effects, 115–116; height effects, 270–273, 281; size effects, 114–115, 281
nest site: inspection, 111–112; properties, 101–108, 281; selection, 112–120
Nolan, Willis J., 148
Nye, William P., 90–91

observation hive, 127–129, 137–138, 150, 158–159, 163–164, 189, 197–198, 201, 205, 213, 216, 223, 238, 240–242
Oldroyd, Benjamin P., 39
Oleksa, Andrzej, 36
Ostwald, Madeleine M., 240–242
Owens, Charles D., 225–226

Paddock, F. B., 87–89
Paenibacillus larvae. See American foulbrood
Page, Robert E., Jr., 46, 158, 172
Parise, Stephen G., 117
Park, O. Wallace, 87–89, 240–242
Peck, David T., 55, 211–214, 257–258
Peer, Donald F., 181–182
Pellett, Frank C., 87–89
Peters, Jacob M., 235–237

Pinto, Alice M., 34
pollen: annual collection, 192–195; basket, 136; collecting, 138, 141, 150, 188–189, 194; collection efficiency, 193–194; consumption of, 94, 98, 137, 144, 190, 193; harvesting by beekeepers, 94, 98, 284, 289; load size, 193; nutritional value, 188–189, 279, 289; storage of, 121, 189–190; storage location in nest, 189, 281–282; substitutes, 283
pollination: commercial, 94–95; value of honey bees, 291
polyandry: benefits to colony functioning, 183–185; definition of, 182; level in wild colonies, 183–186
population biology of wild colonies, 165–169
Pratt, Stephen C., 129–135
Primorsky region of Russia, 43, 245–247
Prokopovich, Peter, 69
propolis, 73, 83–84, 91, 106–107, 115, 120, 123, 135–139, 217, 224, 239, 273–276, 279–280, 287, 289, 291
propolis envelope: construction, 135–139; function, 139, 217, 224, 239, 273–276, 279–287; structure, 135
propolis traps, 138
protandry, 157–158

queen: cells, 146–147, 161; lethal behavior of virgins, 162–164; mating flight range, 38, 179–182; preparation for swarming, 161; rearing, 149–150, 158, 161, 285; shaking by workers, 161; sperm storage in, 182; substance pheromone, 178
queen production, timing of: 149–150, 158

Radcliffe, Robin W., 41, 103–104, 230
Rangel, Juliana S., 175
Ratnieks, Francis L. W., 201
recruitment to food sources, 205–210
Reeve, H. Kern, 175
reproduction of colonies: hermaphroditic, 155–157; timing of, 142, 150–151, 157–158
resin: collection, 135–136; 190–191; handling and use, 123, 136–139, 273–276
resin collectors: ages of, 137; behaviors inside nest, 137–138

Rinderer, Thomas E., 118, 246–247
robbing of honey: by beekeepers, 289; benefits and costs, 210, 268, 280; in forest vs. in apiary, 212; robber behavior, 212–213; role in spreading parasites and pathogens, 87, 210–214, 252, 263, 268, 290
Rothamsted Experimental Station, 150
Rothenbuhler, Walter C., 89
Rutschmann, Benjamin, 38–39, 230
Ruttner, Friedrich, 179–181
Ruttner, Hans, 179–181

sacbrood, 109
scale-hive records, 143–146
Schmidt, Justin O., 118
Scofield, Hailey N., 230
Seeley, Robin H., 21–22
sex attractant pheromone, 178
sex investment ratio, 171–174
Shindagin Hollow State Forest, 40–41, 103–104
Simone-Finstrom, Michael, 139
skeps, 69–72
sleep, 2
small-cell comb. *See* comb, cell size
small hive beetle, 282
Smibert, Thomas, 187
Smith, Michael L., 127, 132, 158–160, 172, 240–242, 264
smoke, effects on bees, 82–85
Sonoran Desert, 46–47
Southwick, Edward E., 227
Spivak, Marla, 274
Starks, Philip T., 222
Sternberg, Karin, 82
stinging behavior, 82–84
survival probabilities: established colonies, 168; newly founded colonies, 168
Swabian Alb Biosphere Reserve, 38–39
swarming: control of, 72, 97; effects on *Varroa* mite infestations, 265–268; optimal swarm fraction, 174–177; probability of after-swarming, 163; probability of swarm survival, 168; seasonal pattern of, 111, 128, 148–152, 158; sequence of events during, 161–162
Szabo, Tibor I., 116, 119

Tarpy, David R., 163–164, 183–185, 253–254
Tautz, Jürgen, 216
Tel Rehov, Israel, 62–63
Teleogryllus oceanicus, 14
temperature: of bees' flight muscles, 218–220, 232; in broodnest, 141, 215–219, 234–237; profile of winter cluster, 217, 225; sensitivity of pupal brood, 216, 221, 234; tolerances of adult bees, 221; in winter cluster, 140, 215, 224–230
thermal conductance: of bee cluster, 226–227, 233; of nest cavity's walls, 227–231
tracheal mite, 46, 89–90, 109, 282
Trans-Siberian Railway, 245
tree beekeeping (*Zeidlerei*), 6–7, 24–25, 66–69, 72
Tribe, Geoff, 82
Twain, Mark, 17

uncapping and recapping cells (for resistance to *Varroa*), 258
Varroa destructor: breeding for resistance to, 89, 289; dispersal between colonies, 214; effects on colonies, 46–47, 52–56; effects on individual bees, 44–45; extraordinary nimbleness, 213–214; general biology, 13–15, 33, 42–45; history of spread on *Apis mellifera*, 42–44, 245–259, 282–283; population dynamics in small-hive vs. large-hive colonies, 265–268
Varroa jacobsoni, 245–246
Varroa-sensitive hygienic (VSH) behavior, 251–252, 258
ventilation. *See* nest ventilation
virgin queens: fates of, 164; fighting among, 162–164
virulence, evolution of, 259
Visscher, P. Kirk, 28–31, 38, 41, 48, 151, 197, 205, 208
von Frisch, Karl, 197, 201

waggle dance, 138, 183, 197–201, 205–208, 216, 238
water: collection, 188–190, 236–242; handling/use, 190, 237–238, 240–242; storage in nest, 240, 242
Watson, Lloyd R., 87–88
wax moth, 244
weight records of colonies, 143–144
Welder Wildlife Refuge, 34–35
Williams, George C., 155
Winston, Mark L., 163

Yale Myers Forest, 202–204